Fundamental Theori

Volume 193

Series editors

Henk van Beijeren, Utrecht, The Netherlands
Philippe Blanchard, Bielefeld, Germany
Paul Busch, York, United Kingdom
Bob Coecke, Oxford, United Kingdom
Dennis Dieks, Utrecht, The Netherlands
Bianca Dittrich, Waterloo, Canada
Detlef Dürr, Munich, Germany
Ruth Durrer, Geneva, Switzerland
Roman Frigg, London, United Kingdom
Christopher Fuchs, Boston, USA
Giancarlo Ghirardi, Trieste, Italy
Domenico J. W. Giulini, Bremen, Germany
Gregg Jaeger, Boston, USA
Claus Kiefer, Cologne, Germany
Nicolaas P. Landsman, Nijmegen, The Netherlands
Christian Maes, Leuven, Belgium
Mio Murao, Bunkyo-ku, Tokyo, Japan
Hermann Nicolai, Potsdam, Germany
Vesselin Petkov, Montreal, Canada
Laura Ruetsche, Ann Arbor, USA
Mairi Sakellariadou, London, UK
Alwyn van der Merwe, Denver, USA
Rainer Verch, Leipzig, Germany
Reinhard Werner, Hannover, Germany
Christian Wüthrich, Geneva, Switzerland
Lai-Sang Young, New York City, USA

The international monograph series "Fundamental Theories of Physics" aims to stretch the boundaries of mainstream physics by clarifying and developing the theoretical and conceptual framework of physics and by applying it to a wide range of interdisciplinary scientific fields. Original contributions in well-established fields such as Quantum Physics, Relativity Theory, Cosmology, Quantum Field Theory, Statistical Mechanics and Nonlinear Dynamics are welcome. The series also provides a forum for non-conventional approaches to these fields. Publications should present new and promising ideas, with prospects for their further development, and carefully show how they connect to conventional views of the topic. Although the aim of this series is to go beyond established mainstream physics, a high profile and open-minded Editorial Board will evaluate all contributions carefully to ensure a high scientific standard.

More information about this series at http://www.springer.com/series/6001

Serguei Krasnikov

Back-in-Time and Faster-than-Light Travel in General Relativity

Springer

Serguei Krasnikov
St. Petersburg
Russia

ISSN 0168-1222 ISSN 2365-6425 (electronic)
Fundamental Theories of Physics
ISBN 978-3-030-10260-9 ISBN 978-3-319-72754-7 (eBook)
https://doi.org/10.1007/978-3-319-72754-7

Softcover re-print of the Hardcover 1st edition 2018

Printed on acid-free paper

This Springer imprint is published by the registered company Springer International Publishing AG part of Springer Nature
The registered company address is: Gewerbestrasse 11, 6330 Cham, Switzerland

Речь идет о том, как поступать с задачей,
которая решения не имеет. Это глубоко
принципиальный вопрос [...]

К. Х. Хунта в [189][1]

[1] "We are speaking of how to deal with a problem that has no solution. This is a matter of deep philosophical principle ..." Cristobal Junta in [189].

To my parents

Preface

Can a cause–effect chain close into a circle? Can a person travel to their past? These questions have been intriguing people for millennia.

Until the last century, however, they were discussed only by philosophers and poets.[2] An adequate language appeared only in the end of the nineteenth century, when due to the efforts of people like Wells and Poincare the concept was developed of time as an attribute of a four-dimensional object now called Minkowski space. This concept formed the foundation of special relativity—logically, if not historically—and brought up the third, almost equivalent to the first two, question: is it possible to travel faster than light?

The next major step in understanding causality[3] was made in a few years, when Einstein developed *general* relativity, according to which the Minkowski space is only an approximation, while the actual large-scale geometry of the universe is much more complex and admits—in principle—different wonders: handles, fractures, closed geodesics, etc. As a result, the question of a maximal speed became much harder, even the concept of distance lost its clarity. The problem of time travel also acquired a new aspect: among the above-mentioned wonders, there may be self-intersecting timelike curves. Such a curve would describe an object—a time traveller—returning to the past in a manner that has no analogues in special relativity. The traveller observes no local 'miracles': photons still overtake it, people do not walk backward, raindrops fall down from clouds, etc. Nevertheless, at some moment they meet their younger/older self. To dismiss the possibility of *such* time trips is much more difficult than those in the Minkowski space. Einstein confessed

[2]Oedipus receives (via an oracle) a signal from the future, which compels him to flee from Corinth. This flight is responsible for his meeting (and, consequently, killing) Laius. But it is the message from the oracle concerning this killing that prompted Oedipus to leave Corinth (where he would have never meet Laius). The story comes full circle making up what is known today as a 'bootstrap paradox', see Chap. 6.

[3]The importance of that step went far beyond physics. It is worth mentioning, for example, that modern philosophers of time use a general relativity based language [38].

that this problem disturbed him already at the time of the building up of the general theory of relativity[4] without his having succeeded in clarifying it [40].

In a next few decades, little progress was made. The finding by Gödel [66] of a cosmological model in which causality breaks down in every point somehow had no developments. And the pithy discussion of tachyons concerned only causality in Minkowski space. True, at that time the concept of wormhole gained popularity. Misner and Wheeler suggested that what we think are charges are actually wormhole mouths [134]. This idea, however, was not turned into a realistic model. At the same time, the specific wormholes proposed by Ellis [44] and Bronnikov [18] required the matter source to be quite exotic (which is unavoidable, as we understand now). As a result, the interest in wormholes waned.

The situation changed drastically after the paper by Morris and Thorne had been published where they noted that the existence of wormholes would apparently imply also the existence of closed timelike curves [136]. This idea aroused considerable interest and gave rise to a wave of publications. In a few years, however, it became clear—somewhat unexpectedly, it seems—that there are no obvious arguments against the possibility of time travel. This ended the period of facile attacks leaving us with the understanding that

1. the problem was harder than it looked;
2. the formation of a time machine may be unstable;
3. both time and faster-than-light journeys (the latter was represented by a flight through a wormhole) require, as a source of the spacetime curvature, matter with *negative* (in a certain frame) energy density.

The results obtained by the early 1990s are summed up—with an emphasis on wormholes—in [172].

The next period was characterized by a shift of accents. In particular, the wormholes, time machines and—after the publication of Alcubierre's paper [3]—superluminal motion, began to be studied on equal footing with other relativistic phenomena (singularities, say), and not as curiosities, whose impossibility, though obvious, still is not convincingly proven, for some reason. The new approach turned out to be more productive and in the last 20 years a number of interesting results have been obtained. The time seems right to gather them into one monograph and thus to spare those interested in the subject from having to work through numerous original papers.

That is how this book came about (see also [187]). Not being a review, it is not intended to be complete: to keep the size of the book reasonable without sacrificing thoroughness, some topics were not included. The criteria of rejection were subjective; it is worth mentioning only that among the omitted topics there are important (e.g. Gott's time machine), fashionable (e.g. wormholes in the 'thin shell approximation') and highly controversial (e.g. Fomalont and Kopeikin's measurement of the speed of gravity) ones.

[4]Maybe, in fact, even earlier [41].

Though some of its parts must be understandable to a reader unfamiliar with mathematics, this book is by no means popular. To understand it *entirely*, including the proofs, a reader should be acquainted with basic notions of general topology (such as connectedness, compactness, etc.) and differential geometry (manifolds, tensors, etc.). It will be helpful also, if the reader is familiar with the basics of relativity. On the other hand, some fundamentals are recapitulated in the first four sections of Chap. 1.

I acknowledge the many people who have helped this book to appear. Although they are too numerous to list here, I wish to thank them *all*. Special gratitude goes to R. R. Zapatrin, the only person with whom I seriously discussed the subject of the book.

St. Petersburg, Russia
Serguei Krasnikov

Contents

Notations

Throughout the book, we use the Planck units: $G = c = \hbar = 1$. The metric signature and other sign conventions for geometric quantities are those adopted in [135]. The neighbourhoods are defined to be open. As a rule,

$A_{B(C)}$	A_B or A_C
$A \leftrightharpoons B$	A is defined to be B, A denotes B
$M,\ U,\ \ldots$	n-dimensional manifolds, spacetimes
Int W, $\overline{W}$ and $\dot{W}$	The interior, closure and boundary of W
$\mathrm{Cl}_U\ W$, $\mathrm{Bd}_U\ W$	The closure and boundary of W in U
$\mathcal{B},\ \mathcal{S}\ \ldots$	$(n-1)$-dimensional surfaces, or (sometimes) their closures
$\Theta,\ \Sigma\ldots$	Sets of lower than $n-1$ dimension
$p,\ q\ldots$	Points
$\gamma,\ \lambda\ldots$	Curves
γ_{pq}	A curve from p to q
$\dot{\gamma}(\chi)$	The velocity of a curve γ at a point x
$\boldsymbol{t},\ \boldsymbol{k},\ \boldsymbol{T}\ldots$	Vectors and tensors
$\boldsymbol{\phi}$	A field operator
$g(\boldsymbol{t},\boldsymbol{k}) \leftrightharpoons t^a k_a,\ g$	A (covariant) metric tensor
g^R	A Riemannian metric tensor
$\chi,\ \phi,\ \psi\ \ldots$	Isometries
$\overset{\sigma}{\sim}$	The equivalence relation generated by an isometry σ
$q \in I^+_U(p)$	q is in the chronological future of a point p in U
$p \prec q$	The same, when $U = M$
$q \in J^+_U(p)$	q is in the causal future of a point p in U
$p \preccurlyeq q$	The same, when $U = M$
$<p,\ q>_U$ and $\leqslant p, q \geqslant_U$	are defined on p. 17
$\mathcal{D}(U)$	The Cauchy domain of a set U
$\mathcal{H}$	A horizon
CG(D)CH	A compactly generated (determined) Cauchy horizon
c	A local condition or property
C	The set of all spacetimes satisfying C

$\mathsf{R}(M)$	is defined on p. 17
$\mathbb{I} \leftrightharpoons (0, 1)$	
$\mathbb{E}^n$	The n-dimensional Euclidean space
$\mathbb{L}^n$	The n-dimensional Minkowski space
$\mathscr{D}(U)$	The space of smooth functions with compact support lying in U
$\overset{\looparrowright}{M}$	The set of all points of M in which the causality condition does not hold
M^{r}	is defined on p. 86
$T_p(M)$	The space tangent to M at the point p
$\Box \leftrightharpoons \partial^a \partial_a$	
c_1, c_2 ...	Constants, the values of which are irrelevant at the moment
$M_\odot$	The solar mass

In a reference to an equation, proposition, etc., the number of the chapter is indicated when and only when it differs from the current. Thus, for example, the proposition formulated at page 88 is referred to as 'proposition 6' throughout Chap. 4, and as 'proposition 4.6' everywhere else.

Part I
Classical Treatment

Chapter 1
Geometrical Introduction

It has become apparent in recent decades that in discussing fundamental properties of space and time the adequate language is Lorentzian geometry. Correspondingly, we start the book with a brief introduction to that discipline. In the first four sections of this chapter some basic notions—such as convexity, causal simplicity, and global hyperbolicity—are defined and some basic facts about them are provided, such as Whitehead's and Geroch's theorems, the Gauss lemma, etc. Of course, this recapitulation cannot substitute a systematic presentation, so wherever possible the proofs of those facts are dropped, which is understood as the reference to classical monographs such as [141] or [76]. Then, in Sect. 5, some *new* objects—perfectly simple sets—appear and the theorem is proved which states that they exist in arbitrary spacetime. We shall use this theorem in proving Theorem 2 in Chap. 5, but, as it seems, it is interesting also by itself (the perfectly simple sets are 'as nice as possible'—they are both convex *and* globally hyperbolic, which makes them exceptionally useful in proving statements).

Finally, in Sect. 6 the 'cut-and-paste' surgery is rigorously described. This is necessary because we shall widely employ that surgery in constructing spacetimes from more simple ones. And experience suggests that even though the method seems simple and pictorial, neglecting some its subtleties may lead to serious mistakes.

Thus, this chapter is technical and a reader can safely skip it or use it as a glossary, if they do not intend to analyse the proofs of the statements formulated in the consequent chapters.

1 Spacetimes

Classical general relativity, as it is understood in this book, is the theory describing the universe by an (inextendible, see below) *spacetime* (M, g), which is a connected Hausdorff manifold M endowed with a smooth Lorentz metric g and a time

S. Krasnikov, *Back-in-Time and Faster-than-Light Travel in General Relativity*, Fundamental Theories of Physics 193, https://doi.org/10.1007/978-3-319-72754-7_1

orientation. Sometimes, when it is obvious or irrelevant what metric g is meant, the manifold M itself will be also called a spacetime.

Comments 1 (1) A rigorous definition of time orientation can be found in [141]. Roughly speaking, it is a choice—made in a smooth manner–of which timelike vectors, see Definition 15, are future- and which are past-directed. (2) M is automatically smooth and paracompact. For the proof of the latter, non-trivial, fact see [61].

For building a full-fledged theory, the geometrical description must be supplemented with postulates determining how matter acts on geometry (usually these are Einstein's equations) and, conversely, how matter is acted upon by non-trivial geometry (typically the 'comma-goes-to-semicolon' rule is adopted, see Sect. 16 of [135]). Those postulates are of secondary importance. For slightly varying them (introducing, say, a small Λ-term), one typically changes the theory inconsiderably (at least at non-cosmological scale). At the same time, none of the above-listed properties defining the spacetime can be dropped without mutilating the theory beyond recognition.

A spacetime does not have to describe *the whole universe*. It is easy to check that *any* open connected subset of a spacetime is a spacetime too. However, as the Minkowski space exemplifies, the converse is false: even a noncompact spacetime may not be a part of a larger one.

Definition 2 A spacetime $M' \neq M$ is called an *extension* of a spacetime M, if the latter is an open subset of the former or is isometric to such a subset. M is *extendible*, if it has an extension and *inextendible*, or *maximal*, otherwise.

Remark 3 An open subset $M \subset M'$ is more than just a spacetime. It is a spacetime *imbedded in* M' by a particular isometry. In considering such an imbedded spacetime, we may be interested both in properties defined by its geometry (we shall call such properties *intrinsic*, see below) *and* those defined by the imbedding. With this in mind, we shall not automatically identify spacetimes only because they are isometric.

It seems natural to interpret an extension of M as describing a larger than M portion of the universe and to consider M an adequate model of the universe only if it is inextendible. In doing so, however, one encounters some technical problems associated with infinities. For example, the regions $t < 0$ and $t < -1$ of the Minkowski space are each other's extensions. To avoid such problems, it is convenient to consider not spacetimes, but *triples* $T = (M, p, \{\boldsymbol{e}_{(i)}\})$, where p is a point of M, and $\{\boldsymbol{e}_{(i)}\}$ is a basis in the space tangent to M in p. A triple T_2 will be called an extension of T_1, if there is an isometry

$$\zeta_{12}\colon \qquad M_1 \to \zeta_{12}(M_1) \subsetneq M_2,$$

sending p_1 to p_2 so that the differential $\mathrm{d}\zeta_{12}$ maps $\{\boldsymbol{e}_{(i)1}\}$ to $\{\boldsymbol{e}_{(i)2}\}$. Such an isometry, if it exists, is unique, which makes the—obviously transitive—relation 'to be an extension of' asymmetric on the set of the triples. Thus, the relation in question is a (strict) partial order [87], which enables one to show [62] that any spacetime has a maximal extension (we shall use the same line of reasoning in Chap. 5). Note that an extendible spacetime generally has *infinitely many* maximal extensions. Correspondingly, one may ask: given the known part of the universe is described by M, which of its maximal extensions describes the whole universe? The answer is very far from being clear as we shall discuss in Chap. 2.

2 Local Geodesic Structure

2.1 Geodesics

Any spacetime, just because there is a metric in it, possesses a distinguished class of curves—the geodesics. Their properties to a great extent determine the whole geometry of a spacetime including its causal structure.

According to the simplest definition, a curve is a geodesic if its velocity vector is parallel translated (or, equivalently, covariantly constant) along that curve. It only remains to recall that if $\boldsymbol{T}$ is a smooth tensor field of type (q, s)—with components $T^{a\dots e}{}_{b\dots f}$—then its covariant derivative $\nabla \boldsymbol{T}$ is defined to be a smooth tensor field of type $(q, s+1)$ with the components defined in the coordinate basis by the formula

$$T^{a\dots e}{}_{b\dots f;c} \leftrightharpoons T^{a\dots e}{}_{b\dots f,c} + \Gamma^a_{ci} T^{i\dots e}_{b\dots f} + \dots - \Gamma^i_{bc} T^{a\dots e}{}_{i\dots f} - \dots ,$$

where

$$\Gamma^a_{bc} \leftrightharpoons \tfrac{1}{2} g^{ai} (g_{ib,c} + g_{ic,b} - g_{bc,i})$$

are the so-called *Christoffel symbols*. And the covariant derivative along a curve $\lambda(\zeta)$, written $\frac{D}{\partial \zeta}\boldsymbol{T}$, is defined to be the projection of $\nabla \boldsymbol{T}$ on $\boldsymbol{v} \leftrightharpoons \partial_\zeta$, that is to be the tensor[1] $T^{a\dots e}{}_{b\dots f;c} v^c$.

Properties 4 *Covariant derivative (in our case, when the connection is Levi-Civita's)*

(*a*) *coincides with the 'usual' derivative for scalar functions* f: $\nabla f = \mathrm{d} f$ *and* $f_{:a} = f_{,a} \leftrightharpoons \partial f / \partial x^a$;
(*b*) *obeys the Leibnizian product rule:* $\nabla(\boldsymbol{T} \otimes \boldsymbol{S}) = \nabla \boldsymbol{T} \otimes \boldsymbol{S} + \boldsymbol{T} \otimes \nabla \boldsymbol{S}$;
(*c*) *is linear and commutes with contraction;*
(*d*) *is compatible with the metric:* $g_{ab;c} = 0$ *(hence the lengths of vectors and angles between them do not change, when the vectors are parallel translated);*
(*e*) *is symmetric:* $\Gamma^c_{ab} = \Gamma^c_{ba}$.

The symmetry immediately results in two important equalities:

[1] For the sake of brevity, a tensor (vector) field is sometimes called a tensor (vector).

$$f_{;ab} = f_{;ba}, \tag{1}$$

for all smooth functions f, and

$$u^a{}_{;b}v^b - v^a{}_{;b}u^b = u^a{}_{,b}\, v^b - v^a{}_{,b}\, u^b, \tag{2}$$

for any smooth vector fields $\boldsymbol{u}$ and $\boldsymbol{v}$. The right-hand side of equality (2) is sometimes called the *commutator* of $\boldsymbol{u}$ and $\boldsymbol{v}$ and is denoted $[\boldsymbol{u}, \boldsymbol{v}]$.

Corollary 5 *Let* $\boldsymbol{u} \leftrightarrows \partial_x$ *and* $\boldsymbol{v} \leftrightarrows \partial_y$*—be the velocity vectors of the coordinate lines of some coordinate system* $\{x, y, \dots\}$*. Then*

$$u^a{}_{;b}v^b - v^a{}_{;b}u^b = u^a{}_{,b}\, v^b - v^a{}_{,b}\, u^b = 0.$$

Proof Indeed, in the coordinate basis

$$\boldsymbol{u} = (1, 0, \dots) \quad \text{and} \quad \boldsymbol{v} = (0, 1, \dots).$$

Clearly, any partial derivatives of these vectors are equal to zero, and hence so is the commutator. But by (2) that latter fact is coordinate independent. □

Now let us define a curve $\gamma(\xi)$ to be geodesic if for some choice of h the following equation holds:

$$\left(\frac{D}{\partial\xi}\frac{\partial}{\partial\xi}\right)^a \equiv v^a_{;b}v^b = h(\xi)v^a, \tag{3}$$

where $\boldsymbol{v} \leftrightarrows \partial_\xi$ and, correspondingly, $v^i \leftrightarrows \mathrm{d}x^i(\xi)/\mathrm{d}\xi$.

Remark 6 In what follows the word 'curve' will mean both a map whose domain is a segment *and*, quite often, the *image* of that map. Thus, for example, we call γ a curve whether it is the map

$$\gamma\colon\ [0, 1] \to \mathbb{R}^2, \qquad \xi \mapsto (x = \cos 2\pi\xi,\ y = \sin 2\pi\xi)$$

or the resulting circle. In rare instances when these two objects *must* be distinguished we shall write, correspondingly, $\gamma(\xi)$ and simply γ. Different maps with the same images will be referred to as a different *parameterization* of the same curve.

The definition based on Eq. (3) is actually equivalent to that given in the beginning of the section, because if (3) holds for *some* h it also holds for $h = 0$ (and an appropriate parameter ξ). The so chosen parameter is called *affine*, any two such parameters being related by an affine transformation $\xi' = c_1\xi + c_2$. In the literature, the term 'geodesic' is assigned sometimes only to $\gamma(\xi)$, where ξ is affine, while the common name for the solutions of (3) is *pregeodesics.*

A geodesic $\gamma(\xi)$ whose affine parameter ξ is bounded above by some value ξ_0 is called *incomplete* in the direction in which ξ grows (i.e. for example, future incomplete or past incomplete). Further, it is called *extendible* in that direction, if there is a geodesic $\bar{\gamma}$

$$\gamma \subset \bar{\gamma}, \qquad \gamma(\xi_0) \in \bar{\gamma}$$

and *inextendible* otherwise.[2] For example, the ray ($y = 0, x < 0$) of the Euclidean (or Minkowskian, does not matter) plane with cartesian coordinates is an extendible geodesic, while the same ray is *in*extendible in the space obtained by removing the origin from the plane. In both cases, the geodesic is incomplete. Below inextendible in one or (more often) both directions, geodesics will be called *maximal*.

Remark 7 Completeness, in contrast to extendibility, is an attribute of a geodesic understood as a function rather than as a set of points. We shall see below that even a *closed* geodesic can be incomplete.

A spacetime M in which all inextendible geodesics are complete is termed *geodesic complete*. A geodesic *in*complete spacetime is considered singular and the singularity is called *irremovable*, if the relevant geodesic remains inextendible in *any* extension of M.

Example 8 Let us introduce coordinates (they are called Cartesian) in the Minkowski plane $\mathbb{L}^2$ so that the metric takes the form

$$\mathrm{d}s^2 = -\mathrm{d}x_0^2 + \mathrm{d}x_1^2.$$

All straight lines are geodesics and on every such line with nonconstant x_1 this parameter is affine. Now, if we remove the ray $x_0 = 0, x_1 \geqslant 0$ from the plane, the thus obtained spacetime M_χ will be singular, because, for example, the maximal geodesic $\gamma : x_0 < 0, x_1 = 1$ is incomplete. The singularity is removable because the geodesic is extendible in $\mathbb{L}^2 \supset M_\chi$. However, after M_χ is extended to, say, the twofold covering of $\mathbb{L}^2 - o$, where o is the origin, the singularity becomes irremovable (loosely speaking the lacking point cannot be glued back into the spacetime. Similarly, one cannot fill in the white circles in Fig. 5b; the case illustrated by that figure is a little more complex, it is obtained by changing $\mathbb{L}^2 - o$ to $\mathbb{L}^3 - \mathbb{S}^1$).

2.2 Normal, Convex and Simple Neighbourhoods

In any point $p \in M$, the vectors tangent to M form the vector space $T_p(M)$ (this is just the space in which the bilinear form g acts). Geodesics enable one to build a canonical bijection between a neighbourhood of p and a neighbourhood of the origin of T_p. The resulting coordinate system is a powerful tool as it casts the expressions for some geometrical quantities into an especially simple form (for example, in these coordinates the Christoffel symbols vanish at p).

---

[2]To avoid confusion, note that 'extendible' and 'extended' are almost antonyms: an extendible geodesic being that which is *not* (fully) extended.

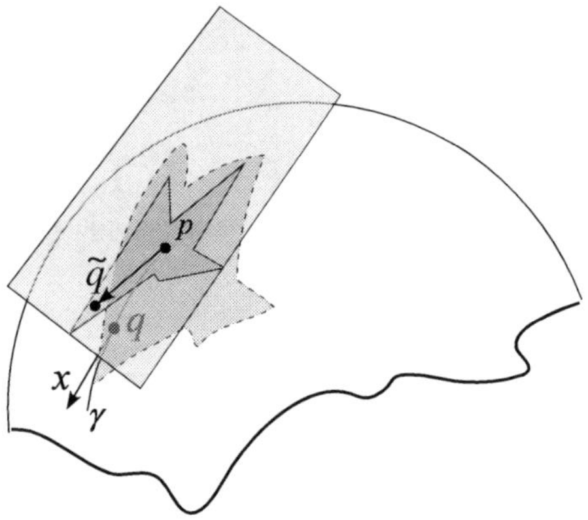

Fig. 1 The dark star-shaped region is $\tilde{N}$. The light region is N

Let $\gamma_{p,t}(\xi)$ be an affinely parameterized geodesic starting at p with the initial velocity $\boldsymbol{t}$ [i.e. $\gamma_{p,t}(0) = p$, $\boldsymbol{v}(0) = \boldsymbol{t}$]. Then, the *exponential map* $\exp_p$ is the function sending to the point $\gamma_{p,t}(1)$ each $\boldsymbol{t}$ for which that point exists, see Fig. 1. The geodesics through p may intersect in some other points too. However, one can expect (rightly, see below) that the exponential map is injective in a sufficiently small neighbourhood of p.

Definition 9 Let $\tilde{N}$ be a neighbourhood of the origin of $T_p(M)$ and be star-shaped (the latter means that $\tilde{N}$ with any vector $\boldsymbol{u}$ contains also all vectors $c\boldsymbol{u}$, where $c \in [0, 1]$; topologically a star-shaped set is just a ball [151]). If the restriction of $\exp_p$ to $\tilde{N}$ is a diffeomorphism, the neighbourhood $N \leftrightharpoons \exp_p \tilde{N}$ of p is called *normal*.

As it is seen from the definition, every point of a normal neighbourhood N is connected to the origin by a (unique) geodesic segment lying entirely in N. It is this fact that makes possible the above-mentioned convenient coordinates. They are built as follows. Pick a basis $\{\boldsymbol{e}_{(i)}\}$ in T_p. The *normal coordinates* of a point $q \in N$ are the components of the vector $\tilde{\boldsymbol{q}} \leftrightharpoons \exp_p^{-1}(q)$ in that basis (note that a different choice of $\{\boldsymbol{e}_{(i)}\}$ would lead to a different—though also normal—coordinate system). Using the definition of the exponential map, we can put it another way: the normal coordinates of a point q are $X^a(q)$ when and only when there is an affinely parameterized geodesic $\gamma_{pq}(\tau)$ in N with the ends

$$\gamma_{pq}(0) = p, \qquad \gamma_{pq}(1) = q,$$

and with the initial velocity

$$\boldsymbol{v}(p) = X^a(q)\boldsymbol{e}_{(a)}, \qquad \boldsymbol{v} \leftrightharpoons \partial_\tau. \tag{4}$$

Now note that any point

$$r \leftrightharpoons \gamma_{pq}(\tau_r), \qquad 0 \leqslant \tau_r \leqslant 1$$

is at the same time an end point of the segment $\gamma_{pr}(\zeta)$ with the ends $\gamma_{pr}(0) = p$, $\gamma_{pr}(1) = r$ and initial velocity $\partial_\zeta(0) = \tau_r \boldsymbol{v}(0)$. Hence, the coordinates of r in the system under discussion are $\tau_r X^a(q)$, which proves that in normal coordinates a radial geodesic γ_{pq} is just a straight line

$$X^a[\gamma_{pq}(\tau)] = \tau X^a(q), \qquad 0 \leqslant \tau \leqslant 1.$$

The components of the tangent vector $\boldsymbol{v}(\tau) = X^a(q)\partial_{X^a}$, in the coordinate basis $\{\partial_{X^a}\}$, do not depend on τ,

$$v^a(\tau) = X^a(q), \qquad a = 1 \dots n \tag{5}$$

[the values of a are listed in order to emphasize that this is *not* a vector equality: the vectors $\boldsymbol{X} \in T_p$ and $\boldsymbol{v}(\tau) = X^a(q)\partial_{X^a} \in T_{\gamma_{pq}(\tau)}$ belong to *different* spaces; therefore, in some other coordinates the equality (5) will not hold]. The vector $\boldsymbol{v}(1) \in T_q$ is of a particular interest, so we give to it the special name, *position vector*, and the special notation, $\boldsymbol{x}(q)$. Clearly,

$$x^a(q) = X^a(q), \qquad a = 1 \dots n \tag{6}$$

[again the right-hand side is a vector of T_p, while the left-hand side is a (written in the coordinate basis) vector of T_q].

Pick an arbitrary point $q \in N$. It follows immediately from (5) that

$$v^a{}_{,b}\, v^b = \frac{\mathrm{d}v^a(1)}{\mathrm{d}\tau} = \frac{\mathrm{d}}{\mathrm{d}\tau} X^a(q) = 0.$$

On the other hand, $\boldsymbol{v} = \partial_\tau$ and τ is an affine parameter. So,

$$v^a{}_{;b} v^b = \left(\frac{D}{\partial\tau}\frac{\partial}{\partial\tau}\right)^a = 0. \tag{7}$$

Comparing these two equalities, we conclude that

$$\Gamma^a_{cb}(q) v^c(q) v^b(q) = 0, \qquad \forall q \in N.$$

But a radial geodesic with initial velocity $\boldsymbol{v}(p) \sim \boldsymbol{u}$ can be found for *any* $\boldsymbol{u}$. Hence,

$$\Gamma^a_{cb}(p) u^c u^b = 0, \qquad \forall \boldsymbol{u} \in T_p.$$

Now pick arbitrary vectors $\boldsymbol{u}, \boldsymbol{w} \in T_p$ and define $\boldsymbol{z} \leftrightharpoons \frac{1}{2}(\boldsymbol{u} + \boldsymbol{w})$, $\boldsymbol{y} \leftrightharpoons \frac{1}{2}(\boldsymbol{u} - \boldsymbol{w})$. Then the just obtained equation in view of the symmetry $\Gamma^a_{cb} = \Gamma^a_{bc}$ gives

$$\Gamma^a_{cb}(p) u^c w^b = \Gamma^a_{cb}(p)(z + y)^c (z - y)^b = \Gamma^a_{cb}(p) z^c z^b + \Gamma^a_{cb}(p) y^c y^b = 0,$$

which implies

$$\Gamma^{a}_{cb}(p) = 0. \tag{8}$$

Another fundamental and highly non-trivial property of the exponential map is the equality

$$x^b(q)g_{ba}(q) = x^b(q)g_{ba}(p), \qquad a = 1\ldots n, \qquad \forall q \in N \tag{9}$$

(which, again, has this form only in the coordinates under consideration). In proving this equality—known as the Gauss lemma—and Eq. (11) below we closely follow the proofs in [141] of, respectively, Lemma 1 of Chap. 5 and Corollary 3 of that lemma.

Proof Pick a vector $\boldsymbol{w} \in T_p$, $\boldsymbol{w} \nparallel \boldsymbol{v}$ (throughout the proof, we shall write $\boldsymbol{v}$ for $\boldsymbol{v}(0)$) and consider a one-parameter family of geodesics $\exp_p \tau(\boldsymbol{v} + s\boldsymbol{w})$, where s ranges over a neighbourhood of zero. These geodesics sweep out a two-dimensional surface, with the coordinates τ and s in it (in particular, q has the coordinates $\tau = 1$, $s = 0$). Two vector fields, ∂_τ and ∂_s, are defined on that surface. By Corollary 5, they commute, which enables us to write

$$\begin{aligned}\frac{\partial}{\partial\tau} g\Big(\frac{\partial}{\partial\tau}, \frac{\partial}{\partial s}\Big) = \frac{D}{\partial\tau} g\Big(\frac{\partial}{\partial\tau}, \frac{\partial}{\partial s}\Big) = g\Big(\frac{\partial}{\partial\tau}, \frac{D}{\partial\tau}\frac{\partial}{\partial s}\Big) = g\Big(\frac{\partial}{\partial\tau}, \frac{D}{\partial s}\frac{\partial}{\partial\tau}\Big) = \\ = \frac{1}{2}\frac{\partial}{\partial s} g\Big(\frac{\partial}{\partial\tau}, \frac{\partial}{\partial\tau}\Big) = \frac{1}{2}\frac{\partial}{\partial s} g(\boldsymbol{v} + s\boldsymbol{w}, \boldsymbol{v} + s\boldsymbol{w}),\end{aligned}$$

where we have used property 4(*a*) (in the first and the last but one equalities) and property 4(*d*) combined with Eq. (7) (in the second equality). Thus, the variation of the scalar product of ∂_τ and ∂_s along a radial geodesic obeys the equation

$$\frac{\partial}{\partial\tau} g\Big(\frac{\partial}{\partial\tau}, \frac{\partial}{\partial s}\Big)\Big|_{s=0} = g(\boldsymbol{v}, \boldsymbol{w}).$$

The initial condition for this equation is the vanishing of ∂_s at $\tau = 0$, that is at p. Integrating the equation, one gets

$$g\Big(\frac{\partial}{\partial\tau}, \frac{\partial}{\partial s}\Big)\Big|_{s=0,\tau=1} = g(\boldsymbol{v}, \boldsymbol{w})\tau\Big|_{\tau=1} = v^b g_{ba}(p) w^a. \tag{$*$}$$

The left-hand side is the scalar product at q of $\frac{\partial}{\partial\tau} = \boldsymbol{v}$ and $\frac{\partial}{\partial s}$, the latter being tangent at $s = 0$ to the curve with the (normal) coordinates $(\boldsymbol{v} + s\boldsymbol{w})$, whence its components in the coordinate basis are w^a. Thus, $(*)$ reduces to

$$v^b g_{ba}(q) w^a = v^b g_{ba}(p) w^a,$$

which proves (9) since $\boldsymbol{w}$ has been chosen arbitrarily and v^b equals $x^b(q)$, see (5), (6). □

Corollary 10 *The function defined by*

$$\sigma(q) \leftrightharpoons g_{ab}(q)x^a(q)x^b(q) \tag{10}$$

(note that all quantities in this definition depend on the choice of the origin of coordinates and are therefore functions of p even though as a rule we shall not indicate this explicitly) obeys the equation

$$\sigma_{,a}(q) = 2x_a(q). \tag{11}$$

Proof $\sigma_{,a}(q) = [x^b(q)g_{bc}(p)x^c(q)]_{,a} = 2x^b(q)g_{ba}(p) = 2x^b(q)g_{ba}(q) = 2x_a(q).$ □

The normal neighbourhoods are remarkably convenient and yet they can be improved further: not all points of such neighbourhoods are equal. Indeed, $\exp_{p'}^{-1}$ does not have to map the neighbourhood to a star-shaped region of $T_{p'}$ when $p' \neq p$.

Definition 11 An open set O is *convex*, if it is a normal neighbourhood of each of its points.

With any two points x, y a convex set O contains also a (unique) geodesic segment γ_{xy} from x to y which does not leave O. The converse is also true: if in a normal spacetime each pair of points can be connected by a single geodesic, then this spacetime is a star-shaped neighbourhood of each of its points and hence is convex.

Definition 12 *([145])* An open convex set O is *simple*,[3] if its closure is a compact subset of some normal neighbourhood.

Moreover, any convex subset of a simple set is simple.

Proposition 13 *Let O_1 be convex. Then any connected component C of the intersection $O_1 \cap O_2$ is convex (simple), if so is O_2.*

Proof The closure of *any* subset of a compactum is compact, so, it suffices to demonstrate that C is a normal neighbourhood of each $p \in C$.

Consider the inverse images $\tilde{O}_{1,2}$ of $O_{1,2}$ under the exponential map

$$\tilde{O}_{1,2} = \exp_p^{-1} O_{1,2}.$$

It follows right from Definition 12 that the restriction of $\exp_p$ to each of them is a diffeomorphism. Hence, so is the restriction of $\exp_p$ to $\tilde{C} \leftrightharpoons \tilde{O}_1 \cap \tilde{O}_2$. Thus, $\tilde{C}$ satisfies the second condition of Definition 9. Suppose now that $\boldsymbol{u}$ lies in $\tilde{C}$. Then any vector $c\boldsymbol{u}$ with $c \in [0, 1]$ must lie both in $\tilde{O}_1$ *and* in $\tilde{O}_2$ and hence in $\tilde{C}$. So, the first condition is fulfilled too and $\exp_p \tilde{C}$ is convex. It remains to check that $\exp_p \tilde{C} = C$. The inclusion $\exp_p \tilde{C} \subset C$ is obvious (being a ball $\exp_p \tilde{C}$ is connected

[3]Not *causally* simple.

of course). But the inclusion $C \subset \exp_p \tilde{C}$ is obvious as well, because $\exp_p \tilde{C}$ is the *entire* $O_1 \cap O_2$ (not only one of its connected components). □

Generally, the intersection of two convex sets need not be convex or even connected (this can be easily verified by considering a couple of convex subsets of a cylinder). However, the situation is different for subsets of a convex space, cf. [141, p. 131].

Corollary 14 *If O_1 is simple, and its subsets O_2, O_3 are convex, then each of the sets*

$$O_2, \quad O_3, \quad O_2 \cap O_3$$

is simple.

Proof The simplicity of $O_{2,3}$ is obvious. Now let $x, y \in (O_2 \cap O_3)$. Since $O_{2,3}$ are convex, there must exist geodesic segments $\gamma_{2,3} \subset O_{2,3}$ connecting x to y. On the other hand, it follows from the convexity of O_1 that the geodesic connecting them without leaving O_1 is unique. So γ_2 coincides with γ_3 and, consequently, lies in $O_2 \cap O_3$. This intersection is therefore connected and hence (by Proposition 13) simple. □

By the Whitehead theorem, see [151], any point of any spacetime has a simple neighbourhood. And since the neighbourhood itself is a spacetime (and since its simple subsets are at the same time simple subsets of the initial 'large' spacetime) the simple sets form a base for the spacetime topology. They do not form a topology, however, because the union of two simple (convex, normal) sets may not be simple (convex, normal).

3 The Time

3.1 The Past and the Future

Everything discussed in the previous section applies equally to Riemannian and to Lorentzian manifolds. However, the latter, actually, have a much richer structure, because the vectors tangent to them are non-equivalent.

Definition 15 A non-zero vector $\boldsymbol{v}$ tangent to M, is called *timelike* [*(non-)spacelike*, *null*], if $g(\boldsymbol{v}, \boldsymbol{v})$ is, respectively, negative [(non-)positive, zero].

The space T_p tangent to M at $p \in M$ is essentially Minkowski space. Therefore, in each such space, the non-spacelike position vectors form two cones with the boundaries generated by null vectors. The cones meet (only) in the common vertex located in the origin of T_p. One of the cones is called *past* and the other is *future*. The vectors constituting the cones are said to be, correspondingly, *past-* and *future-directed*. The division of the non-spacelike vectors into these two classes is continuous (recall that the spacetime is oriented by definition).

Convention 16 The names 'past' and 'future' are arbitrary—there is no 'time's arrow' in general relativity. Thus, the theory is symmetric in the sense that any definition or statement has an (equally correct) 'dual' one, obtained by replacing the words 'past' $\longleftrightarrow$ 'future'. As a rule we shall formulate only one definition or statement, the dual being always implied.

Convention 17 Throughout the book by 'isometry', we actually understand 'isometry preserving the time orientation', thus an isometry sends future-directed vectors to future-directed ones.

Consider a *Riemannian* manifold (M, g^R). Let $\boldsymbol{v}$ be a smooth nowhere zero vector field on M (of course the existence of such a field restricts the choice of M). Now a spacetime can be built by defining on M the *Lorentz* metric

$$g(\boldsymbol{a}, \boldsymbol{b}) \leftrightharpoons g^R(\boldsymbol{a}, \boldsymbol{b}) - 2\frac{g^R(\boldsymbol{a}, \boldsymbol{v})g^R(\boldsymbol{v}, \boldsymbol{b})}{g^R(\boldsymbol{v}, \boldsymbol{v})}, \quad \Leftrightarrow \quad g_{ab} \leftrightharpoons g^R_{ab} - 2\frac{v_a v_b}{g^R_{cd} v_c v_d} \tag{12a}$$

(note that with respect to this Lorentz metric the vectors $\boldsymbol{v}$ are timelike). And, vice versa, the metric

$$\tilde{g}(\boldsymbol{a}, \boldsymbol{b}) \leftrightharpoons g'(\boldsymbol{a}, \boldsymbol{b}) - 2\frac{g'(\boldsymbol{a}, \boldsymbol{v}')g'(\boldsymbol{v}', \boldsymbol{b})}{g'(\boldsymbol{v}', \boldsymbol{v}')}, \tag{12b}$$

defined with the use of a Lorentz metric g' and a timelike vector field $\boldsymbol{v}'$, is Riemannian, i.e. positive defined.

Proof Indeed, in each point of M pick an orthonormal basis diagonalizing g'. Obviously, only one vector of the basis is timelike. The boost sending that vector to $\boldsymbol{v}'$, transforms $\tilde{g}_{ab}$ (the matrix formed by the components of $\tilde{g}$) into the identity matrix. □

The metric $\tilde{g}$ coincides with g^R, when $g' = g$ and $\boldsymbol{v}' = \boldsymbol{v}$ (this is easy to verify in the basis described above, see also Sect. 2.6 in [76]).

It turns out that *any* spacetime can be constructed by this means,

Proposition 18 ([141, Lemma 5.32]) *In any spacetime, there is a smooth future-directed timelike vector field.*

This proposition is important. First, it restricts the possible topology of spacetimes, see, e.g. Sect. 6.4 of [76]. Second, as we shall see, an auxiliary timelike vector field can in itself be a convenient tool. Finally, representing a spacetime as a Riemannian manifold (plus a smooth vector field) is a means of constructing spacetimes with predetermined properties, cf. Example 8 in Chap. 4. Unfortunately, one cannot completely reduce the Lorentzian case to Riemannian, because there is no *preferred* choice of $\boldsymbol{v}$.

Definition 19 A piecewise smooth curve $\lambda(\xi)$ is called (future-directed) *timelike*, *(non-)spacelike* or *null*, if such are all vectors ∂_ξ tangent to it. Sometimes, non-spacelike curves are also called causal.

Convention 20 'Constant' curves, consisting of a single point, are regarded non-spacelike, but not timelike.

On a timelike curve, one can introduce the *natural* parameter ξ defined by the condition $g(\partial_\xi, \partial_\xi) = -1$. In the general case, however, a non-spacelike curve has no preferred parameter and therefore has no property analogous to completeness. On the other hand, the notion of extendibility *can* be generalized to non-geodesic curves.

Definition 21 Let a curve $\lambda(\xi): (a, b) \to M$ be the restriction to the interval (a, b) of a larger curve $\tilde{\lambda}: (A, B) \to M$, where $A < a < b < B$. The points $p_1 = \tilde{\lambda}(a)$ and $p_2 = \tilde{\lambda}(b)$, if exist, are called the *end points* of λ. If λ is future-directed the end points are called *past* and *future*, respectively. A curve is said to be (*past-*, *future-*) *extendible*, if it has a (past, future) end point, and *inextendible* otherwise.

What makes this definition meaningful is the fact that the choice of $\tilde{\lambda}$ is actually immaterial: the end points being just the limit points of all sequences $\lambda(\xi_m)$ at $\xi_m \to a, b$.

Now that timelike and non-spacelike curves are defined we shall use it to introduce some important sets associated with every point of spacetime.

Definitions and notation 22 Let U be a neighbourhood of a point $p \in M$. The set of all points of U which can be reached from p by a future-directed timelike curve lying entirely in U is termed the *chronological future of* p *in* U and denoted by $I^+_U(p)$. When $U = M$, the subscript $_M$ is usually dropped and the set is denoted by $I^+(p)$. For an arbitrary set $P \subset M$ the notation $I^+_U(P)$ means $\bigcup_{p\in P} I^+_U(p)$. Changing in this definition the word 'timelike' to 'non-spacelike', we define the *causal future* of p and P, which are denoted by $J^+_U(p)$ and $J^+_U(P)$, respectively. The chronological and causal *past* of a point p or a set P are defined dually.

The relations $q \in I^+(p)$ and $r \in J^+(s)$ are often written as, respectively, $p \prec q$ and $s \preccurlyeq r$ (or, equivalently, as $q \succ p$ and $r \succcurlyeq s$). Both relations are obviously transitive and the latter one, due to Convention 20, is also reflexive. If $p \preccurlyeq q$ and $p \neq q$, the points p and q are said[4] to be *causally related.*

Through any point p, there is a timelike geodesic and hence (the second membership is a result of Convention 20)

$$p \in \mathrm{Bd} I^\pm(p), \qquad p \in J^\pm(p). \tag{13}$$

Propositions 23 *We shall not prove the following fundamental facts (which are combinations of Lemma 5.33 and Proposition 10.46 of [141] or Propositions 4.5.1 and 4.5.10 of [76]) and restrict ourselves to a few comments.*

(*a*) *The sets* $I^\pm_O(P)$ *are always open.*

(*b*) *If a region* O *contains a timelike geodesic from* p *to* q, *then* $q \in I^\pm_O(p)$. *And for a convex* O *the converse is also true: the points which can be connected by a piecewise smooth timelike curve can be connected also by a timelike geodesic.*

[4]For the reasons discussed in great detail in Sect. 2 in Chap. 2.

The first assertion is trivial, but the second is not. In principle, one could imagine that there is a point q on a spacelike *geodesic through p which can be reached from the latter by a timelike (non-geodesic) curve.*

(c) *If O is convex, then $Bd_O I^{\pm}_O(p)$ is the set of all points that can be reached from p by future-directed (past) null geodesics lying in O. In particular,*

$$Cl_O\left(I^{\pm}_O(p)\right) = J^{\pm}_O(p), \qquad \forall p \in O. \tag{14}$$

Hence, $J^{\pm}_O(p)$ is closed in O. Note that this is not necessarily the case if O is not *normal. Thus, for the point $p = (-1, -1)$ in the Minkowski plane, see Example 8, the relation* (14) *is valid, but it breaks down if we delete the origin.*

(d) *If a* non-spacelike *curve λ from p to q is not a null geodesic, then in any neighbourhood of λ there is a* timelike *curve connecting p and q.*
One of the corollaries is the implication

$$(p \prec q \preccurlyeq r) \ \text{or} \ (p \preccurlyeq q \prec r) \quad \Rightarrow \quad p \prec r. \tag{15}$$

It should be stressed that p and q are not *supposed to lie in a common convex neighbourhood.*

Propositions 23(d) and (a) enable us to formulate already at this stage a (primitive) assertion to the effect that the fastest trip is that made with the speed of light. Consider a timelike curve μ through a point q and interpret it as the world line of an observer. Then either a particle emitted from p and received in q moves on a null geodesic (and therefore with the speed of light), or the observer receives it *later* than some other particle which travelled from p to q with a *sub*luminal speed (because if q is in $I^+(p)$, then the same is true for some points of μ that precede q).

Corollary 24 *The function σ defined in a convex O by the equality* (10) *is negative in $I^{\pm}_O(p)$ and positive outside of $J^{\pm}_O(p)$.*

Corollary 25 *The set $I^+(p)$ is connected and coincides with the interior of its closure: $I^+(p) = \mathrm{Int}\overline{I^+(p)}$.*

Proof The inclusion $I^+(p) \subset \mathrm{Int}\overline{I^+(p)}$ is trivial. To prove the converse, consider an arbitrary point q of the set $\mathrm{Int}\overline{I^+(p)}$. This set is a neighbourhood of q and q lies on the boundary, see (13), of an open, see Proposition 23(a), set $I^-(q)$. Consequently,

$$\mathrm{Int}\overline{I^+(p)} \cap I^-(q) \neq \varnothing.$$

But $I^-(q)$ is open. So, with every point of $\mathrm{Int}\overline{I^+(p)}$ it contains some neighbourhood of that point, and hence some point of the set $I^+(p)$

$$I^+(p) \cap I^-(q) \neq \varnothing,$$

which implies $q \in I^+(p)$. □

The concept of regions (un)bounded in time is formalized by introducing the following notion.

Definition 26 S is a *future set*, if $I^+(S) - S = \varnothing$.

In other words, S is *not* a future set, if and only if there is a point in BdS which can be reached from S by a future-directed timelike curve. A simple example of a future set is $J^+(U)$, where $U \subset M$ is arbitrary.

Proposition 27 *If S is a future set in M, then $I^+(\overline{S}) \subset S$.*

Proof Let $p \in \overline{S}$. We must prove that $p \prec q$ implies $q \in S$. But if an open, by Proposition 23(a), set $I^-(q)$ includes a point $p \in \dot{S}$, then it includes also a point $s \in S$. Thus, $q \in I^+(s) \subset I^+(S) \subset S$. □

Below, we shall also need the following simple criteria.

Criteria 28 Let W be an open future set in a spacetime A. Then

(a) $A - W$ and $A - \overline{W}$ are past sets (such sets are defined dually to the future sets) in A;
(b) If $\mathcal{P}$ is a subset of A, then $W - \mathcal{P}$ is a future set in $A - \mathcal{P}$, when this last set is a spacetime (i.e. when $\mathcal{P}$ is closed and does not separate A);
(c) If B is a subset of an extension of A such that

$$B \cap A \neq \varnothing, \qquad B \cap W = \varnothing,$$

then W is a future set in $B \cup A$.

Proof Criteria 28*(a)* If it were possible to reach a point $p \in \mathrm{Bd}(A - W) = \mathrm{Bd}W$ from $q \in (A - W)$ by a past-directed timelike curve, that would mean that $q \in I^+(W) \subset W$ and hence $W = \varnothing$ (the proof for $A - \overline{W}$ is perfectly analogous).
Criteria 28*(b)* Obvious.
Criteria 28*(c)* If a future-directed timelike curve $\lambda(\tau)$ leaves W, then (since W is open) there must exist the *least* value of its parameter τ_0 at which this takes place, i.e.

$$\exists \tau_0 : \qquad \lambda(\tau) \in W \quad \forall \tau < \tau_0, \qquad \lambda(\tau_0) \notin W.$$

$\lambda(\tau_0)$ cannot lie in B, because the latter is disjoint with W, and hence with its boundary too. But $\lambda(\tau_0)$ cannot lie in A either, by the definition of the future set, see the remark below Definition 26. □

The fact that some region is a future set imposes serious constraints on the region's boundary.

Definition 29 If no two points of a set S can be connected by a timelike (non-spacelike) curve, it is called *achronal* (respectively, *acausal*).

Achronal surfaces should not be confused with *spacelike* ones. The latter are defined to be (sufficiently smooth) surfaces such that all vectors tangent to them are spacelike. At the same time, an achronal surface need not be smooth, and even if the tangent vectors exist they can be null (not spacelike). For example, the surface $t = x$ in the Minkowski space is achronal, but not spacelike. And, vice versa, a spacelike surface does not have to be achronal.

Example 30 Let S be the strip defined in the $(2 + 1)$-dimensional Minkowski space by the following system of equations written in the cylindrical coordinates:

$$\tfrac{9}{10} < \rho < 1, \quad -\pi < \varphi < \pi, \quad t = \alpha\varphi,$$

where α is a small number. The strip is spacelike, because such are the vectors tangent to it (they are linear combination of two spacelike vectors: ∂_ρ and $\partial_\varphi + \alpha\partial_t$). It is, however, not achronal, see Fig. 3a.

Proposition 31 *The boundary of a future set, when non-empty, is a closed, imbedded, achronal C^{1-} hypersurface.*[5]

For a proof see [76, Proposition 6.3.1].

Let us introduce the following convenient notation.

Notation 32 For any two points $p, q \in U$

$$<p, q>_U \rightleftharpoons I^+_U(p) \cap I^-_U(q), \quad \leqslant p, q \geqslant_U \rightleftharpoons J^+_U(p) \cap J^-_U(q)$$

(again, when $U = M$ the subscript $_U$ as a rule will be dropped). Further, for a given spacetime M we denote by $\mathsf{R}(M)$ the family of sets which consists of M and all its subsets of the form $<x, y>_M$.

The obvious fact that for any spacetime U and any two pairs of points $x, y \in U$, $p, q \in <x, y>_U$

$$<p, q>_{<x,y>_U} = <p, q>_U \quad \text{and} \quad \leqslant p, q \geqslant_{<x,y>_U} = \leqslant p, q \geqslant_U, \tag{16}$$

leads to the following implication:

$$U \in \mathsf{R}(M) \quad \Rightarrow \quad \mathsf{R}(U) \subset \mathsf{R}(M). \tag{17}$$

It seems tempting to relate the topological and the causal structures of spacetime using $\mathsf{R}(M)$ as a base of topology. It turns out, however, that generally the result—called the Alexandrov topology—does not coincide with the initial topology the spacetime.

[5] A hypersurface is a submanifold of codimension 1.

3.2 Causality Condition

We shall say that the *causality condition* (causality) holds in a point p of a spacetime M if there are no closed non-spacelike curves through p, i.e. if $J^+_M(p) \cap J^-_M(p) = p$. For any set $U \subset M$, we shall denote by $\overset{\looparrowright}{U}$ the set of the points violating causality in U:

$$\overset{\looparrowright}{U} \rightleftharpoons \{p\colon \ \leqslant p, p \geqslant_U \neq p\}.$$

(In this notation, the causality condition in p is $p \notin \overset{\looparrowright}{M}$).

Definition 33 A spacetime M will be called *causal*, if $\overset{\looparrowright}{M} = \varnothing$, and *non-causal*[6] otherwise.

Remark 34 Whether a point is in $\overset{\looparrowright}{U}$ is defined by the causal structure of U and does not depend on anything in $M - U$. Generally,

$$\overset{\looparrowright}{U} \subsetneq (\overset{\looparrowright}{M} \cap U)$$

(it may happen that causality is violated in a point of U even though $\overset{\looparrowright}{U} = \varnothing$). There is, however, an important exception. If U is a past set in M and ℓ is a closed non-spacelike curve through a point $q \in U$, then by (15) the entire ℓ lies in the chronological past of q and it follows from the definition of a past set that $\ell \subset U$. Thus, in this case

$$\overset{\looparrowright}{U} = \overset{\looparrowright}{M} \cap U.$$

Proposition 35 *The two following statements are, respectively, Propositions 6.4.2 and 6.4.3 of [76]:*

(*a*) *Any compact spacetime is non-causal;*

(*b*) $\overset{\looparrowright}{M}$ *is a union of disjoint sets of the form*

$$J^+(p) \cap J^-(p), \quad p \in M.$$

3.3 Causally Convex and Strongly Causal Sets

Definition 36 An open set $N \subset M$ is *causally convex in* M, or—when M is evident—just *causally convex*, if with any two points p, q it also contains the set

[6] We have to use this awkward term because the word 'acausal' is already taken, see Definition 29.

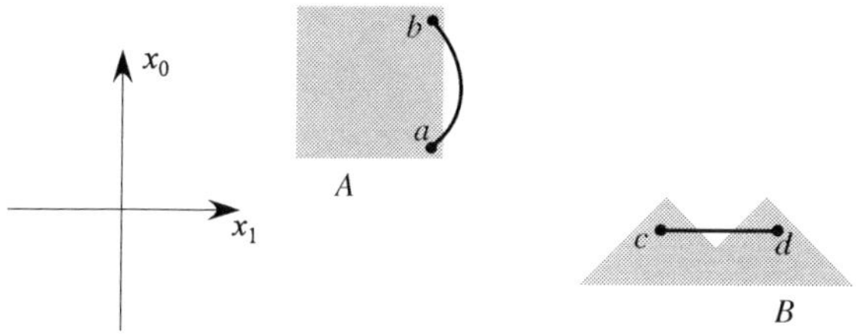

Fig. 2 For a given region C, consider the set $\mathsf{S}(C)$ of the curves which have both end points in C, but do not lie there entirely. Then, C is causally convex if and only if no curve of $\mathsf{S}(C)$ is timelike. Applying this to the grey regions of the Minkowski plane, we conclude that B is causally convex, but A is not, because $\mathsf{S}(A)$ contains, in particular, the curve (ab). Likewise, when M is convex, C is convex too, if and only if no curve of $\mathsf{S}(C)$ is geodesic. Therefore, A is convex, but B is not, because $\mathsf{S}(B) \ni (c, d)$

$<p, q>_M$ or, to put it differently, if

$$\forall p, q \in N \qquad <p, q>_N = <p, q>_M.$$

Clearly, N is causally convex, if and only if every non-closed inextendible timelike curve lying in M intersects N in a connected set, cf. [11]. Note also that changing $<p, q>_M$ to $\leqslant p, q \geqslant_M$ in the definition above, one obtains an equivalent definition. Indeed, consider a pair of points $p', q' \in N$ such that $p, q \in <p', q'>$ [existence of p' and q' is guaranteed by the fact that $I^+(p)$ and $I^+(q)$ are open]. It follows from (15) that any $r \in \leqslant p, q \geqslant$ is also in $<p', q'>$ and hence lies all the same in a causally convex [according to Definition 36] set N. And the converse implication

$$\leqslant p, q \geqslant_M \subset N \quad \Rightarrow \quad <p, q>_M \subset N$$

is obvious.

Convexity and causal convexity are defined by the behaviour of two quite different families of curves—geodesic and timelike. It is not surprising therefore that these two properties are unrelated. For example, if ℓ is a closed timelike curve in a spacetime M, it leaves a sufficiently small convex neighbourhood U of a point $p \in \ell$. So, U is convex, but not causally convex in M. And vice versa the subset $x_0 < |x_1|$ of the Minkowski plane is causally convex but not convex. Two examples more are depicted in Fig. 2.

Proposition 37 *It is easy to check that*

(a) *Any future (past) set is causally convex;*
(b) *If A and B are causally convex, then so is $A \cap B$;*
(c) *For any points p, q the set $<p, q>_M$ is causally convex (this follows from the previous two items);*

(*d*) *If* N *is causally convex, then so is any* $A \in \mathsf{R}(N)$ *(indeed, if* $p, q \in A$*, then* $< p, q >_A = < p, q >_N = < p, q >_M$*; the former equality being a consequence of* (16)*, and the latter—of the causal convexity of* N*, see the definition).*

The causality condition can be strengthened by prohibiting—in addition to *closed* causal curves—'almost closed' ones.

Definition 38 A set N is termed *strongly causal in* M, if each of its points has an arbitrarily small causally convex neighbourhood.

Comments 39 The expression 'an arbitrarily small such-and-such neighbourhood' is a—very convenient, we shall use it throughout the book—abbreviation for 'a such-and-such sub-neighbourhood of an arbitrary neighbourhood'.

Remark 40 Alternatively, one can define strong causality in $p \in M$ to be the following property: each neighbourhood $U \ni p$ contains a sub-neighbourhood V such that any non-spacelike curve with the end points in V lies entirely in U (see 14.10 in [141]). Now a set is called strongly causal if strong causality holds in each of its points. In fact this definition is equivalent to Definition 38. Indeed, if p has arbitrarily small causally convex neighbourhoods, then at least one of them lies in U. Such a neighbourhood satisfies the requirements for V. Conversely, assume strong causality in the sense of [141] holds in p and let U and V be the sets entering the definition. Consider the set $W = \bigcup_{r,q \in V} < r, q >$. Clearly, W is a neighbourhood of p and by construction it is causally convex. On the other hand, each $< r, q >$ by hypothesis lies entirely in U and hence $W \subset U$. This means that p has arbitrarily small causally convex neighbourhoods.

Remark 41 It should be stressed that strong causality is a property not of a spacetime $N \subset M$, but rather of the pair—N *and* its embedding in the ambient space M (technically, this follows from the fact that Definition 38 uses the notion of causal convexity and hence the sets $< p, q >_M$, not just $< p, q >_N$). So, if M_1 and M_2 are two different extensions of N, it may happen that N is a strongly causal subset of M_1, but not of M_2. Therefore, the expression 'a strongly causal spacetime N' is, strictly speaking, meaningless, cf. Remark 3. It is used only when M is evident, mostly when $N = M$ (in which case one speaks of *intrinsic strong causality*, see Definition 46). The same is with causal convexity.

Example 42 Let N be the rhombus $|x_0 \pm x_1| < 1$ with the metric

$$\eta : \qquad \mathrm{d}s^2 = -\mathrm{d}x_0^2 + \mathrm{d}x_1^2.$$

It is easy to check that (N, η) is a strongly causal subset of Minkowski plane $(\mathbb{R}^2, \eta)$. At the same time N can be extended to the cylinder (C_M, η)

$$C_M : \qquad x_{0,1} \in \mathbb{R}^1, \quad x_0 = x_0 + 3.$$

In the latter case, there is a closed non-spacelike curve $x_1 = \mathit{const}$ through every point of N. Thus, N is *not* a strongly causal subset of C_M.

The curves akin to 'almost closed' are those which fail to leave forever a *finite* region.

Definition 43 A future inextendible future-directed curve $\lambda\colon\ \mathbb{I} \to M$ is said to be *totally future imprisoned* in a set K, if for some $\xi_0 \in \mathbb{I}$

$$\lambda(\xi) \in K, \qquad \forall \xi > \xi_0.$$

If a weaker condition

$$\lambda\big|_{\xi > \xi_0} \cap K \neq \varnothing, \qquad \forall \xi_0 \in \mathbb{I},$$

holds, then $\lambda(x)$ is *partially future imprisoned* in K. Past imprisonment is defined dually.

In a strongly causal spacetime, neither type of imprisonment is possible, cf. Proposition [76, 6.4.7] and [141, 14.13], respectively.

Proposition 44 *A future (past) inextendible non-spacelike curve leaves any compact strongly causal set $\mathcal{N}$ never to return.*

Proof Let a future-directed (and automatically non-spacelike) curve $\lambda(s)\colon\ \mathbb{R}_+ \to \mathcal{N}$ be future inextendible. If λ is imprisoned in $\mathcal{N}$, then there is a sequence s_i such that

$$s_i \to \infty, \qquad \lambda(s_i) \to p \in \mathcal{N}. \tag{$*$}$$

By Definition 38, any neighbourhood U of the point p has a causally convex subneighbourhood $W \subset U$. As follows from $(*)$, all $\lambda(s_i)$ beginning from some s_{i_0} are in W. But λ is non-spacelike. Hence, all points of $\lambda\big|_{(s_i, s_{i+1})}$ also lie in the—causally convex—W. Thus, the entire λ beginning from some point lies in W and hence in U. So, p is an end point of λ, which contradicts the inextendibility of the latter. □

4 Global Hyperbolicity

In spite of all their remarkable properties, the strongly causal sets in the general case are still 'insufficiently good' and can have different undesirable pathologies (if, for example, one removes an arbitrary closed set from such a spacetime, it will remain strongly causal). That is why a more important role in general relativity is played by a narrower class of spacetimes.

Definition 45 A strongly causal set $N \subset M$ is called a *globally hyperbolic subset of* M, if for any $p,\ q \in N$ the set $\leqslant p, q \geqslant_M$ is compact and lies in N.

A simple example is provided by a pair of half-planes—$x_0 < 0$ and $x_1 < 0$—in $\mathbb{L}^2$. Only the former is a globally hyperbolic subset even though both are strongly causal.

Before discussing the exceptional simplicity and other merits of globally hyperbolic spacetimes, we have to dwell on a terminological problem (already mentioned above). The point is that global hyperbolicity, as well as strong causality, is a property of a *triple* (M, N, ϕ) and not of N alone, see Remark 41. However, the formally correct expression 'a globally hyperbolic subset of a given spacetime' is too cumbersome and is often shortened. In particular, the set N from Definition 45 is called just 'globally hyperbolic' (we shall also follow this tradition), which may lead to confusion. To avoid it, we introduce a special term [94].

Definition 46 A spacetime which is a globally hyperbolic (or strongly causal) subset of itself will be called *intrinsically* globally hyperbolic (respectively, intrinsically strongly causal).

A globally hyperbolic subset N of a spacetime M is always intrinsically globally hyperbolic. The converse, however, is not always true, as can be seen from Example 42, where the rhombus N being intrinsically globally hyperbolic is nevertheless a non-globally hyperbolic subset of the cylinder C_M. The situation is reversed if N is causally convex.

Proposition 47 *An intrinsically globally hyperbolic spacetime is globally hyperbolic, if and only if it is causally convex.*

Proof Let N be causally convex. Then, the compactness of $\leqslant p, q \geqslant_N$, where p, $q \in N$, implies the compactness of $\leqslant p, q \geqslant_M$ (which is merely the same set). Likewise, the intrinsic strong causality of N, due to the causal convexity thereof, implies the strong causality of N as a subset of M. These two implications prove the 'if' assertion.

Further, if N is *non*-causally convex, then there is a non-spacelike curve which starts at $p \in N$, leaves N, and returns back to terminate in some $q \in N$. Thus, $\leqslant p, q \geqslant_M \not\subset N$ and therefore N is not a globally hyperbolic subset of M. □

Proposition 48 ([76, Proposition 6.6.1]) *If $\mathcal{K}$ is a compact subset (a point, for example) of a globally hyperbolic region N, then both sets $J^{\pm}(\mathcal{K}) \cap N$ are closed in N (this property of N is called 'causal simplicity').*

Corollary 49 ([141, Lemma 14.22]) *Let M be a globally hyperbolic spacetime and let $p_i, q_i \in M$, $i = 1, 2 \dots$ are sequences converging to, respectively, p and q. Then the relation $\forall i \;\, p_i \preccurlyeq q_i$ implies $p \preccurlyeq q$.*

The importance of the intrinsically globally hyperbolic spaces stems mostly from the fact that physics in such spaces can be studied in the customary terms of the Cauchy problem.

Proposition 50 *A spacetime N is intrinsically globally hyperbolic if and only if it possesses a subset—called a* Cauchy surface*—which meets in a single point, every inextendible timelike curve in N.*

The 'if' part of this fundamental proposition is [141, Corollary 14.39]. And the 'only if' part is a half of Geroch's splitting theorem (see [76, Proposition 6.6.8]).

Warning. Our definitions of a Cauchy surface and (below) of a Cauchy domain are those adopted by O'Neil in [141]. Correspondingly, the former differs from that used by Hawking and Ellis in [76] or by Geroch and Horowitz in [64] (in particular, we do not require the surface to be spacelike), and the latter—from that adopted in [64], which is obtained from ours by replacing the word 'non-spacelike' with the word 'timelike', see Definition 53. As a result, Geroch and Horowitz's Cauchy domain is the closure of ours (cf. [76, Proposition 6.5.1]).

It turns out that the geometry of globally hyperbolic spacetimes is quite specific.

Proposition 51 *Let $\mathcal{S}$ be a Cauchy surface of a spacetime M. Then,*

(*a*) *$\mathcal{S}$ is an achronal connected closed topological (i.e. C^0) hypersurface;*
(*b*) *$\mathcal{S}$ is homeomorphic to any other Cauchy surface in M;*
(*c*) *$M = \mathbb{R}^1 \times \mathcal{S}$, with $\{t\} \times \mathcal{S}$ being a Cauchy surface for every t.*

The connectedness of $\mathcal{S}$ is proven in [141, Proposition 14.31], its achronality is a part of the definition, and the remainder of Proposition 51(*a*), cf. [141, Proposition 14.29], is a simple corollary of Proposition 31 [the achronality of $\mathcal{S}$ implies that $I^+(\mathcal{S})$ is disjoint with $I^-(\mathcal{S})$ and hence $\mathcal{S}$ is the boundary of the past set $I^-(\mathcal{S})$]. Finally, Proposition 51*(b)* is [141, Proposition 14.32], while Proposition 51*(c)* is the second half of Geroch's splitting theorem.

Though topologically all Cauchy surfaces are equivalent, they might differ *geometrically*. The existence of sufficiently regular ones is established by the following proposition (acausality—under the assumption of achronality—is proven for spacelike hypersurfaces in [141, Proposition 14.42]).

Proposition 52 ([12]) *In any globally hyperbolic spacetime M, there is a smooth spacelike acausal Cauchy surface $\mathcal{S}$ such that M is diffeomorphic to $\mathbb{R}^1 \times \mathcal{S}$.*

It follows from the last two propositions that in some—well-defined—sense the topology of a globally hyperbolic spacetimes does not change with time. Loosely speaking, a handle will never appear in a globally hyperbolic spacetime born simply connected.

Propositions 47 and 50 provide a simple way of building globally hyperbolic spaces: pick a subset S of a spacetime M and remove from M all inextendible timelike curves which do *not* meet S or meet it more than once. The remaining set N (if non-empty and open) will be a globally hyperbolic subset of M. Indeed, by construction S is a Cauchy surface in N, so N is intrinsically globally hyperbolic. At the same time, N is causally convex in M, because otherwise there would exist a non-spacelike curve λ such that the intersection $\lambda \cup N$ would have more than one connected component. But each of those components would have to meet S (by the definition of a Cauchy surface in N) in contradiction with the achronality thereof. Let us formulate this idea strictly.

Definition 53 *The Cauchy domain* of a set U—written $\mathcal{D}(U)$—is the set of all points p such that every inextendible non-spacelike curve through p meets U.

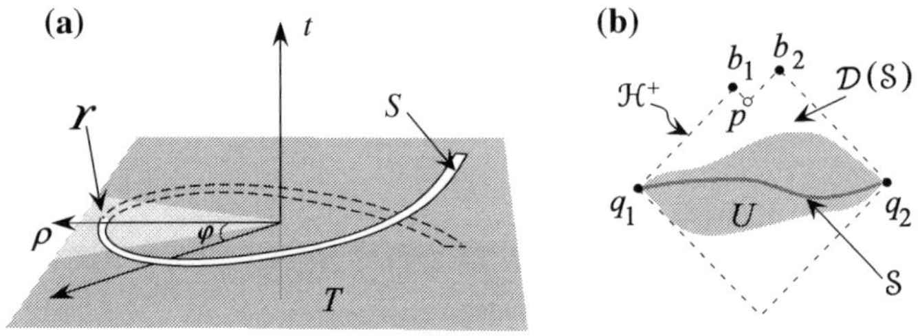

Fig. 3 **a** The strip S is spacelike, but not achronal: its upper and lower ends contain points which can be connected by timelike curves. The Cauchy domain of the strip in M' is intrinsically globally hyperbolic, but it is not a globally hyperbolic subset of M. **b** $\mathcal{H}^+$ is the future Cauchy horizon for the subset U (the grey region in the picture) of the Minkowski plane punctured in the point p. If the points q_1 and q_2 are removed too, the properties of $\mathcal{H}^+$ are governed by Proposition 56. b_1 and b_2 are points in which two generators meet

$\mathcal{D}(U)$ is never empty, because it always contains U. It is possible, however (even if U is achronal), that it contains nothing more: $\mathcal{D}(U) = U$. Such is the case, for example, when U is a null geodesic in the Minkowski space. The following proposition is a combination of Lemmas 14.42 and 14.43 from [141] with [76, Proposition 6.6.3].

Proposition 54 *If S is a spacelike achronal hypersurface in a spacetime M, then $\mathcal{D}(S)$ is an open globally hyperbolic subset of M. And if S is closed and achronal, then Int$\mathcal{D}(S)$ is globally hyperbolic, when non-empty.*

Note that if S is a Cauchy surface of a globally hyperbolic $N \subset M$, then $\mathcal{D}(S)$ does not need to coincide with N, though the inclusion $N \subset \mathcal{D}(S)$ is always true, of course. Thus, $\mathcal{D}(S)$ is the *maximal* (by inclusion) globally hyperbolic subset of M whose Cauchy surface is S.

Example 55 (*[94]*) Let S be the strip considered in Example 30, T be the plane $t = 0$, and Υ (shown in light grey in Fig. 3a) be a sector $\varphi \in (-c, c)$ in that plane. The constant c is chosen so that none of non-spacelike curves which have both end points in S meet Υ:

$$\Upsilon \cap J^-(S) \cap J^+(S) = \varnothing.$$

Denote by N the (clearly non-empty) Cauchy domain of S in the spacetime $M' \leftrightharpoons M - (T - \Upsilon)$. M' differs from M in that S is achronal in the former. Since S is also spacelike and achronal, N is a globally hyperbolic subset of M'. Therefore, N is intrinsically globally hyperbolic. At the same time, N is *not* a globally hyperbolic subset of M (not being causally convex there).

With every globally hyperbolic region $U \subset M$, one can put in correspondence a set $\mathcal{H}^+$ called the *future Cauchy horizon*:

$$\mathcal{H}^+(U) \leftrightharpoons \mathrm{Bd}\mathcal{D}(\mathcal{S}) - I^-(\mathcal{D}(\mathcal{S})), \quad \text{where } \mathcal{S} \text{ is a Cauchy surface of } U \qquad (18)$$

[$\mathcal{H}^+(U)$ will be written also as $\mathcal{H}^+(\mathcal{S})$, or simply $\mathcal{H}$]. The dually defined *past Cauchy horizon* is, naturally, denoted $\mathcal{H}^-$, while $\mathcal{H}$ is their union $\mathcal{H} = \mathcal{H}^+ \cup \mathcal{H}^-$. These definitions seem to differ from the standard ones (which at this moment are less convenient for us), see [141, Definition 14.49], but actually are equivalent to them. Note that in defining $\mathcal{H}^+$ both $\mathcal{S}$ and U are *auxiliary* in a sense. Instead of $\mathcal{S}$, one could take any other Cauchy surface of U and instead of U any other globally hyperbolic subset of M with the same Cauchy surface. $\mathcal{H}^+$ would not change.

Proposition 56 *Assume, the surface $\mathcal{S}$ entering definition* (18) *is closed. Then $\mathcal{H}^+$ is a closed achronal topological hypersurface and through each of its points there is a past inextendible null geodesic which lies entirely in $\mathcal{H}^+$.*

The proof can be found in [141, Proposition 14.53], for an illustration see Fig. 3b.

Definition 57 The just mentioned geodesic, unless it is a portion of a larger one of the same kind, is termed a *generator* of $\mathcal{H}^+$.

Corollary 58 *If two points of $\mathcal{H}^+(\mathcal{S})$, where $\mathcal{S}$ is closed, are causally related, they belong to the same generator. It follows, in particular, that a point common to two generators is the future end point of either.*

Proof Suppose $p, q \in \mathcal{H}^+$ are, respectively, the past and the future end points of a non-spacelike curve γ which is *not* a horizon generator. Then by Proposition 23(*d*) the broken line consisting of γ and the segment of a generator which connects p to some $r \in \mathcal{H}^+$, $r \preccurlyeq p$ (the existence of r is guaranteed by Proposition 56) can be deformed into a past-directed timelike curve from q to r. But this contradicts the achronality of $\mathcal{H}^+$.

Now suppose that generators α_1 and α_2 meet in a point b and one of them—let it be α_2 for definiteness—extends (in the future direction, obviously) beyond b to some $c \in \mathcal{H}^+$. Then any point $a \in \alpha_1$ can be connected to c by a non-spacelike curve which is not a horizon generator (specifically, by the segment of α_1 from a to b combined with the segment of α_2 from b to c). But we just have established that this is impossible. □

5 Perfectly Simple Spacetimes

Now we intend to relate the causal and the geodesic structures. That relation depends ultimately on Proposition 23(*b*) and the first result of this kind will be [145, Proposition 4.10].

Proposition 59 *Any convex spacetime is intrinsically strongly causal.*

Proof Let O be a convex spacetime and $W \subset O$ be a region with compact closure. To prove the proposition, we only have to demonstrate that W (however, 'small' it is) contains a causally convex neighbourhood of any $q \in W$.

Pick two sequences $p_m, r_m \in W$, $m = 1, 2 \ldots$ such that

$$p_i, r_i \xrightarrow[i\to\infty]{} q, \qquad q \in <p_m, r_m>, \quad \forall m$$

and consider a timelike curve from p_m to r_m. To derive a contradiction, suppose it leaves W. Then there must be a point

$$x_m \in (<p_m, r_m> \cap \operatorname{Bd}W)$$

and [as follows from Proposition 23(*b*)] two geodesics: a future-directed $\lambda_{x_m r_m}$ from x_m to r_m and a past-directed $\lambda_{x_m p_m}$ from x_m to p_m. If such x_m existed for infinitely many m, there would exist a subsequence $\{x_i\}$ converging to some $x \in \operatorname{Bd}W$ and sequences $\{\lambda_{x_j r_j}\}$ and $\{\lambda_{x_j p_j}\}$ converging to the same (non-zero, because $q \notin \operatorname{Bd}W$) geodesic λ_{xq}, which thus would be at the same time future- and past-directed: a contradiction. □

Remark 60 Generally speaking, convergence of curves is quite a subtle matter, see [11], but, fortunately, not in the case of geodesics in a convex spacetime. Choose a normal coordinate system in W and for each $\lambda_{x_m r_m}$ fix its affine parameter τ by the requirement that $v^0_{(m)} = -1$, where $\boldsymbol{v}_{(m)} = \partial_\tau(x_m)$ and the components are found in the coordinate basis. Being thus normalized the set of all pairs (x, v^a) is obviously compact[7], and hence there exists a vector $\boldsymbol{v}(x)$ to which $\{\boldsymbol{v}_{(j)}\}$ converge componentwise. That vector must be tangent to λ_{xq} (which must exist by the convexity of M) because geodesics are solutions of the relevant system of ordinary differential equations and therefore depend smoothly on the initial conditions. So, λ_{xq} is, indeed, non-spacelike and future-directed.

The utility of Proposition 59 is restricted by the following fact. Strongly causal sets, as mentioned above, can be quite pathological. In particular, they do not need to be convex. Thus, the Whitehead theorem guarantees the existence of arbitrarily small (in the sense of Comment 39) convex neighbourhoods of any point. Proposition 59 in turn guarantees that each of those neighbourhoods contains a strongly causal sub-neighbourhood. But it is not clear whether a given point has a neighbourhood which is strongly causal and convex *at the same time*. Our next goal is to show, see Theorem 65, that such a neighbourhood (and even a 'better' one) does always exist [97].

Definition 61 A spacetime V is called *perfectly simple* if each $A \in \mathsf{R}(V)$ (see p. 17 for notation) (1) is convex, (2) is intrinsically globally hyperbolic and (3) with any two points p, q contains also a pair of points r, s such that $p, q \in <r, s>_A$.

Remarks 62 (*a*). Perfect simplicity, in contrast to simplicity, is an intrinsic property: it does not depend on how (if at all) V is embedded in a larger spacetime. (*b*). It follows from (17) that if V is perfectly simple, then so is every $A \in \mathsf{R}(V)$. (*c*).

[7] In the tangent bundle $T(M)$, of course.

For any compact subset $\mathcal{K}$ of a perfectly simple V, there are points r, s such that $\mathcal{K} \subset \lt r, s\gt_V$.

Lemma 63 *Any spacetime point has an arbitrarily small simple neighbourhood O such that for every $p \in O$ the sets $I^{\pm}_O(p)$ are also simple.*

Let O' be a simple neighbourhood of a point s, and $\boldsymbol{e}_{(i)}$, $i = 0\dots3$ be an orthonormal tetrad in O', that is a set of four smooth vector fields subject to the relations

$$e_{(m)}{}^a e_{(k)a} = \delta_{mk} - 2\delta_{m0}\delta_{k0}$$

in each point (due to the normality of O' such fields always exist [141]). For every point $p \in O'$, the corresponding tetrad defines a basis—it is the set $\{\boldsymbol{e}_{(n)}(p)\}$—in T_p and, consequently, a normal coordinate system, see Sect. 2.2. The coordinates of a point r in that system will be denoted by $X^a(p; r)$. The functions $X^a(p; r)$ vanish at $r = p$ and depend smoothly on r and p. So, for any positive number δ there is a simple sub-neighbourhood $O_\delta \ni s$ such that the coordinates of all of its points are bounded by δ:

$$\overline{O_\delta} \subset O' \qquad \text{and} \qquad |X^a(p; r)| < \delta \quad \forall\, p, r \in O_\delta, \qquad a = 0\dots3. \tag{19}$$

Thus, the lemma will be proven once we demonstrate that at some δ it is true irrespective of p that

> a geodesic segment with both end points in $I^+_{O_\delta}(p)$ lies in that set entirely $\quad (\star)$

(such a segment exists and is unique in O' owing to the normality thereof). Indeed, this would prove that $I^+_{O_\delta}(p)$ (for the sake of definiteness from now on we discuss $I^+_{O_\delta}(p)$; the properties of $I^-_{O_\delta}(p)$ are obviously the same) is convex and even (being a subset of the simple set O_δ) simple. So, O_δ possesses all properties of the sought-for O.

Proof of the Lemma Pick a δ. In the corresponding O_δ, see (19), choose a normal coordinate system with the origin at $p \in O_\delta$ and define the function $\sigma : O_\delta \to \mathbb{R}$ by formula (10).

Now suppose $(\star)$ is false. Then there is a geodesic segment $\mu(\xi)$, $\xi \in [0, 1]$ such that

$$\mu(0), \mu(1) \in O_\delta, \qquad \mu \not\subset O_\delta,$$

see Fig. 4. By Corollary 24, this implies $\sigma < 0$ in the end points of μ and $\sigma \geqslant 0$ in some inner point. Clearly, this is impossible when μ is sufficiently short and our strategy will be to show that it remains impossible as long as μ fits in O_δ with sufficiently small δ. To this end, we shall consider a homotopy (with one fixed end) interpolating between a 'sufficiently short' μ and that under discussion. We focus upon the point r_0 at which σ ceases to be negative for the first time. At r_0 the geodesic

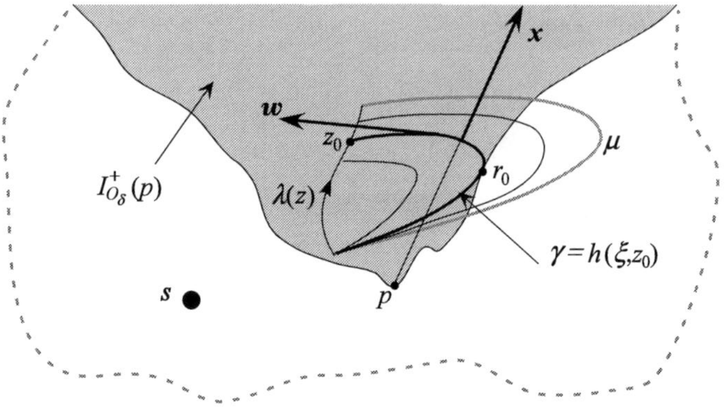

Fig. 4 The situation which, as we argue, is impossible at a sufficiently small δ. The dashed line is the boundary of the region O_δ. Both ends of μ are in the future of p, but are not causally related to each other

μ, as we shall see, must touch the cone $\mathrm{Bd}I^-_{O_\delta}(p)$ from inside, but the cone is 'too convex' for that.

Let us connect $\mu(0)$ to $\mu(1)$ by a curve $\lambda(z)$, $z \in [0, 1]$ lying entirely (in contrast to μ) in $I^+_{O_\delta}(p)$, see Fig. 4, and define a surface $h(\xi, z) \subset O_\delta$ by the equalities

$$h(\xi, 0) = \mu(0), \qquad h(1, z) = \lambda(z)$$

and the requirement that the curve $h(\xi)$ at each (constant) $z \neq 0$ be an affinely parameterized geodesic. Consider the maximal value of σ on each of those geodesics. Since σ is smooth, the function $\sigma_m(z) \leftrightharpoons \max_\xi \sigma[h(\xi, z)]$ must be continuous. But $\sigma_m(0) < 0$ and $\sigma_m(1) \geqslant 0$, so $\sigma_m(z_0) = 0$ at some z_0. Thus, one can see that the violation of ($\star$) would imply the existence of a geodesic segment $\gamma(\xi) \leftrightharpoons h(\xi, z_0)$ such that the function $\bar\sigma(\xi) \leftrightharpoons \sigma \circ \gamma$ is negative on the boundary of its domain but has a maximum $\bar\sigma(\xi_0) = 0$.

Now to prove the lemma it would suffice to show the non-existence—at sufficiently small δ—of such $\gamma(\xi)$, which we shall do by establishing the implication

$$\bar\sigma(\xi_0) = 0,\ \bar\sigma'(\xi_0) = 0 \quad\Rightarrow\quad \bar\sigma''(\xi_0) > 0. \qquad (\star\star)$$

(it says that a geodesic in O_δ can touch the null cone only from outside). In doing so, we shall regard $\gamma(\xi)$ spacelike

$$w^a w_a = 1, \qquad \boldsymbol{w} \leftrightharpoons \partial_\xi, \qquad (20)$$

because for timelike geodesics ($\star$) and, in particular, ($\star\star$) follow directly from the definition of $I^+_{O_\delta}(p)$ and for null ones—from the same definition in combination with Proposition 23(d). For future use, it is also convenient to write down the first two derivatives of σ in terms of $\boldsymbol{w}$:

$$\bar{\sigma}' = \sigma,_a w^a = 2x_a w^a, \qquad \bar{\sigma}'' = (2x_b w^b),_a w^a = 2x_{b:a} w^b w^a \tag{21}$$

(in the first chain of equalities, relation (11) is used and in the second chain—the fact that γ is a geodesic).

Consider now a point $r_0 \leftrightharpoons \gamma(\xi_0)$ in which γ is tangent to the future null cone of p, that is in which the left-hand side of (⋆⋆) holds. Denote by $T_\perp$ the subspace of T_{r_0} orthogonal to the position vector[8] $\boldsymbol{x}_0 \leftrightharpoons \boldsymbol{x}(r_0)$. By the definition of σ the equality $\bar{\sigma}(\xi_0) = 0$ implies that $\boldsymbol{x}_0$ is null and, correspondingly $\boldsymbol{x}_0 \in T_\perp$. We use this fact to construct a basis $\{\boldsymbol{x}_0, \boldsymbol{d}_{(1)}, \boldsymbol{d}_{(2)}\}$ in $T_\perp$ defined (non-uniquely, of course) by the following relations:

$$\boldsymbol{d}_{(1,2)} \in T_\perp, \quad d_{(i)}{}^a d_{(j)a} = \delta_{ij}, \quad d_{(i)}{}^a e_{(0)a} = 0 \qquad i, j = 1, 2. \tag{22}$$

[we have chosen in addition to $\boldsymbol{x}_0$ a pair of unit spacelike vectors which are orthogonal to each other, to $\boldsymbol{x}_0$, and to $\boldsymbol{e}_{(0)}(r_0)$]. Since $\bar{\sigma}'(\xi_0) = 0$, it follows from the first equation in (21) that $\boldsymbol{w}(\xi_0) \in T_\perp$ and so can be decomposed as

$$\boldsymbol{w}(\xi_0) = w^{(1)}\boldsymbol{d}_{(1)} + w^{(2)}\boldsymbol{d}_{(2)} + w^{(3)}\boldsymbol{x}_0, \qquad \text{where} \quad {w^{(1)}}^2 + {w^{(2)}}^2 = 1 \tag{23}$$

[the second equality follows from the normalization condition (20)].

Our next step is to estimate $\bar{\sigma}''$ using (23). To this end, substitute this equation into (21) and note that the term $w^{(3)}\boldsymbol{x}_0$ does not contribute to $\bar{\sigma}''$. Indeed, the contribution is $(x_{b:a}x^a y_1^b + x_{b:a}x^b y_2^a)\big|_{r_0}$, where $\boldsymbol{y}_{1,2} \in T_\perp$. But $x_{b:a}x^a$ is proportional to x_b (since $\boldsymbol{x}$ is tangent to a radial geodesic), and hence the whole term $x_{b:a}x^a y_1^b$ is proportional to $x_{0b} y_1^b = 0$. The second term, in view of (11), vanishes too: $\frac{1}{2}(x_b x^b)_{:a} y_2^a = \frac{1}{4}\sigma,_a y_2^a = \frac{1}{2}x_a y_2^a = 0$. Thus, in r_0

$$\bar{\sigma}''(r_0) = 2w^{(i)}w^{(j)}d^b_{(i)}d^a_{(j)}x_{b:a} = 2w^{(i)}w^{(j)}d^b_{(i)}d^a_{(j)}(g_{ab} + \Gamma_{b.ac}x^c), \tag{24}$$

where $\Gamma_{b.ac} \leftrightharpoons \frac{1}{2}(g_{ab,c} + g_{bc,a} - g_{ac,b})$, and an obvious equality $x^c{}_{,d} = \delta^c_d$ is used. Now, taking into consideration (19), (22) and (23) one gets

$$|\bar{\sigma}''(r_0) - 2| \leqslant 4\Gamma\delta, \qquad \Gamma \leftrightharpoons \Big|w^{(i)}w^{(j)}d^b_{(i)}d^a_{(j)} \sup_{\substack{b,a,c \\ r,p \in O'}} \Gamma_{b.ac}\Big|.$$

But $w^{(i)}$ are bounded, see (23), as well as $d^b_{(i)}$ (since these are components of unit vectors in the space $\mathbb{E}^3$ spanned by the unit vectors $\{\mathbf{e}_{(k)}\}$ $k = 1, 2, 3$, and therefore are bounded being continuous functions on the compact $\overline{O_\delta}$). Thus, Γ is finite and hence $\bar{\sigma}''(r_0)$ is positive, if δ was chosen sufficiently small. This proves (⋆⋆) and, consequently, the whole lemma. □

---

[8]Whenever a coordinate-dependent entities—such as σ, position vector or Christoffel symbols (below)—is mentioned in this proof, it is understood that the coordinates $X^a(p; r)$ are used.

Corollary 64 *Any point s of a strongly causal spacetime has an arbitrarily small convex causally convex neighbourhood with compact closure.*[9]

Proof It is easy to check that all the listed properties are possessed by sets of the form $V \leftrightharpoons <pr>_O$, where $p \in I^-_O(s)$, $r \in I^+_O(s)$ and O, which is the neighbourhood appearing in the lemma, is chosen so as to lie inside a causally convex neighbourhood of s. Indeed, V is convex by Lemma 63 and Proposition 13, while its causal convexity follows immediately from its definition, when (16) is taken into account. □

Theorem 65 *Every spacetime point has an arbitrarily small perfectly simple neighbourhood with compact closure.*

Proof Let O and V be that from Corollary 64 (this time, though, we do not require M to be strongly causal) and let A be a spacetime from $\mathsf{R}(V)$. Clearly, A is a (connected, of course) intersection of two simple (by Lemma 63) sets. Consequently, by Proposition 13, it is simple too. This means in particular that, by Proposition 59, A is intrinsically strongly causal. On the other hand, for any $a, b \in A$ the set $\leqslant a, b \geqslant_A$ is compact, as can be seen from the fact that by (16)

$$\leqslant a, b \geqslant_A = \leqslant a, b \geqslant_O,$$

the right-hand side of which is a closed [see (14)] subset of the compactum $\overline{O}$. Thus, A is intrinsically globally hyperbolic. And, finally, the fulfilment of the requirement 3 in Definition 61 is also evident. So, V is perfectly simple. □

To exemplify the utility of the just proven theorem, we formulate the following test (it will be needed later).

Test 66 If a spacetime M' is an extension of a spacetime M, then the latter contains inextendible geodesics which are extendible in the former, cf. Exercise 5.15c in [141]. There are (infinitely many if $n > 2$) such geodesics of each of the three characters—null, timelike and spacelike.

Proof The proof consists in the construction of a null geodesic segment γ which starts in M and ends in $(M' - M)$. The existence of this geodesic would prove our assertion because
(a) the geodesic $\tilde{\gamma} \cap M$, where $\tilde{\gamma}$ is the maximal extension of γ, has the required properties (it is extendible in M', but not in M);
(b) the same reasoning applies to every geodesic segment that starts at the same point as γ and has an initial velocity sufficiently close to that of γ. Among such segments, there are infinitely many timelike, spacelike and (at $n > 2$) null ones.

Consider a perfectly simple neighbourhood O of a point $p \in \mathrm{Bd}_{M'} M$ and a point $r \in I^+_O(p)$. Let for definiteness $r \in M$ [the case $r \in (M' - M)$ is similar, one only

---

[9]In [76], such neighbourhoods are called *local causality neighbourhoods* and this corollary is accepted without proof. Note in this connection that in [145] by 'local causality neighbourhoods' *different* (not necessarily convex) sets are meant.

has to replace the point p by some $p' \in I^-_O(r) \cap M$]. Then, all we need is to find a point d that would be connected by null geodesic segments (dr) and (dp) with, respectively, r and p: clearly, either (dr) or (dp) has its end points one in M and the other in M' and thus can serve as γ.

Let μ be a past-directed null geodesic trough r. Since $\langle r, p \rangle_O$ is compact (recall that O is intrinsically globally hyperbolic), μ must leave it by Proposition 44. Hence, a point $x \in (\mu \cap O)$ can be found such that the geodesic ν_{px} connecting p with x is spacelike. At the same time, the geodesic ν_{pr} is timelike. Hence, by continuity there must be a point in μ such that the geodesic which connects it to p is null. It is this point that we take as d. □

6 'Cutting and Pasting' Spacetimes

In this section, we consider in detail a convenient and pictorial way of describing spacetimes: one spacetime is represented as a result of some surgery applied to another.

6.1 Gluing Spacetimes Together

Let (N_1, g_1) and (N_2, g_2) be spacetimes such that two of their regions $U_i \subsetneq N_i$, $i = 1, 2$ are related by a time-orientation-preserving isometry ψ. These spacetimes can be *glued along U* or, equivalently, *glued by the isometry* ψ, see Fig. 5a, which means that they are combined into a single spacetime (M, g) as follows. Consider the map π—called *natural projection*—which leaves all points of $N_i - U_i$, $i = 1, 2$ intact and identifies the points of each pair p_1, p_2, where $p_1 \in U_1$, $p_2 = \psi(p_1)$. It is the image $\pi(N_1 \cup N_2)$ endowed with the quotient topology that we take as M:

$$M \leftrightharpoons N_1 \cup_\psi N_2. \tag{25}$$

The restrictions of π to N_i will be denoted by π_i and the images $\pi_i(N_i) = \pi(N_i)$ by M_i (we speak of *images*, because according to our convention $M_i \neq N_i$, see Remark 3). Thus,

$$M = M_1 \cup M_2$$

(note the disappearance of the subscript ψ).

The projection π induces both a smoothness and a smooth metric on M (it is to ensure that π_1 and π_2 induce the *same* metric that we required ψ to be an isometry),

Terminology 67 Whenever we shall mention identification (joining, gluing together, etc.) of isometric open sets, we shall imply that the smoothness and the metric on the resulting space are those induced by the natural projection.

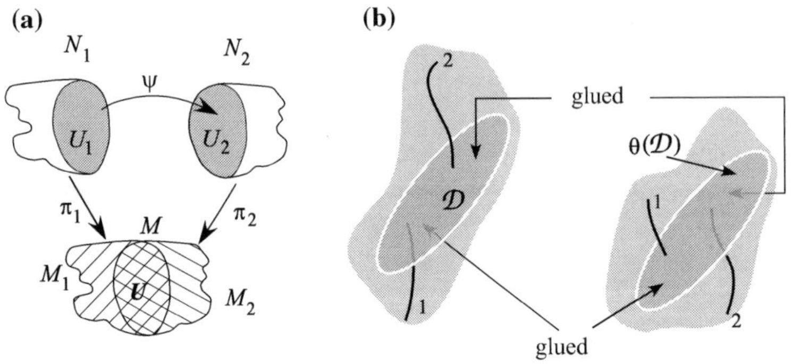

Fig. 5 **a** M is the result of 'gluing N_2 to N_1 by the isometry ψ'. **b** The spacetimes are joined so that curves 1 and 2 are continuous. The white circles represent the—missing—edges of discs Ξ and $\theta(\Xi)$

Thus, M is a smooth connected pseudo-Riemannian time-oriented (owing to the connectedness of U_i) manifold. However, it does not have to be a spacetime: we do not know whether it is Hausdorff.

Test 68 M is a spacetime (and thus an extension of either of N_i), if and only if for any converging sequence of points $b_k \in U_1$, $k = 1, 2\ldots$ the following implication holds:

$$\psi(b_k) \to p \qquad \Rightarrow \qquad p \in U_2. \tag{26}$$

In other words, M is *not* a spacetime, if and only if there is a converging sequence in N_1 such that its image in N_2 has a limit point on the *boundary* of U_2.

The straightforward proof consists in elementary but tedious search through variants. A simpler way is to represent M as the quotient space $N/\overset{\psi}{\sim}$, where $N \leftrightharpoons N_1 \cup N_2$, and $\overset{\psi}{\sim}$ is a (clearly, open) equivalence relation

$$p \overset{\psi}{\sim} q \quad \Leftrightarrow \quad p = q, \psi(q), \text{ or } \psi^{-1}(q).$$

The closedness in $N \times N$ of the graph of this relation is equivalent to the validity of the implication (26). Therefore, the proof of our assertion follows from Proposition [186, III 1.6].

Corollary 69 *If $Cl_{N_i} U_i$ is compact, then the result of gluing together N_i along U is* not *a spacetime [indeed in this case* any *sequence converging to a point on the boundary of U_1 provides a counterexample to* (26)*].*

Any extension M of a spacetime N_1 has the form (25).

Corollary 70 *Any compact spacetime is maximal.*

By definition, spacetimes are Hausdorff and pseudo-Riemannian. Note that the former property is essential to the corollary, but the latter is not.

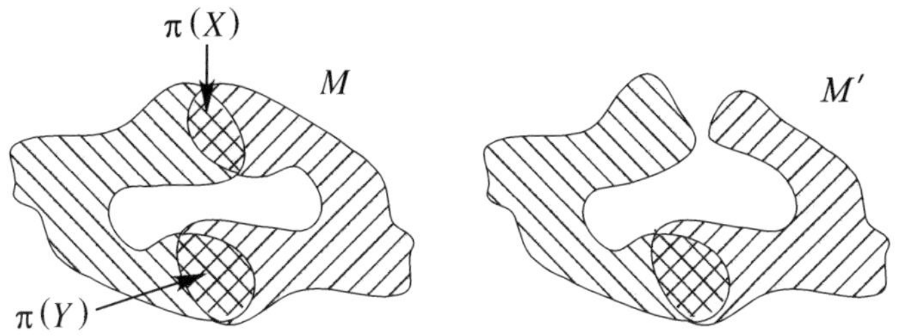

Fig. 6 M' is the result of ungluing M along $\pi(X)$

6.2 Ungluing Spacetimes

Consider a spacetime M of the form $M = N_1 \cup_\psi N_2$. This time, however, in contrast to the case considered in the previous subsection, U_1 is not connected: it is a union of two disjoint non-empty open sets X and Y, see Fig. 6. Define a new spacetime

$$M' \leftrightharpoons N_1 \cup_{\psi_Y} N_2,$$

where ψ_Y is the restriction of ψ to Y.

Assertion 71 M' is a spacetime.

The just described procedure of obtaining M' from M will be called [*partial*][10] *ungluing* the latter [along $\pi(X)$]. Note that while the local properties are the same in isometric regions, the *global* properties of M and M' (such as causality, extendibility, etc.) may differ considerably.

Not any spacetime can be unglued. It is easy to see, for example, that two curves, one of which connects a point $p \in Y$ to $q \in X$ and the other—$\psi(p)$ to $\psi(q)$—are projected by π to a pair of curves that have the same end points, but that are not fixed-end-point homotopic. So, to be partially unglued a spacetime must be non-simply connected. There is, however, a quite similar surgery which can be applied to *any* spacetime. Let us consider a simple example.

Denote by B, Ξ, $\mathfrak{D}$, and $B^\pm$, respectively, a coordinate ball, its equator, the $(n-1)$-dimensional disc bounded by Ξ and the halves of the balls separated by $\mathfrak{D}$. To put it differently, these are the sets that are defined, in some coordinates $\{z_j\}$, $j = 1, \dots n$ covering the whole $\overline{B}$, by the following expressions:

$$B \leftrightharpoons \{p \in M\colon\ z_1^2(p) + z_2^2(p) + \dots z_n^2(p) < 1\},$$
$$B^\pm \leftrightharpoons \{p \in B\colon\ z_1(p) \gtrless 0\}, \qquad \mathfrak{D} \leftrightharpoons \{p \in B\colon\ z_1 = 0\}, \qquad \Xi \leftrightharpoons \overline{\mathfrak{D}} - \mathfrak{D}.$$

The space $M - \Xi$ is non-simply connected and can be easily unglued along B^-: it suffices to take $M - \overline{\mathfrak{D}}$, B, B^+ and B^- as, respectively, M_1, M_2, $\psi(X)$ and $\psi(Y)$,

[10]It would be complete, if Y were empty and M' were the union of two disjoint spacetimes.

that is to represent $M - \Xi$ as the result of filling with B the slit obtained by deleting the disc $\overline{\mathcal{D}}$. The resulting space is used as an intermediate in constructing maximal spacetimes such as 'dihedral wormholes' [172], 'Deutsch–Politzer time machines', see below, 'stringlike singularities' [101] etc., which are built as follows.

Let θ be an isometry such that

$$\theta(B) \in M, \qquad \overline{B} \cap \overline{\theta(B)} = \varnothing,$$

where $B \in M$ is the aforementioned ball. Remove from M both Ξ and $\theta(\Xi)$. Then, unglue the remaining spacetime along B^- and along $\theta(B^+)$. Finally, use θ to glue $B^\pm$ and $\theta(B^\mp)$ together. By Test 68 the result is a spacetime. To visualize it, remove two discs from M, see Fig. 5b, and glue the left edge of either slit to the right edge of the other.

Remark 72 Throughout the book, we shall speak of gluing together surfaces, not regions bounded by them, because this language is more traditional (it is used, for example, in complex analysis, in visualizing Riemann surfaces). Note, however, that it may be somewhat misleading. In particular, it may happen that there are non-equivalent ways of identifying the surfaces, see examples below and in [101]: B can be mapped to $\theta(B)$ and $\mathcal{D}$—to $\theta(\mathcal{D})$ by more than one isometry.

Example 73 (Deutsch–Politzer (DP) spacetime [34, 150]) Choose M to be a Minkowski space $\mathbb{L}^n$, that is $\mathbb{R}^n$ with a flat metric η on it

$$\eta: \qquad ds^2 = -dx_0^2 + dx_1^2 + \cdots + dx_{n-1}^2.$$

The Deutsch–Politzer space (M_{DP}, η) is obtained from it by, first, making two 'horizontal' (i.e. spacelike) cuts and then gluing the upper bank of either slit to the lower bank of the other, see Fig. 7a. More accurately, it is described as the result of the surgery under discussion with B, $\mathcal{D}$ and θ being, respectively, a coordinate ellipsoid

$$4(x_0 + 1)^2 + \sum_{j=1}^{n-1} x_j^2 = 1$$

(depicted by the lower grey ellipse in Fig. 7a), the disc $\sum_{j=1}^{n-1} x_j^2 < 1$, $x_0 = -1$, and the translation $x_0 \mapsto x_0 + 2$.

For future use, note that the DP space is an extension of the region $\mathbb{L}^n - \overline{\mathcal{D}} - \theta(\overline{\mathcal{D}})$; therefore, one can use there the same coordinates $\{x_k\}$ as in the initial Minkowski space keeping in mind, though, that now they do not cover the *whole* manifold, that is, their values are restricted by the following conditions:

$$x_k \in \mathbb{R} \quad \text{at } x_0 \neq \pm 1, \qquad \sum_{k=1}^{n-1} x_k^2 > 1 \quad \text{at } x_0 = \pm 1.$$

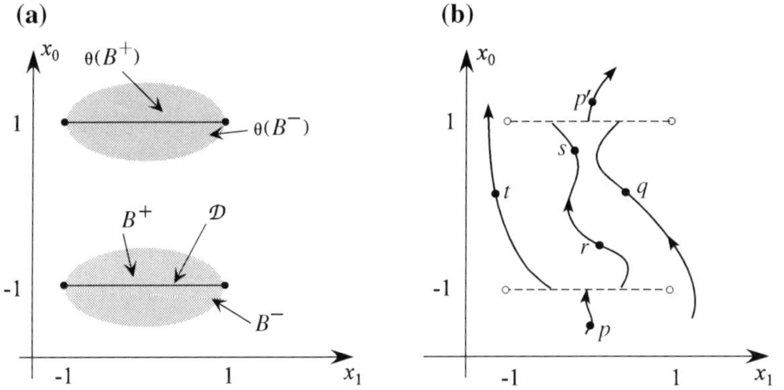

Fig. 7 **a** The initial Minkowski plane. **b** The (twisted) Deutsch–Politzer space as it looks in the coordinates inherited from the Minkowski plane. Some smooth curves *look* discontinuous: after reaching a 'seam' (a former cut shown by a dashed line) such a curve continues from the other—seemingly distant—bank of the cut. Actually, however, the arrowed segments make two, if the depicted space is $M_{\rm DP}$, or three, in the case of $M_{\rm TDP}$, continuous curves. These are (p, p') and $(qrst)$ in the former case and (p, p'), (qt) and—closed—(rs) in the latter

The sets *not* covered by the coordinates $\{x_k\}$ are shown by the dashed lines in Fig. 7b. These are the—glued together in pairs—former banks of the cuts. It is convenient to formulate the difference between the initial Minkowski space and the DP space obtained from it as a difference in rules determining which curves are smooth, see Fig. 7b.

Example 74 (*Twisted DP space*) There are *many* isometries in the Minkowski space sending B to a region disjoint with it. One can choose, for example, θ to be a superposition of the translation and the reflection $x_i \mapsto -x_i$, where $i = 1$ in the two-dimensional case and $i = i_1, i_2 \neq 0$ when $n > 2$. Now before gluing the inner edges together, one of them is rotated by 180°. The thus obtained spacetime $M_{\rm TDP}$, see Fig. 7 and the beginning of Sect. 3.2 in Chap. 6, will be called the *twisted* DP space. In the two-dimensional case such a space is necessarily non-orientable (though time-orientable, of course), but in four dimensions it is orientable, even though its subset (x_0, x_{i_1}) is a two-dimensional DP space.

A remarkable property of M is the 'lack of some points' (for example, those denoted by the white circles in Fig. 5). Indeed, we started the construction of $M_{\rm DP}$ from deleting the points of Ξ. And we cannot return them back (lest the Hausdorff condition be violated, see Test 68), that is, we cannot attach them as end points to curves which initially terminated at Ξ. Thus, $M_{\rm DP}$ is singular.[11] Whether such singularities exist in reality is still an open question (we shall return to it in Sect. 3 in Chap. 2). But anyway it should be stressed that, contrary to a widespread opinion, the surgery used for building $M_{\rm DP}$ is merely a convenient pictorial *description* of $M_{\rm DP}$, by no

[11] There are different definitions of singularity, see [62], for example. In this case, apparently the difference is immaterial.

means it indicates any deficiency of the spacetime. It *is* 'constructed by cutting and gluing together decent space-times' [43], but this is not a property of a particular spacetime—exactly the same can be said of *any* spacetime.

One more way of obtaining a spacetime from another one—denote it M—is the transition from M to its covering $\tilde{M}$ (which is meaningful, of course, only when M is non-simply connected). The reverse procedure is taking the quotient of a spacetime $\tilde{M}$ by a group of isometries G acting on it [17].

Consider again the equivalence relation

$$p \overset{G}{\sim} q \iff p = \varsigma(q), \quad \varsigma \in G.$$

Suppose, G is properly discontinuous and acts freely on $\tilde{M}$, i.e. (the following definition is that used in [141]):
(1) Each point of $\tilde{M}$ has a neighbourhood V such that for any $\varsigma \in G,\ \varsigma \neq \mathrm{id}$

$$\varsigma(V) \cap V = \varnothing, \tag{27a}$$

(2) Points $p, p' \in \tilde{M},\ p \overset{G}{\not\sim} p'$ have neighbourhoods $V \ni p$ and $V' \ni p'$ such that

$$\varsigma(V) \cap V' = \varnothing, \quad \forall \varsigma \in G. \tag{27b}$$

Then, $M \leftrightharpoons \tilde{M}/\overset{G}{\sim}$ (sometimes $\tilde{M}/G$ is written instead of $\tilde{M}/\overset{G}{\sim}$) is a spacetime (the smoothness and the metric are again understood to be inherited from $\tilde{M}$, cf. Corollary 7.12 in [141]) and the natural projection $\pi \colon \tilde{M} \to M$ is a covering.

The same spacetime can be obtained by the 'cut-and-paste' surgery. To this end, pick a fundamental region $\mathcal{F}$ of the group G and in a neighbourhood $U \supset \mathcal{F}$ identify the regions related by isometries from G. If, for example, $\tilde{M}$ is the Minkowski space and G is generated by the translation $x_0 \mapsto x_0 + 3$, then $\tilde{M}/\overset{G}{\sim}$ is the cylinder C_M from Example 42. The same cylinder is the result of gluing together the regions $x_0 \in (-1, 0)$ and $x_0 \in (2, 3)$ in the strip $x_0 \in (-1, 3)$. A more complex example (the Misner space) will be considered in Sect. 2 in Chap. 4.

Chapter 2
Physical Predilections

The discussion in this book is based on classical general relativity. But this science is still young, each of its components—from the action to the smoothness requirements—being questioned now and then. So, it is important that there is almost a consensus on the validity of a few meta principles that restrict the arbitrariness of possible modifications. Sometimes these principles are explicitly formulated—differently by each author—but much more often they manifest themselves in casual remarks like, 'This violates causality and is therefore impossible'. Perhaps, a too thorough analysis of these principles would be redundant, but on the other hand, in discussing the subject matter of this book the traditional half-poetic approach is definitely insufficient.

Our aim in this chapter is to briefly clarify what can be called 'locality', 'causality', etc., and what kind of prohibitions are associated with these concepts. Unfortunately, the mentioned terms are so heavily burdened with their rich prehistory that we have to take a radical approach. Specifically, instead of comparing the already proposed formulations and adopting the most suitable one, we analyze the matter ab ovo. Meanwhile, the reader is requested to keep in mind that this is a *theoretical physics* and *not a philosophy* book.

1 Locality

One of important tasks encountered within relativity is the comparison of different worlds. The worlds are assumed to differ *geometrically* only, the laws governing their matter content being the same. A question that immediately arises is how to formulate the relevant condition rigorously, how to capture the idea of sameness? Fortunately, this is possible in the most important—though special—case of *local geometrical* laws. The emphasized words are understood as follows. According to a universally accepted view only 'coordinate independent' quantities are physically meaningful.

S. Krasnikov, *Back-in-Time and Faster-than-Light Travel in General Relativity*,
Fundamental Theories of Physics 193, https://doi.org/10.1007/978-3-319-72754-7_2

Fig. 1 Vertices $V_{2,3}$ are related to V_1 by Lorentz transformations

Fields, particles and their evolution are described in terms of mathematical objects (e. g. tensors or spinors) that transform in a certain prescribed way in response to coordinate transformation. Thus any isometry $\psi\colon\ U \to U'$ induces a transformation $f \to f'$ of fields in U and we call *geometric*[1] laws of motion such that a field f' satisfies them in the spacetime U' if and only if so does f in U. We shall regard this property as obviously mandatory and consider no exceptions.

Further, we call a law of motion (of some field, say) *local* if (1) the question of whether this law holds at a particular *point* is meaningful and (2) the answer is fully determined by the values or the field in an arbitrarily small neighbourhood of the point (the rigorous definition is given in the next subsection). Note that locality is something totally different from causality. By saying that a pair of events p and q are related by a local interaction one does not imply that p is a cause (or an effect) of q or, say, that p and q cannot happen simultaneously. One only states that these events are not related *directly*, the interaction somehow *propagates* from one event to the other. Clearly, all 'usual' dynamical laws are local (in particular, this is automatically true for any laws formulated as differential equations). So, locality is not a too restrictive requirement. Still, it is violation of locality (implicit, as a rule) that leads to serious problems in studying some subtle points of spacetime evolution (see, for example [83, 94]; we shall return to this question on p. 207).

Example 1 Suppose, a theory of pointlike particles in the Minkowski space admits interaction shown by the vertex V_1 in Fig. 1. Then to be local and geometrical the theory *must* admit all vertices—$V_{2,3}$, for instance—obtained from V_1 by a Lorentz transformation, since any such collision has a neighbourhood related to the corresponding neighbourhood of V_1 by the aforementioned transformation.

1.1 **C**-*Spaces*

The notion of locality turns out to be useful in discussing both matter and gravitational fields. Suppose a spacetime M can be represented as a union

$$M = \bigcup_{\alpha} V_{\alpha},$$

[1]This definition is not common by any means. Fortunately, we shall use it only a couple of times.

where V_α are some suitable sets (balls, convex or perfectly simple neighbourhoods, etc.). It is easy to see that properties of M fall into two categories: to establish some of them (for example, convexity) one needs to know, how *exactly* V_α are united to form M. At the same time, other properties (for example, flatness) can be deduced from examining each of V_α separately. It is these later properties that will be called local.

Definition 2 Let C denote the set of all spacetimes, satisfying a condition c (or possessing a property c). Then the condition (respectively, property) c is *local*, if for any open covering $\{V_\alpha\}$ of an arbitrary spacetime M the following equivalence is true[2]

$$M \in \mathsf{C} \quad \Leftrightarrow \quad V_\alpha \in \mathsf{C} \ \ \forall \alpha.$$

From now on the elements of C will be called c-*spaces* for brevity.

We shall need the following evident fact

Fact 3 If c is local, then the following implications are true:

(a) $V \subset W, W \in \mathsf{C} \quad \Rightarrow \quad V \in \mathsf{C};$
(b) $O_1, O_2 \in \mathsf{C} \quad \Rightarrow \quad (O_1 \cup O_2), (O_1 \cap O_2) \in \mathsf{C};$
(c) $N \in \mathsf{C}$ and N' is locally isometric to $N \quad \Rightarrow \quad N' \in \mathsf{C}.$

Example 4 The property 'to satisfy the Einstein equations'

$$G_{ab} = 8\pi T_{ab} \tag{1}$$

is local, when the stress–energy tensor T_{ab} is determined in each point by the value of the fields (and its derivatives) in that point. The properties 'to be geodesically complete' or 'to be b-complete', see Sect. A.3, clearly do not satisfy the $\Rightarrow$ part of the equivalence. So, these properties are *non*local (this is one of the reasons why the study of singularities is so hard). Extendibility and causality are also non-local: both violate the $\Leftarrow$ part. For the same reason staticity also is non-local, as is demonstrated by the Misner space, see Sect. 2.1 in Chap. 4. At the same time, 'local staticity', that is the property to have a unique (up to a constant factor) surface orthogonal timelike Killing vector field in each simply connected region [55] is, in agreement with its name, local.

Definition 2 is formulated in purely topological terms, so, if desired, one could change the word 'spacetime' in it to 'manifold' or even to 'topological space'. Then one would discover that, for example smoothness is local, while connectedness is not. We do not need such generality. A law will be called local, if—irrespective of the choice of an open covering $M = \cup_\alpha V_\alpha$—it holds in the entire spacetime M, when and only when it holds in each element V_α.

---

[2] Cf. the 'three particularly interesting conditions' considered in Appendix B of [62].

Remark. The definition of locality must be given with certain caution. For example, one could think that Definition 2 without much detriment can be changed to the following one: C is local, if it is true that

$$M \in \mathsf{C} \quad \Leftrightarrow \quad U \in \mathsf{C} \qquad \forall U: \quad U \text{ is open and } U \subsetneq M.$$

In fact, however, this would lead to quite a different concept. For example, in the Riemannian case the latter definition is satisfied by the—non-local according to the former one—property 'to have diameter not greater than 1'. We adopt Definition 2 because it is closer to the intuitive notion of locality: in particular, it makes the implications (3)b and (3)c valid.

The difference between the two definitions is essential in this book, too. The requirement of locality enters the formulation of Theorem 2 in Chap. 5 and if we adopted the 'modified' definition of locality, as was declared (erroneously) in [97], the theorem would be false or, at least, unproven [122, 126].

Dealing with certain relativistic problems, such as the lack of predictability in relativity, see Sect. 3, one may be tempted to modify the definition of spacetime by adding to it some supplementary requirements. Such requirements are presumed to be local, so it is worth noting that:

1. There is no rational argument (to my knowledge) for that presumption;
2. Some of the standard conditions (in particular, Hausdorffness and inextendibility) are *not* local;
3. The imposing of an additional local condition C may make it necessary to revise the concept of inextendibility. This will happen if an extendible C-space is found which has no extensions satisfying C. An example will be given in a moment.

Definition 5 Let C be a local property. M' is called a C-*extension* of a C-space M, if $M \subsetneq M'$ and M' is a C-space too. A spacetime is C-*extendible*, if it has a C-extension, and C-*maximal* otherwise.

The difference between extendibility and C-extendibility is demonstrated by the example of region IV (this is one of the two regions with $r > 2m$) of the Schwarzschild space, see Sect. 2.1 in Chap. 9. If C is the property 'to be locally static' the said region is a C-space. And it is extendible (for example, to the Kruskal space). However, in each of its extensions the Killing field at the boundary of the region becomes lightlike and hence the extension does not have the property C. Correspondingly, the region is C-maximal. Thus, the requirement that a spacetime satisfy an additional (local) condition, may well come into conflict with the requirement that it be inextendible.

1.2 The Weak Energy Condition (WEC)

An important class of local conditions are those imposed on the Einstein tensor G_{ab} or, insofar as we accept the Einstein equations (1), on the stress–energy tensor T_{ab}. Such is, in particular, the weak energy condition

$$T_{ab}v^a v^b \geqslant 0 \qquad \forall \text{ non-spacelike } \boldsymbol{v}. \tag{2}$$

The importance of the WEC lies in the fact that it relates a (quite pictorial) geometric characteristic of spacetime to a widespread feature of its matter content.

As a property of the matter source, the WEC is the condition that the energy density measured in the proper frame of an arbitrary observer be positive. There is evidence that the WEC may break down (the Casimir effect [14, 70] was observed *experimentally* [110, 160]). Moreover, it is not impossible that the WEC violating matter constitutes the dominant bulk of the universe [118]. Still, it is generally believed that as long as we ignore quantum effects (see Chap. 8) the weak energy condition must hold.[3]

To understand the *geometrical* meaning of the WEC consider a point p and a two-dimensional oriented spacelike surface Ξ through p. Pick an (as small as necessary) neighbourhood $O \ni p$ and let $\mathcal{H}$ be the set—it is a three-dimensional surface called also beam, or congruence—formed by the null geodesics emanating from every point of $O \cap \Xi$ in the direction normal to Ξ [strictly speaking there are *two* such directions in each point of Ξ; we choose *one* of them by
(1) specifying which of the two such geodesics passing through p lies in $\mathcal{H}$, that geodesic will be denoted γ_p;
(2) requiring that the choice be continuous on Ξ[4]].
Let ζ be the affine parameter on γ_p fixed, up to a constant factor, by the condition $\zeta(p) = 0$, and let $A(\zeta)$ be the area of the cross section of the beam in the point $\gamma_p(\zeta)$. Now $\mathcal{H}$ is characterized by the expansion $\theta(p)$—the quantity proportional to $A^{-1}\frac{\mathrm{d}}{\mathrm{d}\zeta}A\big|_{\zeta=0}$ in the limit of infinitely narrow beam. Clearly, θ is positive, if the beam diverges and negative otherwise. It is essential that θ obeys also the inequality

$$\frac{\mathrm{d}}{\mathrm{d}\zeta}\theta \leqslant -\tfrac{1}{2}\theta^2 - (\partial_\zeta)^a R_{ab}(\partial_\zeta)^b, \tag{3}$$

see, for example Chap. 4 of [76]. This inequality is a purely geometrical fact stemming from the properties of geodesics, it holds for *any* thus constructed congruence (i. e. it does not depend on the choice of Ξ). As we see,

---

[3] For an alternative point of view see [8, 174].

[4] It is this step in constructing $\mathcal{H}$ where we need Ξ to be oriented.

$$\frac{\mathrm{d}}{\mathrm{d}\zeta}\theta \leqslant -\tfrac{1}{2}\theta^2 \qquad \text{when the WEC holds.} \tag{4}$$

Thus, the curvature caused by the presence of ordinary (WEC-respecting) matter can focus light rays ('gravitational lensing'), but not disperse them: an initially converging beam keeps converging and its cross section shrinks to a point within a finite parameter distance. Indeed, (4) is easily integrated to give

$$\theta(\zeta) = \big(\zeta/2 + 1/\theta(0) + f\big)^{-1},$$

where f is a non-decreasing function vanishing at $\zeta = 0$. This proves the following version of [76, Proposition 4.4.6].

Proposition 6 *Provided that the WEC is satisfied and $\theta(0) < 0$, there is a number ζ_* such that $\theta(\zeta_*) = -\infty$, if γ_p can be extended to the point $p_* \leftrightharpoons \gamma_p(\zeta_*)$.*

In the point p_* the area of the beam's cross section vanishes, which means that γ_p meets there (or 'almost meets', see [141, Example 10.31]) infinitesimally neighbouring geodesics of $\mathcal{H}$. The importance of p_*—such points are called *conjugate to* Ξ *along* γ_p—lies in the fact that any point in γ_p beyond p_* can be reached from Ξ by a timelike curve (see [76, Proposition 4.5.14]; when some of the geodesics *do* intersect in p_*, this is a simple corollary of Proposition 23(d)). Thus, Proposition 6 establishes a connection between the maximality of the speed of light and positivity of energy that can be formulated as follows.

Statement 7 Suppose a geodesically complete spacetime satisfies the weak energy condition. Then a null geodesic emanating from p reaches a sufficiently distant destination *later* than some other non-spacelike curve starting at the same two-dimensional oriented spacelike surface Ξ if this surface is chosen so that $\theta(p) < 0$.

Below, we shall also use the following corollary of Propositions 6 and [76, 4.5.14]

Corollary 8 *If the null geodesics that form $\mathcal{H}$ are future (or past) complete and in a point $p \in \mathcal{H}$ the inequality $\theta(p) < 0$ [respectively, $\theta(p) > 0$] holds, then $\mathcal{H}$ is not achronal.*

Remark 9 We shall deal only with congruences of *null* geodesics. So we could change the WEC to the *null energy condition* or (which is the same when the Einstein equations are satisfied) to the *null convergence condition*, which are the requirements that, respectively, $T_{ab}l^al^b$ and $R_{ab}l^al^b$ are non-negative for any null $\boldsymbol{l}$. However, these conditions do not have such transparent physical meaning. Other local energy conditions can be found in [76].

2 The Maximal Speed and Causality

> Бросая в воду камешки, смотри на круги, ими образуемые; иначе такое бросание будет пустою забавою.
>
> Козьма Прутков [188].[5]

It is a matter of common knowledge that 'according to relativity, nothing can travel faster than light' [75]. And it is of vital importance to this book that, if taken literally, the quoted simple, neat, and plausible assertion is *wrong*. A counterexample is provided by light spots and shadows[6] which well can move faster than light without coming into conflict with relativity. But is the fallacy serious? Can't it be corrected by a slight reformulation? It turns out that the answer is negative. Experience shows, in particular, that simple replacements of 'nothing' by 'no material object', or 'no field perturbation' (it is implied that perturbations travel with the group velocity of the corresponding field), etc., do not solve the question. It is this problem in formulation of what one might think as a fundamental of relativity that makes this section necessary. Its goal is to find out:

(1) exactly what speed limit is set by relativity?
(2) *how* is it enforced? If such a limit is a property of the corresponding fields and particles, then what has it to do with relativity? On the other hand, if it is a property of spacetime, then where is it hidden in the definition thereof?
(3) how is the speed limit in question related to locality and causality often mentioned in this context?

Our primary interest is the 'speed of gravity'. As for the matter fields and particles, we content ourselves with clarifying the logical status of the principle of causality. No attempts are made to develop a rigorous detailed theory.

2.1 *The Speed to Be Restricted*

In the Minkowski space, consider a pointlike particle with non-zero mass m. The world line of the particle is a curve $x^i(t)$, where $t \leftrightharpoons x^0$ and x^i, $i = 1, 2, 3$ are the standard Cartesian coordinates, see Example 73 in Chap. 1. Define the *3-velocity* v of this particle to be $v \leftrightharpoons \sqrt{(\mathrm{d}x^1/\mathrm{d}t)^2 + (\mathrm{d}x^2/\mathrm{d}t)^2 + (\mathrm{d}x^3/\mathrm{d}t)^2}$ and assume that the particle is *sub*luminal, i. e. that $v < 1$. Then the particle's energy as measured in the coordinate basis is $m/\sqrt{1 - v^2}$. This expression diverges at $v \to 1$ and we conclude

[5]Throwing pebbles into the water, look at the ripples they form on the surface. Otherwise this activity will be an empty amusement. Kozma Prutkov [152].

[6]Other popular counterexamples can be found in good textbooks in special relativity, see, for instance [169].

that a massive initially subluminal particle cannot be accelerated to superluminal velocities. Put differently, the world line of such a particle is timelike.

It is this clear and important fact that one would like to generalize to more complex situations like field theories. In doing so, however, one encounters the problem of deciding exactly what speed is bounded.

Example 10 Imagine a closely packed row of y-oriented electric dipoles aligned along the x-direction. Every dipole is attached to a device able to flip it at an appointed moment. Consider now two cases:

(a) Every dipole flips (back and forth) at the moment $t(x) = x/v_a$, where x is the coordinate of the dipole and v_a is a constant.
(b) *Only* the dipole with $x = 0$ flips and *only* at the moment $t = 0$.

Consider the y-component of the electric field in either case. Obviously, $E_y(x, t)$ at positive x is a pulse running in the positive direction. Its speed[7] is v_a in the first case and a certain v_b in the second.

Clearly, the Maxwell equations (*not* relativity) guarantee that $v_b \leqslant 1$, but no reason is seen for v_a not to be arbitrarily large (examples of superluminal electromagnetic waves of that kind can be found in [22, 155]). It is natural to ask, what is so different between the two pulses that some fundamental restriction applies only to one of them? A self-suggesting answer would be that the pulse from 10(a) is not 'real'. Roughly speaking, it is, rather, a set of independent 'swellings', each appears in its own place and stays put. Its evolution reduces to changes in size and shape (at first it grows, then shrinks, and, finally, vanishes), but not in location. Correspondingly, its velocity is zero and the resemblance between the evolution of the set of such swellings and the propagation of a pulse is nothing more than an illusion.

On the other hand, this illusion is so persistent that one would like to have a formal criterion for discriminating between 'real' and 'illusory' pulses. Such a criterion must be based, as it seems, on the difference between *independent* and *causally related* events. It is this difference that is the subject matter of the present section, its goal being to outline a way of introducing such a vague concept as the 'cause–effect relation' in such a rigorous science as classical relativity.

Remark 11 Actually, the aforesaid differentiation is a very old problem. Consider, for example what appears to be an arrow with the world line $\alpha(t)$. Should we speak of an arrow flying with the speed $\dot{\alpha}$ or of a set of motionless arrows, each of which exists for a single moment in a point of the curve α? This, seemingly philosophical question [185] becomes physical (to some extent) in discussing causality.

The connection of the cause–effect relation to the 'light speed barrier' is well exemplified by pointlike superluminal particles, *tachyons*. Are they forbidden by relativity? At first glance the positive answer is dictated by the following thought experiment going back to Einstein [41], see also [167]. Let an observer $\mathscr{A}$ at the point a emit a tachyon τ_1, see Fig. 2a. The tachyon is received at the point b by observer $\mathscr{B}$ who

[7]The pulse will also deform in the course of propagation. But we neglect this irrelevant effect.

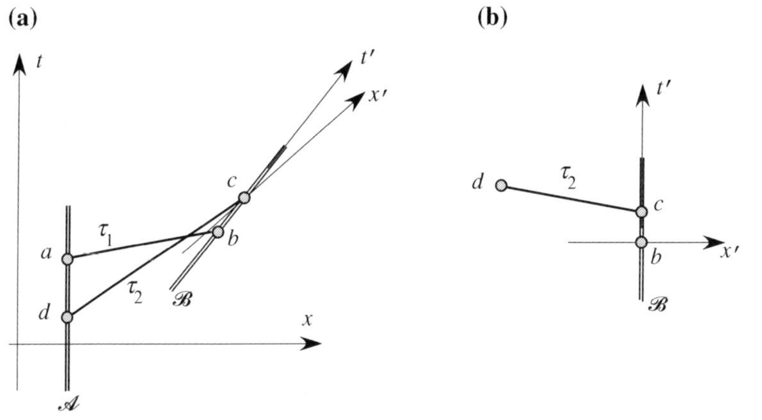

Fig. 2 A thought experiment with tachyons leading to an (apparent?) contradiction with causality. It stems from the fact that the segment dc looks future-directed for observer $\mathscr{A}$, but past-directed for observer $\mathscr{B}$

re-emits it at the point c, see Fig. 2b. Then the re-emitted tachyon—it is denoted τ_2—is absorbed by $\mathscr{A}$ at d. Since the world lines of the tachyons are spacelike the parameters of the experiment (the velocity of $\mathscr{B}$, the delay between the reception of τ_1 and emission of τ_2, etc.) can be chosen so that d will *precede* a on the world line of $\mathscr{A}$. In other words, $\mathscr{A}$ in this experiment manages to send a signal (from a) to their own past. And this, as one[8] might think, gives rise to a variety of paradoxes, see Chap. 6.

Since 1970s, the view has prevailed that no signals to the past are sent by the observer and the whole problem is purely interpretational [13]. Indeed, b happens later than a and c is later than d only in the sense that

$$t(a) < t(b), \qquad t'(c) < t'(d), \tag{$*$}$$

where t and t' are the time coordinates in the proper coordinate systems of the observers $\mathscr{A}$ and $\mathscr{B}$, respectively. (These relations *resemble* the relations $a \preccurlyeq b$, $c \preccurlyeq d$, but—just when the events are spacelike separated—do not coincide with them.) The interpretation of a and c as the emission and b and d as the absorption of tachyons (or, which is the same, a and c as the causes of, respectively, b and d) is based solely on the inequalities ($*$). But this substantiation (*post hoc, ergo propter hoc*) is notorious for its inconsistency. And indeed, the comparison of the second inequality with the—also correct—relation $t(d) < t(c)$ shows that ($*$) is unsuitable for establishing cause–effect relations. On the other hand, abandoning that interpretation and declaring the events c and d causally unrelated (the 'reinterpretation principle' [13]), we immediately get rid of problems with 'paradoxes': tachyon τ_2 is not a signal any more ($\mathscr{A}$ in point d does not receive any information about $\mathscr{B}$), etc.

[8]Not Einstein, see [41]!.

However, solving one problem this approach at the same time gives rise to another one, equally hard, cf. [33]. The point is that if all events on a curve (on the segment *cd* in this case) are causally unrelated, the curve can hardly be referred to as the world line of a particle. Obviously, the reinterpretation principle being consistently applied transforms the tachyon from a pointlike particle into an analog of a light spot or a succession of swellings, see above. Thus, we can conclude that relativity does forbid superluminal particles insofar as we regard sending a signal to the past impossible.

2.2 Cause–Effect Relation

Our goal is to formalize somehow the notions of cause and effect within the framework of general relativity. In this subsection, we restrict ourselves to matter fields on a fixed background whose geometry is independent of those fields. The consideration of the subtler and more important question of what 'the speed of gravity' is will be postponed to Sect. 2.4.

Consider a classical theory of a (not necessarily scalar) matter field f. Normally, such a theory consists of a spacetime M and a convention on which functions f in M are 'physical', i. e. are regarded appropriate, describing reality. Typically, that convention consists of an equation governing the dynamics of the field, (the 'equation of motion') and a set B of constraints on admissible solutions. For example, in electrodynamics, f is the vector potential A^{μ}, its equations of motion are the Maxwell equations, and B comprises, first, the requirements on the field's behavior at infinity (usually it is required to decay sufficiently fast) and, second, some of the laws of motion determining the evolution of the sources (which are also a kind of boundary conditions), cf. Example 12(c).

Example 12 Let M be the Minkowski plane and f be the scalar field ϕ obeying the equation

$$(\partial_t^2 - \partial_x^2)\phi = 0. \tag{5}$$

Consider different choices of B:

(a) Denote by B_a the requirement that ϕ be smooth and that at each t its derivatives be square integrable on the x-axis. Then the choice $\mathsf{B} = \mathsf{B}_a$ means, physically, that we regard as legitimate any field configuration with finite energy and energy density.
(b) Imagine that *always*

$$\phi(t_0, x_0 + 1) = \phi(t_0, x_0)$$

or, put differently, that in nature ϕ exists only as waves of length 1. This property seems quite exotic, but what matters is only that it does not contradict (5). And were it experimentally discovered, one would have to adopt $\mathsf{B} \supset \mathsf{B}_b$, where B_b is the periodicity condition.

Now let M be the four-dimensional Minkowski space and

(c) B_c be the requirement that

(i) ϕ as classical function be defined on $M - \lambda$, where λ is a curve $\boldsymbol{r}(t)$ and $\boldsymbol{r}$ is the position vector in the three-dimensional space $t = const$;

(ii) ϕ be a (distributional) solution of the equation $\Box\phi = \delta\Big(\boldsymbol{r} - \boldsymbol{r}(t)\Big)$.

Clearly, by setting $\mathsf{B} = \mathsf{B}_c$ we incorporate pointlike sources of the field into consideration. These sources are not provided by specific dynamical laws [otherwise we would have to treat them on exactly the same basis as (5)], but one can impose one or another condition on λ (on its velocity $\boldsymbol{v} = \mathrm{d}\boldsymbol{r}/\mathrm{d}t$, in particular).

Note that constraints constituting B may be non-local (such as, for example the integrability condition in the first example). It is also noteworthy that additional conditions entering into B in one theory, may follow from the laws of motion in another—broader—theory. That is what happens, for example with the aforementioned constrains on the motion of sources, if one transits from the theory of free electromagnetic field to magnetohydrodynamics.

Definition 13 Pick an event (point) $p \in M$ and suppose that there is a closed set $\mathcal{A} \subset M$ such that $p \notin \mathcal{A}$ but nevertheless $f(p)$ is uniquely determined by the values of f and its derivatives in the points of $\mathcal{A}$. Suppose, further, that no closed proper subset of $\mathcal{A}$ inherits that property. Then we shall say that $\mathcal{A}$ *determines* p. Also we shall write $q \leftrightarrows p$ for 'q is in a set determining p'.

Below the relation $\leftrightarrows$ will be used for introducing the cause–effect connection, so an important point to emphasize is that $\leftrightarrows$ depends on the choice of B, which is an *independent* element of theory by no means deducible from equations of motion. This distinguishes the speed of signal propagation from the phase velocity, the speed of energy transmission and the like.

Example 14 Consider two segments in $\mathbb{L}^2$:

$$\Sigma_1: \quad \{t = 1,\ x \in [-1, 1]\} \qquad \text{and} \qquad \Sigma_2: \quad \{t = 1,\ x \in [0, 1]\}.$$

By fixing at Σ_1 the initial conditions for Eq. (5) we, of course, uniquely determine $\phi(o)$, where o is the origin of coordinates. On the other hand, neither Σ_2, nor any closed proper subset of Σ_1 have this property when $\mathsf{B} = \mathsf{B}_a$. Thus under the assumptions of Example 12(a), Σ_1 determines o, while Σ_2 does not. And in Example 12(b) owing to the periodicity of ϕ the value $\phi(o)$ is uniquely defined by the data fixed at Σ_2 (but not at any of its subsets). Hence, now, conversely, Σ_2 determines o, and Σ_1 does not. Likewise, it is easy to check that, for example, $p \not\leftrightarrows o$ in the first case and $p \leftrightarrows o$ in the second, where p stands for the point $\{t = 1,\ x = 2\}$. Finally, when $\mathsf{B} = \mathsf{B}_c$, the relation between p and o depends on conditions (if any) imposed on λ. For instance, the condition $|\boldsymbol{v}| < 1$ implies $p \not\leftrightarrows o$.

It seems absolutely natural to regard events p and q causally unrelated

$$q \not\leftrightharpoons p, \qquad p \not\leftrightharpoons q,$$

if what happens in p (i. e. the values of f and its derivatives) does not affect the situation in q (as long as those values remain constant in all points of some set determining q) and vice versa. So, the relation $\leftrightharpoons$ is quite close to the cause–effect relation. But not quite, yet. In particular, it does not distinguish between the cause and effect: one cannot infer from $q \leftrightharpoons p$ whether q is a cause or an effect of p. This was to be expected. Indeed, choosing different functions $f_{\mathcal{A}}$ (restrictions of f to a set $\mathcal{A} \ni q$ determining p), one would discover that $f(p)$ varies accordingly, irrespective of—admissible—values taken by f *outside* $\mathcal{A}$. This suggests that p must be understood as effect. However, exactly as well one could choose different $f(p)$ and state that each time this gives rise to a new $f_{\mathcal{A}}$ (irrespective, again, on which admissible f are fixed in other points of $M - p - \mathcal{A}$). Then one would find it equally natural to interpret p as a cause of q. Another drawback of $\leftrightharpoons$ is that it does not—automatically—take into account the possibility of indirect influence, when p causes q by affecting some cause of q instead of q itself.

To remedy the situation we resort to two as yet unused properties of causality as it is understood intuitively. Namely, (1) a cause of a cause is a cause and (2) events a and b are the same if and only if a is a cause of b and b is a cause of a.

Definition 15 $\trianglelefteqslant$ ('is a cause of', or 'can affect') is a reflexive (i. e. $a \trianglelefteqslant a$), anti-symmetric (i. e. from $a \trianglelefteqslant b$, $b \trianglelefteqslant a$ it follows that $a = b$), and transitive (i. e. from $a \trianglelefteqslant b$, $b \trianglelefteqslant c$ it follows that $a \trianglelefteqslant c$) relation[9] such that

$$a \trianglelefteqslant b, \qquad \Rightarrow \qquad \begin{array}{l} a \leftrightarrow b, \text{ or} \\ \exists c: \quad a \leftrightarrow c, \quad c \leftrightarrow b \quad \text{and} \quad a \trianglelefteqslant c, \quad c \trianglelefteqslant b, \end{array}$$

where $q \leftrightarrow p$ is an abbreviation for '$q \leftrightharpoons p$ or $p \leftrightharpoons q$'.

Remark 16 The so-defined $\trianglelefteqslant$ is still non-unique. For example, one might expect that the relation $\trianglerighteqslant$ ('is an effect') defined by the equivalence

$$p \trianglerighteqslant q \quad \Leftrightarrow \quad q \trianglelefteqslant p,$$

typically will satisfy Definition 15. One way to fix this arbitrariness is to relate the difference between effects and causes with the difference between the past and the future by imposing, for example the following additional condition

$$p \trianglelefteqslant q \quad \Rightarrow \quad p \notin J^+(q)$$

[9] Such relations are called partial orders.

(a cause must not lie in the future of its effect). It is this requirement that is often called 'the principle of causality'. We shall reserve this term for a more meaningful[10] condition.

Thus, we have formalized the notions of cause and effect as attributes of a matter field theory. The theory is understood to include, in addition to dynamics laws, a set of requirements on admissible solutions. Causality in this approach characterizes not a specific solution, but the theory *as whole*. Loosely speaking, it tells one what is considered freely specifiable within that theory and what is to be found. As a next step let us make the consideration 'more local' by introducing the notion of *signal.*

Definition 17 A signal from $\alpha(0)$ to $\alpha(1)$ is a curve $\alpha(\xi)\colon\ [0,1] \to M$ such that

$$\alpha(\xi_1) \trianglelefteq \alpha(\xi_2), \quad \forall \xi_1 < \xi_2.$$

The vector tangent to α is, correspondingly, the *velocity of the signal.*

It must be emphasized that one should be cautious in identifying the velocity of a signal with the velocity of some particular carrier (particles or waves).

Example 18 Suppose, a certain $\mathcal{O}$ lives 4 km from the laboratory where an experimenter $\mathcal{E}$ every noon tosses a coin, this event will be denoted p. If $\mathcal{E}$ gets tails, he stays at laboratory, but if the coin comes up heads, $\mathcal{E}$ walks to $\mathcal{O}$ and tells him the outcome of the experiment. When $\mathcal{E}$ has to make this trip on a weekday he always moves on some stipulated geodesic γ (being a curve in the four-dimensional world γ is different every day), but on holidays he is allowed to choose his way at haphazard. In both cases their meeting with $\mathcal{O}$—it will be denoted q—happens strictly at 1:00 pm. The world line of $\mathcal{E}$ between the noon and 1:00 pm will be denoted β (on a weekday, if $\mathcal{E}$ gets heads, $\beta = \gamma$).

Let us pick a particular day and analyze the causal relations between the relevant events. Taking the approach developed above we shall regard p a cause of an event e, if the observation in e (that is, in fact, the checking of whether $\mathcal{E}$ is present in e) allows one to learn the result of the tossing (because p in such a case is obviously in a set determining e).

For a start, let the day in consideration be one of those weekdays when the coin shows heads and, respectively, $\mathcal{E}$ walks from the laboratory to $\mathcal{O}$'s residence.Then an observer in every point of β learns (just from meeting $\mathcal{E}$ there) the outcome of the tossing and thus gets affected by p. So, β is a signal and, in particular, $p \trianglelefteq q$. The speed of the signal is equal to the speed of $\mathcal{E}$ and everything looks rather trivial.

Let us, however, consider a weekday in which the coin comes up tails. This time $\mathcal{O}$ does not receive $\mathcal{E}$'s report. Nevertheless, the non-appearance of $\mathcal{E}$ will *by itself* let $\mathcal{O}$ know how the coin landed. So, $p \trianglelefteq q$ again. Moreover, the same reasoning applies to any point of γ, hence γ is a signal. The speed of this signal is obviously 4 km/h. Thus we arrive at a rather counter-intuitive situation: there is a signal 'propagating'

[10]Though in pre-relativistic times the difference might seem inessential.

along a specific curve at a specific speed, even though there is nothing *material*—be it a particle or a light spot—the motion of which could be associated with the signal.

Suppose, finally, that the day when $\mathscr{E}$ stayed at laboratory happened to be a *holiday*. For $\mathscr{O}$ this fact is immaterial, so p still affects q. However, this time an observer between p and q, not knowing β, gains no knowledge from the non-observation of $\mathscr{E}$. We thus obtain one more counter-intuitive result: p and q are causally related even though *there is no signals* from one to the other.

2.3 The Principle of Causality

So far we have discussed the cause and effect as notions pertaining to the combination of a field theory *and* the geometry of the background. To use these notions in formulating non-trivial statements one can tie them to another—purely geometrical, say—relation defined on the relevant spacetime. One way to do so is to require that $\trianglelefteq$ be not stronger than $\preccurlyeq$.

Condition 19 'The principle of causality'. An event p can affect an event q, only if there is a future-directed non-spacelike curve from p to q:

$$p \trianglelefteq q \qquad \Rightarrow \qquad p \preccurlyeq q.$$

This condition can be made 'more local', though at the cost of some weakening:

Condition 20 'Light barrier'. The speed of a signal cannot exceed the speed of light.

Remark 21 One also could require the existence of an open covering of M such that in each of its elements, when it is considered as a spacetime by itself, Condition 19 holds. If the conditions in B are local (e. g. the solutions only must be smooth) and M is sufficiently 'nice' (strongly causal, for example), then this requirement[11] is apparently equivalent to the Condition 19. *Generally*, however, this is not the case. Both the causality principle and the boundedness of the speed of signal propagation can break down in spite of local causality. This takes place, in particular, in Example 12(b). Another example is the field from Example 12(a), if the condition of integrability is dropped and M is taken to be the cylinder C_M from Example 42 of Chap. 1.

As the Kirchhoff formula [175] shows, the field

$$\left(\Box - \tfrac{n-2}{4(n-1)} R\right) \phi = 0, \qquad \mathsf{B} = \mathsf{B}_a \tag{6}$$

in the Minkowski space (the term with R will be needed later) allows one to choose $\preccurlyeq$ as $\trianglelefteq$ [the point (t_0, x_0, y_0, z_0), for example is determined by the unit 3-sphere lying in the plane $t = t_0 - 1$ and centered at the point (x_0, y_0, z_0)], so the causality

---

[11] It is, essentially, what is called *local causality* in [76].

principle is satisfied. The same is true for the electromagnetic and, in fact, for any other field which is governed by a hyperbolic system of linear differential equations with non-spacelike characteristics. Actually, the whole everyday physics (including mechanics, quantum field theory, etc., insofar as they are considered in globally hyperbolic spaces) admits the choice '$\trianglelefteq$' = '$\preccurlyeq$' and hence can be subordinated to principles 19, 20.

It is the universality of the principle of causality that makes it so important and explains why the non-spacelike vectors and curves are called causal, the sets $J^{\pm}$ are called the causal future/past, and, generally, the structure associated with the relation $\preccurlyeq$ is often referred to as 'causal' (beyond the present section we also adhere to this tradition).

At the same time, situations are imaginable in which the principle of causality does not hold. The one most pertinent to this book is 'causality violating' spacetimes, see Definition 33 in Chap. 1. In these spacetimes the aforementioned principle cannot hold *by definition* (because $\preccurlyeq$ is not antisymmetric there, contrary to what is required of $\trianglelefteq$). In Chap. 6 we shall discuss the 'paradoxes' ensuing from the absence of a global cause–effect relation, see Chap. 6, and now let us turn to 'more local' causality violations. A simple example is provided by the field governed by the equation $(\partial_t^2 - 2\partial_x^2)\phi = 0$ in the Minkowski space, but it is more interesting to check that the same may happen also with fields obeying geometric (see the beginning of Sect. 1) equations of motion.

Example 22 In a Friedmann universe (M, g)

$$g: \qquad ds^2 = -dt^2 + a^2(t)[d\chi^2 + \chi^2(d\vartheta^2 + \sin^2\vartheta\, d\varphi^2)]$$

consider the field theory (6). Since the metric is conformally flat, the solutions of the equations of motion will differ from the corresponding solutions in the Minkowski space only by a non-zero factor, see Sect. 2.1 in Chap. 7. Hence, in this theory (1) it is possible to choose '$\trianglelefteq$' = '$\preccurlyeq$' [see the reasoning under (6)] and thus to satisfy the principle of causality and (2) the signals exist propagating with the speed of light. Denote by λ one of such signals:

$$\lambda \leftrightharpoons x^c(\xi): \qquad g_{cd}\frac{dx^c}{d\xi}\frac{dx^d}{d\xi} = 0.$$

Next, define on M a *new* metric tensor $\tilde{g}$ by the equation

$$\tilde{g}_{cd} = g_{cd} + \tfrac{1}{2}\tau_c\tau_d, \qquad \boldsymbol{\tau} \leftrightharpoons \partial_t$$

(note that the vector field $\boldsymbol{\tau}$ can be defined in a coordinate independent way as the unit vector field which in every point is normal to the three-surface of constant curvature through that point [76]). $\boldsymbol{\tau}$ have a clear physical meaning: they are the velocities of the flow lines of the fluid whose stress–energy tensor generates, through the Einstein equations, the metric g. In the spacetime $(M, \tilde{g})$, which, incidentally, is also a Friedmann space, though with different factor $a(t)$, consider a theory with the same B and the same equation of motion (specifically Eq. (6), not the one obtained from it by replacing $g \to \tilde{g}$). Clearly, in this new theory exactly the same events are causally related as in the original one. In particular, λ is still a signal. Its velocity, however, is now *greater* than the speed of light:

$$\tilde{g}_{cd} \frac{\mathrm{d}x^c}{\mathrm{d}\xi} \frac{\mathrm{d}x^d}{\mathrm{d}\xi} = \frac{1}{2} \left(\frac{\mathrm{d}x^c}{\mathrm{d}\xi} \tau_c \right)^2 > 0.$$

Thus, we have constructed a 'physically reasonable' theory[12] in which the principle of causality is violated and which nevertheless harbors neither 'logical inconsistencies', nor 'mind-boggling paradoxes'.

In summary, a theory describing matter fields and particles includes dynamic equations and, independently, a set of conditions imposed on their solutions. This set determines (non-uniquely, perhaps) what is freely specifiable in the theory ('initial data') and what is to be found. This division gives rise to relations which we call causal. The principle of causality is the following requirement placed upon both the theory and the geometry of the background: the relation 'to be a (possible) cause of' must imply the relation 'to be connected by a future-directed non-spacelike curve to'. This principle holds in all usual theories, but its violations do not necessarily lead to a catastrophe. Thus, whether to adopt it or not is a matter of taste. Being adopted it is this principle (and not relativity, or locality. or local causality, etc.) that forbids the velocity of a signal (not of a particle, or of the crest—or the front—of a wave, etc.) to exceed the speed of light. We shall neither develop this theory any further, nor make it more rigorous. The role of the principle of causality seems to be greatly exaggerated: if a theory is inconsistent with experimental data, or is self-contradictory, it will be rejected anyway. But if a theory contains no contradictions, than the fact by itself that the principle of causality does not hold in it, is a mere curiosity.

2.4 'The Speed of Gravity' [108]

So far we have been discussing signalling by matter fields. It is reasonable to ask now what the speed of a signal carried by gravity is. Is that speed restricted in any way? The question may seem meaningless, because the universe according to relativity is a 'motionless', 'unchanging' 4-dimensional object ('block universe') and gravity is just the shape of this object. But what can be called a speed of shape? What is the 'speed of being a ball'? Yet, a meaningful question concerning whether 'gravity is faster than light' *can* be formulated [108]. It takes some preliminary work though.

[12]In fact, it is a version of what was proposed in [5].

Suppose an experimenter $\mathscr{E}$ plans to throw a rock.[13] An observer $\mathcal{O}$ located at some distance from $\mathscr{E}$ does not know when exactly this event—let us denote it s—will happen. But if their apparatus is good enough, $\mathcal{O}$ will be able to register changes in the gravitational field of the rock (the event at which the first change is registered we denote by q) and infer that the rock is thrown. It seems quite natural to call s the cause and q the effect. And it is equally natural to ask: How soon after s can q happen? Can, in particular, q happen *before* the first photon from s will reach $\mathcal{O}$?

Remark 23 The positive answer to the last question would mean—among other things—that an outside observer can receive signals from inside the event horizon of a black hole.

The above-formulated questions are quite similar to those discussed in the previous subsections. The reason why a special treatment is required in the case of gravity is this. If $\mathscr{E}$ signalled by a matter field f (for example, by the Coulomb field of the same rock), the statement 's affects q' would mean that the value $f(q)$ depends on the decision of $\mathcal{E}$: it is $f(q) = f_1$, if $\mathscr{E}$ at s throws the rock, and $f(q) = f_2$, otherwise [recall that in our experiment $\mathscr{E}$ plays the role of a source, their behavior is not fully fixed within the theory adopted by $\mathscr{O}$ and the latter has to take into consideration *all* possibilities consistent with B, cf. Example 12(c)]. But if we tried to apply the same approach to gravity we would have to compare the values at q of *the metric*—not of f—in the two cases. And this, generally speaking, is *impossible*. There simply cannot be different metrics—g_1 and g_2, say—in the same *point of a spacetime*: if $g_1 \neq g_2$, then (M_1, g_1) and (M_2, g_2) are different spacetimes *by definition* (irrespective of whether the manifolds M_1 and M_2 are equal). Hence, the metric tensors are compared not in the same point q, but in two different ones (as soon as they belong to different spacetimes): $q_1 \in (M_1, g_1)$ and $q_2 \in (M_2, g_2)$. But to justify out interpretation (according to which the non-equality $g_1(q_1) \neq g_2(q_2)$ signifies that some experiment ended up differently) we must somehow know that q_1 and q_2 correspond to the 'same' (in some sense) time and place. There is no way to establish such a correspondence, so, generally, there is no way to assign a meaning to the statement 'the metric at this point has changed'.

Example 24 It is hard (if possible) to define consistently the speed of gravity in terms of characteristics of Einstein's equation [21, 49]. Indeed, the typical reasoning is: 'The solution [to the Einstein equations] depends, at a point x, only on the initial data within the hypercone of light rays [...] with vertex x, that is, on the relativistic past of that point. This result confirms the relativistic causality principle as well as the fact that gravitation propagates with the speed of light' [21]. The flaw in this argument is that—for the reasons just discussed—in the case of gravity there is no analogue of 'determining sets' (see Sect. 2.2), so 'is fixed as a solution of a differential equation by the data within a set S' and 'is caused only by points of S' is not the same, the latter statement being meaningless.

[13] We have to invent a special thought experiment, because the common question 'If the Sun disappeared, how soon we would find that out' is hard to answer due to its tension with energy conservation.

We shall get around this problem by considering the points in which, vice versa, the metric does *not* change and which therefore can be declared *independent* of s.

Remark 25 In fact, we could do the same with matter fields, too (though it would have been an unnecessary complication). Denote by $N(s)$ the set of all points of M which can*not* be affected by s:

$$N(s) = \{p \in M : \quad s \not\preccurlyeq p\}.$$

Then the principle of causality is the requirement that

$$\big(M - N(s)\big) = J^+(s) \quad \forall s.$$

We plan to capture the idea that M_1 and M_2 describe different developments of the same prehistory (which includes the world line of $\mathcal{E}$ up to s), the next step being the formulation of an analogue to the principle of causality as the exclusion of certain pairs M_1, M_2. To that end we require the existence of a pair of isometric regions $N_k^* \subset M_k$ $k = 1, 2$; it is these regions that describe the aforesaid prehistory. We require further that N_k^* be past sets (clearly the portions—even isometric—of different spacetimes do *not* describe the same region of the universe, if the remembrances of their inhabitants differ).

Definition 26 A pair (M_k, g_k, s_k), where $k = 1, 2$ and s_k is a point of a maximal spacetime (M_k, g_k) is called an *alternative*, if there is a pair of connected open past sets $N_k \supset \Big(J^-(s_k) - s_k\Big)$ and an isometry ϕ sending N_1 to N_2 and $J^-(s_1) - s_1$ to $J^-(s_2) - s_2$.

For a given alternative the pair N_1, ϕ does not have to be unique, there may exist a whole family $\{N_1^\alpha, \phi^\alpha\}$ of such pairs. By (N_1^*, ϕ^*) we denote a *maximal*, by inclusion, element of this family, that is an element such that the family contains no 'greater' element:

$$\nexists \alpha_0 : \qquad N_1^* \subsetneq N_1^{\alpha_0}, \quad \phi^* = \phi^{\alpha_0}\big|_{N_1^*}$$

Zorn's lemma [87] guarantees the existence of such an element, since the regions of M_1 and M_2 are partially ordered by inclusion[14]

$$A \leqslant B \Leftrightarrow A \subset B$$

and with respect to this order every chain $\ldots \leqslant A_1 \leqslant A_2 \leqslant \ldots$ has a supremum $\cup_i A_i]$. Correspondingly, $N_2^* \leftrightharpoons \phi^*(N_1^*)$.

Definition 27 The sets $\mathcal{N}_k \leftrightharpoons \mathrm{Bd}\, N_k^*$, $k = 1, 2$ will be termed *fronts*. A front $\mathcal{N}_k$ is *superluminal*, if $\mathcal{N}_k \not\subset \overline{J^+(s_k)}$.

---

[14]This is because regions are not just spacetimes, but spacetimes plus their embeddings in $M_{1,2}$, cf. the discussion on p. 4.

At either k the front $\mathcal{N}_k$ bounds the set N_k^* (note, incidentally, that since N_k^* is a past set, $\mathcal{N}_k$ by Proposition 31 is a closed achronal embedded C^{1-} hypersurface), which according to our plan is to be interpreted as the common past of the two worlds. This set is maximal and hence in any point of $M_k - N_k^*$ an observer can remember some results of their gravitational experiments that differ their experience from that of any observer from $M_l, l \neq k$. It is such a remembrance that we interpret as a signal from s_k. It is natural to regard the signal superluminal if it is received out of $\overline{J^+(s_k)}$, hence the name of the corresponding front.

The alternative is quite a rough concept. For example, generally, it does not allow one to define the 'speed of a gravitational signal', if the signal is understood to be a front. Even the source of a signal cannot be pinpointed because the same pair of spacetimes can satisfy the definition of alternative for different choices of s. Still, it makes possible the formulation of a *necessary* condition for the speed of gravity—in a particular theory—to be declared superluminal. This is done by ascertaining which alternatives are *admissible* in that theory, i. e. in which cases both spacetimes M_1 and M_2 are legitimate solutions, while the difference between them is assignable to an event s as opposed to the *initial* difference. If none of the admissible alternatives has a superluminal front, we apparently have a theory with the speed of gravity bounded by the speed of light. Note that only admissible alternatives matter and admissibility is established in each theory individually, so we could have defined the alternative just as a pair of pointed spacetimes. The remaining requirements (the existence of the sets N_k^*) in such a case would have formed an additional criterion of admissibility.

Example 28 Let both (M_1, g_1) and (M_2, g_2) be globally hyperbolic. And let $\mathcal{S}_k$, $k = 1, 2$ be Cauchy surfaces through s_k related by the equation $\mathcal{S}_2 - s_2 = \phi^*(\mathcal{S}_1 - s_1)$. An alternative is regarded as admissible if the values of the matter fields (and their derivatives if necessary) in each point of $\mathcal{S}_1$ are equal to their values at the corresponding points of $\mathcal{S}_2$. This criterion does not look far-fetched. Indeed, in the absence of such surfaces no reason is seen to assign the difference between M_1 and M_2 to s and its consequences, cf. [121]; rather such universes should be considered as different *from the outset* and the alternative (M_k, g_k, s_k) as inadmissable. But by assumption $M_{1,2}$ are globally hyperbolic, whence

$$M_k - J^+(s_k) = [J^-(s_k) - s_k] \cup \mathcal{D}(\mathcal{S}_k - s_k)$$

(each point in M_k is connected by every inextendible non-spacelike curve with $\mathcal{S}_k$, so, unless that point is in $J^+(s_k) \cup J^-(s_k)$, all inextendible causal curves through it meet the 'remainder' of $\mathcal{S}_k$, i. e. the set $\mathcal{S}_k - s_k$). By the existence and uniqueness theorem (for the Einstein equations it is proved under some assumptions that we shall discuss in a moment) the equality of the data fixed at three surfaces implies the isometry of the corresponding Cauchy domains. So, the sets $\mathcal{D}(\mathcal{S}_k - s_k)$ are isometric. Hence N_k^* (which includes $J^-(s_k) - s_k$, anyway) contains the entire $M_k - J^+(s_k)$. Thus, neither of the fronts is superluminal. In this sense general relativity,[15] as one would expect

[15] How to adapt this approach to two more exotic theories is briefly discussed in [108, remark 6].

[49], forbids superluminal communication: under the above mentioned assumptions *the speed of gravity is bounded by the speed of light.*

It is important that the theorems of uniqueness for the solutions of Einstein's equations are proved in some 'physically reasonable' assumptions on the properties of their right-hand sides. A possible list of such assumptions can be found, for example, in [76]. One of them is the principle of local causality (see Remark 21), another one is a stability requirement, and the third restricts the stress–energy tensors to expressions polynomial in g^{ab} (the corresponding restriction in [177] allows also the SET to contain first derivatives of the metric). This last assumption is *known* to fail in many physically interesting situations. In particular, vacuum polarization *typically* leads to the appearance of terms with second derivatives of the metric (like the Ricci tensor) in the right-hand side of the Einstein equations, see (7.11) for example. This suggests that superluminal gravitational phenomena, if exist, may take place in situations where quantum effects are strong—in the early universe, on the horizons of black holes, etc.

2.5 The 'Semi-superluminal' Velocity

The fact that a single event is associated with two fronts, each in its own spacetime, has a quite non-trivial consequence because they do not need to be superluminal both *at once*.

Definition 29 An alternative is called *superluminal*, if both its fronts are superluminal, and *semi-superluminal*, if only one is.

Suppose, in a world M_1 a photon (or another *test* particle) is sent from the Earth (we denote this event s_1) to arrive at a distant star at some moment τ_1 by the clock of that star. Let, further, M_2 be the world that was initially the same as M_1 (there are no *objective* criteria, so whether we consider the worlds as initially the same, depends, in particular, on what theory we are using), but in which instead of the photon a mighty spaceship is sent (the start of the spaceship is s_2). On its way to the star the spaceship warps and tears the spacetime by exploding passing stars, merging binary black holes and triggering other imaginable powerful processes. We assume that no superluminal ('tachyonic') matter is involved, so, in spite of all this, the spaceship arrives at the star later than the photon emitted in s_2. It is imaginable, however, that the spaceship's arrival time τ_2 is *less* than τ_1. Thus, the speed of the spaceship in one world (M_2) would exceed the speed of light in another (M_1), which would not contradict the non-tachyonic nature of the spaceship. Nor would such a flight break the 'light barrier' in M_1: the inequality $\tau_2 < \tau_1$ does imply that the front $\mathcal{F}_1$ is superluminal, but no *matter signal* in M_1 corresponds to that front. In particular, there is no spaceship in *that* spacetime associated with $\mathcal{F}_1$. It is such a pair of worlds $M_{1,2}$ that we call a semi-superluminal alternative. A theory admitting such alternatives allows superluminal signalling without tachyons.[16]

[16]One more way to understand the term 'superluminal' is discussed in Chap. 3.

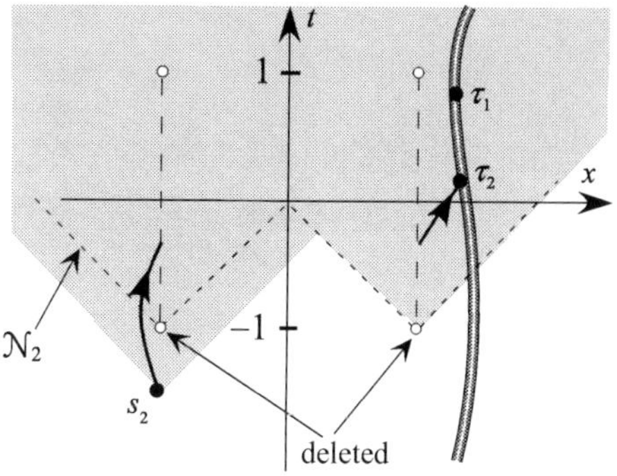

Fig. 3 Spacetime M_2. The curve starting from s_2 is the world line of the spaceship. The grey region is the causal future of s_2 and the dashed broken line is the front $\mathcal{N}_2$ bounding N_2^*

Example 30 Let M_1 be a Minkowski plane and s_1 be its point with coordinates $t = -3/2$, $x = -1$. Let, further, M_2 be the spacetime obtained by removing the segments $t \in [-1, 1]$, $x = \pm 1$ from another Minkowski plane, see Fig. 3, and glueing the left/right edge of either cut to the right/left edge of the other one. (M_2 is an analogue of the DP space, see Example 73 in Chap. 1 with $n = 2$ and a slight change in notation: $x_1 \to t$, $x_0 \to x$). The differences between M_1 and M_2 are confined, in a sense, to the future of the points $t = -1$, $x = -1$ and $t = -1$, $x = 1$, see Fig. 3. Speaking more formally, N_1^* is the complement to the union of two future cones with the vertices at those two points. That N_1^* is maximal, indeed, is clear from the fact that any larger past set would contain a past-directed timelike curve λ terminating at one of the mentioned vertices, while $\phi(\lambda)$ cannot have a past end point (because of the singularity).

Clearly, $\mathcal{N}_1 \not\subset \overline{J^+_{M_1}(s_1)}$, which means that $\mathcal{N}_1$ is superluminal. Its superluminal character *does not contradict* the principle of causality in M_1, because, as discussed above, in the space M_1, which is merely the Minkowski plane, the surface $\mathcal{N}_1$ does not correspond to any signal. The front $\mathcal{N}_2$ *is not* superluminal, so the alternative (M_k, g_k, s_k) is *semi*-superluminal. This name agrees with the fact that, on the one hand, the spaceship reaches the destination at a moment preceding the arrival of any photon emitted in s_1, but, on the other hand, no tachyons are involved: while the photons are in M_1, the spaceship belongs to the universe M_2, where its trajectory is timelike.

From the practical point of view, the alternative from Example 30 has a serious disadvantage: the nature of the difference between M_1 and M_2 is too exotic. It is unknown at the moment whether such alternatives (if they are possible at all) are admissible, i. e. whether something that takes place in $s_{1,2}$ can make 'the topology change' in the required way. Unfortunately, the same is true in the general case: we are going to prove that in *any* semi-superluminal alternative the constituent spacetimes have some causal pathologies.

Proposition 31 *The spacetimes M_1 and M_2 of a semi-superluminal alternative (M_k, g_k, s_k) cannot both be globally hyperbolic.*

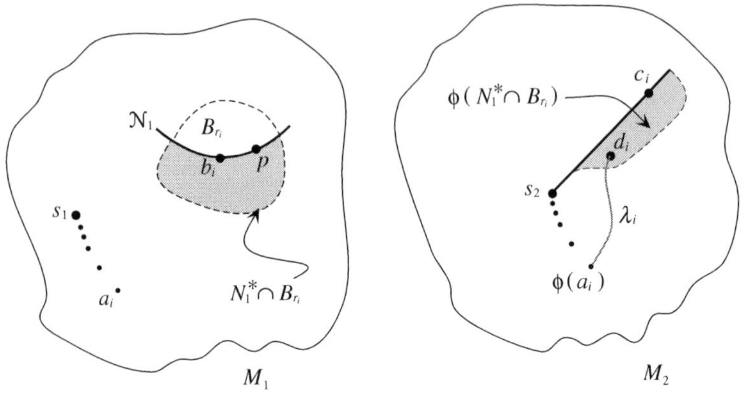

Fig. 4 The ball B_{r_i} bounded by the dashed line lies, by assumption, outside $J^+(s_1)$. But this contradicts the fact that the curves $\phi^{-1}(\lambda_i)$ must converge to a future-directed non-spacelike curve from s_1 to p

Remark 32 This statement differs substantially from that proven in Example 28 (to the effect that at least one of the spacetimes of a *super*luminal alternative must be non-globally hyperbolic). The latter depends on the Einstein equations, while Proposition 31 concerns a purely kinematical fact. Essentially, it is just another property of globally hyperbolic spacetimes.

Proof Let the front $\mathcal{N}_1$ be superluminal. Then some points of $\mathcal{N}_1$ must be separated from the—closed by the global hyperbolicity of M_1, see Proposition 48 in Chap. 1—set $J^+(s_1)$, that is there must be a point p, see Fig. 4, such that

$$p \in \mathcal{N}_1, \qquad B_r \cap \overline{J^+(s_1)} = \varnothing \quad \forall r < \bar{r},$$

where $\bar{r}$ is a constant, and B_r is a coordinate ball of radius r centered at p [i. e. the set of points q which satisfy the inequality $\sum \left(x^j(q)\right)^2 < r^2$ in some—fixed from now on—coordinate system $\{x^j\}$ with the origin at p]. In $I^-(s_1)$ pick a sequence a_i, $i = 1, 2, \dots$ converging to s_1. Our plan is to demonstrate that, unless $\mathcal{N}_2$ is superluminal, there is a timelike curve μ_i from a_i to B_{r_i} for any i and any $r_i < \bar{r}$. This would prove the proposition, because r_i can be chosen so as to tend to zero. In such a case the future end points of μ_i will converge to p, which—due to the global hyperbolicity of M_1, see Corollary 49 in Chap. 1—implies $p \in J^+(s_1)$ in contradiction to the choice of p.

To build for a given i a curve μ_i of the just mentioned type, pick a pair of points

$$b_i \in \left(\mathcal{N}_1 \cap B_{r_i}\right) \quad \text{and} \quad c_i \in \mathcal{N}_2,$$

such that for arbitrary neighbourhoods $U \supset b_i$ and $V \supset c_i$ it is true that

$$\phi\left(N_1^* \cap U\right) \cap V \neq \varnothing.$$

Such pairs always exist, because otherwise the maximal, by hypothesis, spacetime M_2 would have, by Proposition 68 in Chap. 1, an extension $B_{r_i} \cup_{\phi'} M_2$, where ϕ' is the restriction of ϕ to a connected component of $B_{r_i} \cap N_1^*$. Now assume that $\mathcal{N}_2$ is *not* superluminal. Then c_i, being a point of $\mathcal{N}_2$ must lie in $\overline{J^+(s_2)}$ and hence in the (closed, see Proposition 48 in Chap. 1) set $J^+(s_2)$ too. Thus (recall that $a_i \prec s_1$, whence $\phi(a_i) \prec s_2$) a pair a_i, c_i can be found such that $\phi(a_i) \prec s_2 \preccurlyeq c_i$. From Definition 15 in Chap. 1 it follows that $\phi(a_i) \prec c_i$. Hence, there is a neighbourhood of c_i which lies entirely in the (open) set $I^+(\phi(a_i))$. And according to (∗) that neighbourhood contains also points of $\phi(N_1^* \cap B_{r_i})$. So there also must exist points d_i:

$$\phi(a_i) \prec d_i, \qquad d_i \in \phi(N_1^* \cap B_{r_i}) \subset N_2^*.$$

The latter inclusion coupled with the fact that N_2^* is a past set means that any timelike curve λ_i from $\phi(a_i)$ to d_i lies entirely in N_2^* and thus defines the curve $\mu_i \leftrightharpoons \phi^{-1}(\lambda_i)$. This curve possesses all the desired properties: it is timelike, it starts in a_i, and it ends in B_{r_i}.

□

3 Evolutionary Picture

One of the central problems in physics is the determination of the *evolution* of a system. In most general terms it can be formulated as follows: a system (some field configuration, or a set of pointlike particles, etc.) at some moment of time (i. e. at some spacelike surface) is in such-and-such state. Given the system is governed by some known laws of motion, what will be its state at some later moment of time? The laws of motion which are usually considered in this context are such that the said problem has a unique solution in non-relativistic and special relativistic physics.

The situation with general relativity is more complex. The Cauchy problem here is naturally associated with the ('presentist') picture, in which the system—it is the spacetime itself and its matter content—is the result of some evolution processing the 'not-yet-happened future' into the 'already passed past'. Unfortunately, this convenient and intuitive picture so deeply contradicts the very foundations of relativity (with its 'block universe' concept) that, to my knowledge, no way has been found so far to give it a rigorous meaning. Suppose, however, for a moment that one can restrict oneself exclusively to the globally hyperbolic spacetimes. Then the universe would be presented as the product

$$M = \mathbb{L}^1 \times \mathcal{S}, \qquad (\star)$$

see Proposition 51(c) in Chap. 1, and the coordinate t parameterizing the first factor would take the rôle of time in the special relativistic problem. The fact that the spacetime is curved would affect the laws of motion, of course, but the world lines of particles still would be transverse to the surfaces of constant time (as long as we do not consider tachyons, see the previous section) and fields still would be described by hyperbolic differential equations. The Cauchy problem for them would be (typically) well posed, when the initial state were fixed at a Cauchy surface $\{t_0\} \times \mathcal{S}$, and would have a unique solution.[17] After the topology of $\mathcal{S}$ is fixed the geometry of M is fully described by the (local) quantity, its metric g, and the metric can be dealt with as any other field: one fixes g and its derivatives at $\mathcal{S}$ and the Einstein equations give us the metric in any point of M, see Chap. 7 in [76]. General relativity thus would become an ordinary field theory differing from electrodynamics, say, only by details of the field equations.

The questions which we (very briefly) discuss in this section are:

(1) Is the universe globally hyperbolic?
(2) If so, how to derive this fact theoretically?
(3) And if not, how to describe matter (let alone geometry) evolution?

3.1 Is the Universe Globally Hyperbolic?

Unfortunately, in answering the question in the title we cannot be guided by empirical evidence. We do not know how the loss of global hyperbolicity would manifest itself observationally. Assume, for example, that a topology change like that in the spacetime M_2, see Fig. 3, occurs once a year in every ball of the Solar system size. Then the topology change [a process inconsistent with global hyperbolicity, see ($\star$)] would be one of the most common phenomena. And, nonetheless, the assumption contradicts, apparently, no observations. Then again, nothing seems to be inconsistent with the opposite assumption, that our universe *is* globally hyperbolic.

As for *theoretical* analysis, Penrose conjectured that the answer is positive and formulated the principle, see [146] and references therein, that presumably agrees with classical general relativity.

***Conjecture* (*'Strong cosmic censorship'.*)** A 'generic' inextendible solution of the Einstein equations with a 'physically reasonable' right-hand side, is globally hyperbolic.[18]

The words 'generic' and 'physically reasonable' are intentionally left undefined until appropriate theorem comes along. Formally, this makes it impossible to discuss the conjecture in terms of validity: for *some* understanding of those words it is certainly true. Still, the conjecture is not meaningless, because there is more or less clear *intuitive* understanding of those words. 'Generic' essentially means 'that which does not evolve from the initial data so special that it would be physically impossible to achieve' [176]. The requirement that spacetimes be 'physically reasonable' means that their matter content must not be so perverted as to lead to singularities in the

[17] Which, of course, is not a coincidence: originally global hyperbolicity was introduced by Leray [113] as a property guaranteeing the existence and uniqueness of these solution [76].

[18] This is a retelling, not a quotation.

Minkowski space. Perhaps, it will also include some energy conditions (such as the dominant energy condition).

Penrose put forward some arguments in favour of cosmic censorship, see, e. g., [146], but in the later publication [147] he characterized them as 'rather vague' and doubted the existence of *any at all* plausible general lines of argument aimed in this direction. Below we mostly discuss the uniqueness of the relevant solutions to the Einstein equations, but serious difficulties emerge already with proving their existence. Indeed, a lot of inextendible solutions with physically reasonable sources are known to be globally hyperbolic. But this does not bring us any closer to the aim, because as a rule these spacetimes are highly symmetric and thus are by no means 'generic'. A possible way to support the conjecture would be to show that small perturbations of the initial state leave the solution globally hyperbolic. This, however, is quite hard mathematically and, to my knowledge, has been done so far only for a handful of empty spacetimes, see [4, 156] for reviews.

The 'intuitive clarity' of the conjecture should not be overestimated. Note, in particular, that it leans on the concept of 'initial data', and we do need something of this kind: fixing only the (type of) the right-hand side in the Einstein equations by no means fixes their solution and its geometry, in particular. Typically, there will be *infinitely many* different spacetimes corresponding to the same type of the matter source. And most of inextendible ones—in fact, by the uniqueness theorem, there may be at most one exception [23]—will be *non*-globally hyperbolic.

Example 33 Let M_1 be a globally hyperbolic inextendible solution of the Einstein equations with a physically reasonable (whatever this means) matter source. Remove a sphere $\mathbb{S}^{n-2}$ from M_1 and denote by M the double covering of the resulting spacetime $M_1 - \mathbb{S}^{n-2}$, see Fig. 1.5. Obviously, M is exactly as generic and inextendible[19] as M_1 and solves Einstein's equations with the same source.[20] And yet, it is non-globally hyperbolic.

So, the idea suggests itself to exclude the redundant solutions by imposing initial conditions. It is hard, however, even on the intuitive level, to give a sense to the notion of 'initial conditions' in the general non-globally hyperbolic case. Suppose, for example we wish to consider some globally hyperbolic spacetime as a solution of the Einstein equations satisfying such-and-such initial conditions. It is natural to take these conditions to be the geometry in a neighbourhood U of some Cauchy surface $\mathcal{S}$ (for simplicity we consider the vacuum case, otherwise we would have to require also the equality of the values of the matter fields). But which spaces should be regarded then as having *the same* initial conditions (we must be able to answer this question in order to find out whether they all are globally hyperbolic as the conjecture asserts)? If these are *all* spacetimes containing a region $\vartheta(U)$, where ϑ is an isometry, than the uniqueness of the solution satisfying the specified initial conditions is *out of the question even in the class of globally hyperbolic spaces*.

[19] This (hopefully obvious) fact can be rigorously proven much as Proposition 14 in Chap. 5, cf. Corollary 15.

[20] This is important: the problem in discussion is not local, one does not expect anything like '*D*-specialisation' of the geometry [27] in the vicinity of the removed sphere.

Example 34 The set $\mathcal{S}\colon\ x_0 = 0$ is a Cauchy surface of the Minkowski plane, see Example 8 in Chap. 1. The local isometry

$$\vartheta\colon \qquad (x_0, x_1) \mapsto (x_0', x_1'), \quad \text{where } x_0' \leftrightharpoons \mathrm{ch}1\, x_0 + \mathrm{sh}1\, x_1, \quad x_1' \leftrightharpoons \mathrm{sh}1\, x_0 + \mathrm{ch}1\, x_1$$

maps $\mathcal{S}$ to a spiral in the—globally hyperbolic—flat cylinder obtained from that plane by the identification

$$(x_0, x_1) = (x_0, x_1 + 2\pi), \quad \forall x_{1,2}.$$

ϑ sends some region $U \supset \mathcal{S}$ to a neighbourhood of the spiral. In doing so, however, it does not preserve global properties. In particular, the aforementioned spiral (not being achronal) *is not a Cauchy surface*.

Thus, one apparently will have to understand the term 'initial conditions' as a set of conditions that distinguish a particular spacetime from the infinite number of solutions of Einstein's equations with given right-hand side, but that do not reduce to the Cauchy data. *Exactly what* are those conditions is not clear at the moment. This is not that important for relativity itself, but is crucial for the concept of evolving universe: in order to use it, one must be able to answer questions like 'How will the Minkowski half-plane $x_0 < 0$ end up? Will it evolve into the Minkowski space, or, say, into the DP space?'[21]

3.2 *Should General Relativity Be Slightly Modified?*

As is discussed in the previous subsection, to overcome the lack of predictive power of relativity one must have a means to exclude some spacetimes without appealing to initial conditions. One possibility is to prohibit the redundant spacetimes by fiat, i. e. to *modify* general relativity by adding a new postulate to it. To find an appropriate postulate is a highly non-trivial task, but we can try, at least, to formulate (perhaps somewhat vaguely) a list of its desired properties:

1. The sought-for postulate must not be 'too global'. For example, it makes little sense to postulate directly that the spacetime must be globally hyperbolic: a requirement fixing the structure of the entire spacetime from the infinite past to the infinite future, does not well fit in with the evolutionary picture;
2. It must be restrictive enough for determining uniquely the extension of a given space in, at least, simplest cases like the Minkowski half-plane mentioned in the previous subsection;
3. On the other hand, it must be mild enough so as

[21] In this relation one can often read that the Einstein equations do not specify the topology. This is truth, of course, but not the whole truth. Thus, the DP space differs *geometrically* from M_2 of Example 30 even though they have the same topology and both solve the Einstein equations with the same right-hand side and the same initial condition (by which this time we understand that they coincide at $t < -1$).

(a) not to prohibit Minkowski, Schwarzschild, de Sitter, etc. spaces;
(b) not to come into conflict with another postulate of general relativity, according to which the spacetime describing our universe is inextendible. In particular, it must not exclude *all* extensions of some 'admissible' extendible spacetime [as would be the case if we required, for example spacetimes to be static, see the discussion below Definition 5].

For example, one could require of the spacetime[22] that it would remain globally hyperbolic 'as long as possible', cf. [37]. Mathematically this could be formulated as follows: the spacetime must have a closed surface $\mathcal{R}$ such that the region $\mathcal{D}(\mathcal{R})$, see Definition 53 in Chap. 1, has no globally hyperbolic extensions. An advantage of this choice is that—at least for the vacuum solutions of the Einstein equations—$\mathcal{D}(\mathcal{R})$ is unique for a given pair $(\mathcal{R}, g_{\mathcal{R}})$, see Sect. 7.6 in [76]. Yet, this does not solve the entire problem of the non-uniqueness of evolution, because there exist extendible globally hyperbolic spaces, all extensions of which are non-globally hyperbolic, see Sect. 2 in Chap. 4, for example. The postulate in question, contrary to our wish 2, does not allow one to determine which evolution will be preferred by such a spacetime.

To exclude the aforesaid non-uniqueness one might want to prohibit the 'unphysical' singularities like that appearing in Example 33. Thus, Hawking and Ellis proposed [76] to postulate that our universe is described only by *locally inextendible* spacetimes (a spacetime M is called locally extendible, if it contains an open set U with noncompact closure in M, and U has an extension—obviously different from M—in which its closure *is* compact).

Yet another possible postulate was proposed by Geroch who required the spacetime to be 'hole-free', which means the following.

Notation 35 For an arbitrary set $\mathcal{R} \subset M$ denote by $\mathcal{D}^+(\mathcal{R})$ the set of all points p such that every past inextendible causal curve through p meets $\mathcal{R}$.

Clearly, if $\mathcal{D}^-$ is defined dually, i. e. by changing 'past' to 'future', then $\mathcal{D}^+ \cup \mathcal{D}^- = \mathcal{D}$.

Definition 36 [23]A space-time M is *hole-free*, if for any achronal surface $\mathcal{R} \subset M$, and any embedding $\theta\colon U \to M'$, where U is a neighbourhood of $\overline{\mathcal{D}^+(\mathcal{R})}$, it is true that

$$\vartheta(\mathcal{D}^+(\mathcal{R})) = \mathcal{D}^+(\vartheta(\mathcal{R})).$$

(roughly speaking, in a hole-free spacetime each achronal surface has the greatest possible Cauchy development).

It turned out, however, that both postulates suffer from a common drawback—either of them excludes even the Minkowski space, see [11] and [104], respectively.

A few more candidate postulates are obtained by modifying the definition of the hole. One of them was proposed by Clarke [28], see [129] for a critical analysis.

---

[22] In fact, as it seems, this postulate *is* implicitly adopted by relativists.

[23] This is a slightly refined version [104] of the formulation given in [63].

Another one is due to Geroch [125], who altered his initial definition by requiring that $\vartheta(\mathcal{R})$ be acausal. Finally, a postulate different from those two is discussed by Minguzzi in [129]. It seems to satisfy wish 3*a*, but not 3*b*.

3.3 Matter in Non-globally Hyperbolic Spacetimes

So far we have concentrated on evolution of the *geometry* of a spacetime. But the situation with matter is, in a sense, even worse. Indeed, even if we found a way to describe a non-globally hyperbolic spacetime in terms of evolution, this still would not automatically enable us to describe its matter content.

As a simplest example consider a flat strip U

$$\mathrm{d}s^2 = -\mathrm{d}x_0^2 + \mathrm{d}x_1^2, \qquad x_1 \in \mathbb{R}^1, \quad x_0 \in (-1, 1).$$

in an inextendible spacetime M. Suppose, at the moment $x_0 = 0$ (or, put differently, on the Cauchy surface $\mathcal{S}\colon\ x_0 = 0$ of the spacetime U) there is exactly one particle $\mathscr{A}$. Where will the particle find itself in the (proper) time $\Delta\tau = 1$? If we somehow knew that the strip will evolve into a Minkowski space there would be no problems with answering this elementary question. But in the general, non-globally hyperbolic, case $\mathcal{S}$ is of course *not* a Cauchy surface. As a consequence, the question may have infinitely many answers or none at all. Assume, for instance, that M is the DP space. Then there are the following possibilities:

(ai) It may happen that the particle will cease to exist by that time. It will just vanish in the singularity as is shown in the lower part of Fig. 5a.

(aii) It may also happen that the answer simply does not exist, because the specified initial state corresponds to *none* of possible evolutions (i. e. to none of solutions of equations of motion, or, in terms of [39], to the solution of 'multiplicity zero'). For instance, in the theory of pointlike perfectly elastic (identical) particles any solution must contain the *whole* geodesic (composed, perhaps, from segments of the world lines of different particles), if it contains its part. So, if a solution existed with initial state shown in the upper part of Fig. 5a, it would contain the dashed segment with the end points in $\mathcal{S}$. Hence there would be also the particle $\mathcal{A}'$ at the moment $\mathcal{S}$, contrary to the assumption that $\mathcal{A}$ is the only particle there.

(b) Consider, finally, the case—shown in Fig. 5b and in the lower part of Fig. 5a—in which $\mathcal{S}$ is embedded in M so that $\mathcal{D}(\mathcal{S}) \neq M$. This arrangement guarantees the existence of past inextendible causal curves which remain forever in $M - \overline{\mathcal{D}(\mathcal{S})}$ that is beyond the Cauchy horizon. These may be, in particular, closed curves like $\mathfrak{d}_3$, or curves appearing 'from nowhere' like $\mathfrak{d}_1$ and $\mathfrak{d}_2$. Along such curves some unpredictable (for an observer on *this* side of the Cauchy horizon) information enters the spacetime [74]. In particular, these curves may be the world lines of some objects (when this is consistent with the laws of motion thereof) that neither existed in $\mathcal{D}(\mathcal{S})$, nor originated from only objects which existed there. The unusual origin of these objects—for reasons that will become clear in a moment they will be called

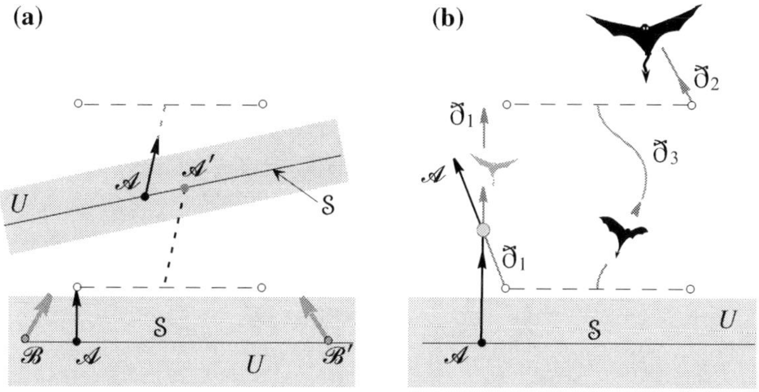

Fig. 5 **a** Two different extensions of U to a DP space. **b** $\eth_i$—are the world lines of Cauchy's demons. In particular, $\eth_3$ is a closed curve (demons of this kind are called jinns in [127]). Varying the initial velocity of $\eth_1$ one obtains infinitely many different evolutions of $\mathscr{A}$ (in [39] such a trajectory is said to have 'infinite multiplicity')

Cauchy demons—does not make them any less physical than the conventional ones: the latter, after all, also appear only from either infinity, or a singularity. So we must take both kinds equally serious (which is especially clear from the 'block universe' viewpoint). This is how the unpredictability of evolution manifests itself: what happens beyond the horizon is affected by factors—the presence of demons and their behaviour—which cannot be taken into account *before* the horizon.

Being legitimate particles, demons, nevertheless, may have quite bizarre interrelation with causality. Let $\mathscr{A}$ be a system whose initial state is fixed at $\mathcal{S}$. Suppose, that in the absence of demons $\mathcal{A}$ has *no* evolution compatible with that state, but such an evolution does exist, if there is a demon $\eth$ in the right place at the right time (for example, this is the case [91] if—instead of $\mathscr{A}$—particles $\mathscr{B}$ and $\mathscr{B}'$ are prepared at $\mathcal{S}$ in the state shown in Fig. 5a, see also examples in Chap. 6). It is natural to ask what will happen if we prepare such a state in a real experiment. Clearly—since we managed to do so—$\eth$ *will* appear where it is needed. But our preparations take place at $\mathcal{S}$, while the world line of $\eth$ does not meet it. So, this demon is brought to existence by quite a wonderful mechanism.

Remark 37 The situation with the Cauchy problem for *fields* is approximately the same. However, *some* spaces were found, where appropriately fixed initial data provide the uniqueness of solution of the wave equation, see [6, 52] and references therein. Also a prescription was formulated [81] which could play the role of the missing data in solving the Cauchy problem for the scalar field in static non-globally hyperbolic spacetimes.

Chapter 3
Shortcuts

The question of interstellar travel under present conditions of physical theory is ... uh ... vague.

Alfred Lanning in [1]

1 The Concept of Shortcut

Signal propagation considered in the previous chapter does not exhaust all types of motion which one might call superluminal. Imagine, for example, that we plan an expedition to Deneb. The distance d from the Sun to Deneb, determined from the parallax, is about 1500 ly. So, a pessimist could think that the expedition, if it is scheduled to start in $t(s) = 2100\,\mathrm{CE}$, will reach the destination no sooner than in $t(f) = 3600\,\mathrm{CE}$ and its report will be received on the Earth no sooner than in $t(w) = 5100\,\mathrm{CE}$, which makes the whole enterprise meaningless.

Notation The Sun, Deneb and the light rays used in determining the distance between them are all supposed to lie in a common simply connected (practically) flat region U, so we use the standard Cartesian coordinates in which t is the time coordinate and in which the Sun and Deneb are at rest. $\overset{\leftrightarrow}{T}$ will denote the time taken by the whole round trip and s, f and w refer, respectively, to the start of the expedition, its arrival at Deneb and its return to home.

Suppose now that the successfully completed expedition returns to the home port at

$$w': \qquad t(w') = 2016\,\mathrm{CE}. \tag{$\star$}$$

Undoubtedly, such a trip would deserve the name 'superluminal'.

The subject of this chapter is the situations in which ($\star$) holds even though (1) no tachyons are involved (in contrast to what was considered in Sect. 1 in Chap. 2) and (2) the expedition takes place in the same world M in which the distance between the Sun and Deneb was found. A simple example of such situations is the flight through an appropriate wormhole, see the following text.

S. Krasnikov, *Back-in-Time and Faster-than-Light Travel in General Relativity*, Fundamental Theories of Physics 193, https://doi.org/10.1007/978-3-319-72754-7_3

The second condition means that the trips under consideration do not reduce to (semi)-superluminal gravitational signals, see Sect. 2.4 in Chap. 2. Yet, there is some similarity between them: two different worlds are compared (the real and a fictitious ones; in this chapter these are, respectively, M and $\mathbb{L}^4$) and a trip made by a spaceship in the former is recognized as superluminal when it is faster than the corresponding trip made by a photon in the *latter*. This time, however, we do not attribute the difference between the two worlds to (consequences of) a particular event s. The wormhole might appear long ago and be known to the pilot even before the start. In the general case that would mean that the points $s, f, w, w' \in M$ have no counterparts in $\mathbb{L}^4$ and we cannot compare the trips in question, see Sect. 2.4 in Chap. 2. We get around this problem by requiring all these points to lie in U, which is isometric to the corresponding region of $\mathbb{L}^4$.

It is the spacetimes in which superluminal, in this sense, trips are possible that will be called shortcuts. Before proceeding further and giving a rigorous definition let us dwell on the question of how the phenomenon is at all possible. Why can a simple bound $\overleftrightarrow{T} > 3000$ yr based on the naive inequality

$$\overleftrightarrow{T} \leftrightharpoons t(w') - t(s) = [t(w') - t(f)] + [t(f) - t(s)] > 2d \tag{1}$$

fail? The point, as we shall see from specific examples, is that, first, the relevant properties are not quite local: a straight segment is guaranteed to be the shortest connection between its end points, *if* it lies in $\mathbb{E}^n$. Belonging to *a* flat region is insufficient, see Example 2. Even if the universe is curved somewhere away from the Sun–Deneb route, the distance between the stars may turn out to be much less than d. Second, in the *pseudo*-Riemannian case it is often hard at all to define a quantity resembling distance.

Example 1 What is the distance d from the horizon of a Schwarzschild black hole of mass m to an observer with $r = r_0 > 2m$? The metric in the relevant region is

$$\mathrm{d}s^2 = -(1 - \tfrac{2m}{r})\mathrm{d}t^2 + (1 - \tfrac{2m}{r})^{-1}\mathrm{d}r^2 + r^2\mathrm{d}\Omega^2,$$

so it might seem natural [111] to define the sought-for distance by the formula

$$d \leftrightharpoons \int_{2m}^{r_0} (1 - \tfrac{2m}{r})^{-1/2}\mathrm{d}r.$$

But the thus defined distance is constant, while the horizon is a sphere moving in each its point with the speed of light.

The idea that changing the geometry of the space *outside* a flat region W we can change the distance between the end points of a segment $\gamma \subset W$ might seem surprising. This happens, however, even in the two-dimensional Riemann case, when the distance between two points is defined, as is customary in the Riemann spaces, to

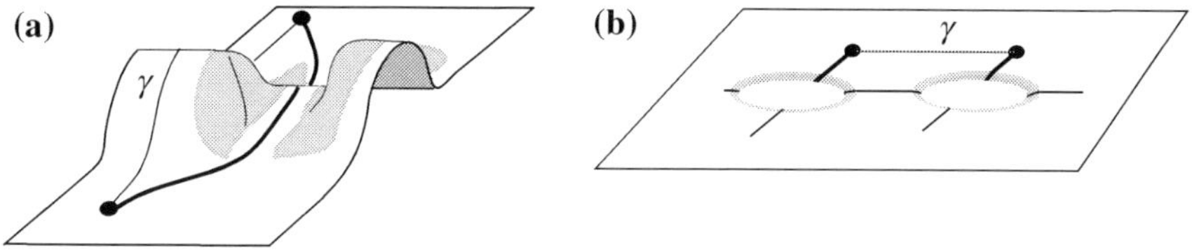

Fig. 1 Either surface differs from $\mathbb{E}^2$ only in the compact regions shaded in grey (in the case of a plane with handle, depicted at the right, it should be borne in mind that the inner parts of the gray annuluses are identified). In both cases, γ lies in a Euclidean region. Nevertheless, γ is *not* the shortest curve with these end points. The latter is depicted as a thick line

be the length of the shortest curve γ_s connecting them. The reason is that the shortest may be a geodesic *leaving* W, see Fig. 1.

Example 2 Consider a plane with metric

$$\mathrm{d}s^2 = \Omega^2(r)(\mathrm{d}x^2 + \mathrm{d}y^2), \qquad r \leftrightharpoons \sqrt{x^2 + y^2}, \tag{2a}$$

where Ω is a monotone function such that

$$\Omega\big|_{r<(1-\delta)r_0} = \Omega_0, \qquad \Omega\big|_{r>r_0} = 1, \tag{2b}$$

$\delta, \Omega_0 \ll 1$, and r_0 being some positive constants. The plane is flat except in a thin annulus $1 - \delta < r/r_0 < 1$, and beyond the annulus, it is indistinguishable from the Euclidean plane. Nevertheless, as is easily checked, the point $(x = -r_0, \ y = 0)$ is *much closer* to the diametrically opposite point $(x = r_0, \ y = 0)$, than if the *whole* space were flat ($\approx 2\Omega_0 r_0$ against $2r_0$).

Example 3 Consider a Riemannian space (M, g^R) depicted in Fig. 1b. It is obtained by removing two equal circles from a plane and deforming some neighbourhoods of the resulting holes so as to glue them together on the next step. To put it rigorously pick a plane and introduce two polar coordinate systems in it: (r, φ) and (r', φ'). Their origins are denoted o and o', respectively, and the angles φ, φ' are measured (clockwise) from the direction of $\overrightarrow{oo'}$. Remove from the plane two equal discs

$$\overline{\mathcal{B}} \leftrightharpoons \{p\colon \ r(p) \leqslant (1 - \delta)r_0\}, \qquad \overline{\mathcal{B}'} \leftrightharpoons \{p\colon \ r'(p) \leqslant (1 - \delta)r_0\},$$

where $0 < \delta \ll 1$ and the constant r_0 is so small that the discs are disjoint. The next step would be to enclose either of the just obtained holes by a ring and to glue these rings together. To do so, however, we need an isometry ψ, see Sect. 6 in Chap. 1, such that after gluing the rings together, by ψ, one obtains a Hausdorff space. But, as follows from Test 68 in Chap. 1, there is no such isometry on $M \leftrightharpoons \mathbb{E}^2 - \overline{\mathcal{B}} - \overline{\mathcal{B}'}$. So, we proceed by deforming the rings, that is by defining a new metric g^R in M and requiring it to satisfy the condition

$$ds^2 = \begin{cases} dr^2 + r^2\, d\varphi^2, & \text{at } r > (1+2\delta)r_0, \\ dr^2 + [(r-r_0)^2 + r_0^2]\, d\varphi^2, & \text{at } (1-\delta)r_0 < r < (1+\delta)r_0, \end{cases}$$

and the same condition with $r \dashrightarrow r'$. Now the space is non-flat in the narrow annuli $r^{(\prime)}/r_0 < (1-\delta, 1+2\delta)$ shown shaded in Fig. 1b, and their inner parts $r^{(\prime)}/r_0 < (1-\delta, 1+\delta)$ are related by the isometry

$$r'[\psi(p)] = 2r_0 - r(p), \qquad \varphi'[\psi(p)] = -\varphi(p). \tag{3}$$

By Test 68 in Chap. 1 the use of ψ in gluing the rings together results in the appearance of a Hausdorff Riemannian space. Topologically it is a plane with handle, see Fig. 3a.

Now note that the segment γ lies entirely in the flat region. Nevertheless, its ends can be connected by a geodesic running through the handle and that geodesic will be shorter than γ (if the distance between the balls $\mathcal{B}$ and $\mathcal{B}'$ was sufficiently large).

Remark 4 The isometry ψ is non-unique. We could use as well any composition of that defined in (3) with a rotation $\varphi \mapsto \varphi + \varphi_0$ or reflection $\varphi \mapsto -\varphi$ (if we don't mind non-orientable surfaces). Each time we would obtain a different space even though the corresponding simply connected regions of all these spaces are isometric.

It is by generalizing in the simplest way the just considered examples to the four-dimensional Lorentzian case that we come to the idea of the shortcut.

Definition 5 Let C be a timelike cylinder in the Minkowski space $\mathbb{L}^4$ (i.e. a set of the form $\mathbb{L}^1 \times \mathcal{B}$, where $\mathcal{B}$ is a spacelike open three-dimensional disc). A region U of a globally hyperbolic spacetime M and also M itself are called *shortcuts*, if there are an isometry $\chi\colon\ (M - \overline{U}) \to (\mathbb{L}^4 - \overline{C})$ and a pair of points $p, q \in (M - \overline{U})$ such that

$$p \preccurlyeq q, \qquad \chi(p) \not\preccurlyeq \chi(q).$$

In other words, we call M a shortcut if it can be obtained from the Minkowski space by replacing a timelike cylinder C with something else (viz. U) so that a pair of initially spacelike separated points become causally related. The trip from p to q is 'faster-than-light' in the sense that the traveller reaches the destination faster than a photon (be the spacetime $\mathbb{L}^4$) would.

Remark 6 An observer in M indifferent to what takes place in U may think that they live in the Minkowski space. The presence of the shortcut would show up only in the 'tachyonlike' behaviour of ordinary particles, which can now travel between spacelike connected points. So, one may conjecture [93] that, by analogy with a pair of tachyons, see Sect. 2.1 in Chap. 2, a combination of two shortcuts can be converted into a time machine. This conjecture was confirmed in special cases [46, 47, 136].

2 Wormholes

2.1 What Are Wormholes?

The concept of wormholes ('bridges') is almost as old as relativity itself [42]. Still there is no commonly accepted rigorous definition of wormhole as of today. The reason probably is that one usually needs such definitions only in proving general theorems, while theorems concerning wormholes are very few.

To visualize the spacetime in question imagine two Euclidean spaces $\mathbb{E}^3$, from each of which a ball is removed. Connect smoothly—with a cylinder $\mathbb{S}^2 \times \mathbb{I}$—the boundaries of the holes left by the missing balls (the difference with what was done in Example 3 is in that the balls there lied initially in the *same* space). The resulting space, see Fig. 2a, is called an *inter-universe wormhole* [172]. A vicinity of the 'seam'—shown in grey—is called the *throat* and the (somewhat indefinite) neighbourhoods of the former holes are *mouths*. When the space under discussion is viewed as a spacelike section of a spacetime, the latter is called a wormhole, too. Also, the same name is sometimes given to spaces *resembling* that described above (we shall consider a couple of examples), or even to a throat alone. In particular, there is a 'local' approach, within which the defining property of the wormhole is the presence of a throat, which is understood as a closed two-surface of the minimal—with respect to some infinitesimal deformations—area [80]. In this approach the space depicted in Fig. 2b is a wormhole.

A classic example of a wormhole is the Schwarzschild space (though in this case the spaces connected by the throat are only *asymptotically* flat), see Chap. 31 and especially Fig. 31.5 in [137]. Another typical wormhole is the Morris–Thorne spacetime. This is an (obviously static and spherically symmetric) space with metric

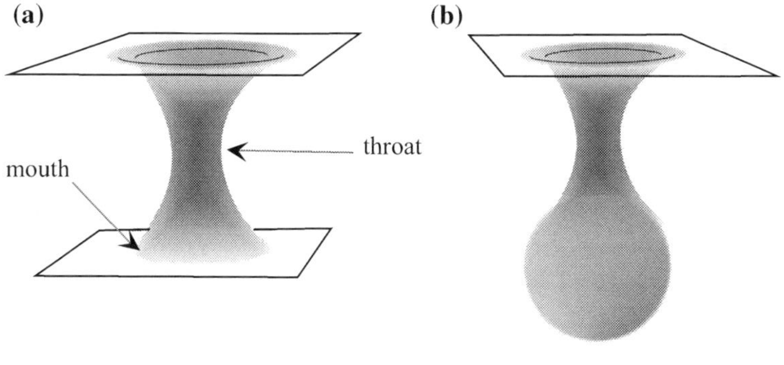

Fig. 2 As a rule, by a wormhole one understands a three-dimensional analogue of the space (**a**), or the four-dimensional result of its evolution. Sometimes this name is extended to spaces like **b** [172]

$$ds^2 = -e^{2\Phi(x)}dt^2 + dx^2 + r^2(x)(d\vartheta^2 + \sin^2\vartheta\, d\varphi^2), \qquad x \in \mathbb{R}, \quad r > 0, \quad (4)$$

where r and Φ are smooth functions tending at $x \to \pm\infty$, respectively, to $|x|+c_1 \pm c_2$ and to $\pm c_3$. In the region, where $dr/dx \neq 0$, this metric is often written [136] as

$$ds^2 = -e^{2\Phi}dt^2 + \frac{dr^2}{1 - b/r} + r^2(d\vartheta^2 + \sin^2\vartheta\, d\varphi^2), \qquad r \neq b(r).$$

The functions $\Phi(x)$ and $b(x)$ in this representation are called, respectively, the shape and the redshift functions. If the space outside of a Morris–Thorne wormhole is empty, Φ and b by Birkhoff's theorem must take the same form as in the Schwarzschild space. This allows one to talk about the 'mass' of a wormhole's mouth (note that the masses of the mouths do not have to be equal). It is worthy of note that there are no reasons to expect such a mass to be positive either.

Example 7 Consider an *externally flat* wormhole, which is defined to be a Morris–Thorne space W_{X}, with

$$\Phi = 0, \quad r = |x| + \mathfrak{p} \qquad \text{at } x \notin (-\mathfrak{z}, \mathfrak{z}),$$

where $\mathfrak{p}$ and $\mathfrak{z}$ are positive constants. As is easily seen, the wormhole consists of three regions: the throat 'of length $2\mathfrak{z}$'—it is the set of the points with $|x| \leqslant \mathfrak{z}$—and two Minkowski spaces, in either of which the set $r \leqslant \mathfrak{p} + \mathfrak{z}$ is missing. The masses of both mouths of this wormhole are zero.

Remark 8 We shall consider many spherically symmetric spacetimes. These, formally speaking, are spacetimes of the form $\mathbb{L}^1 \times \mathbb{S}$ in each section $\{t\} \times \mathbb{S}$, $t \in \mathbb{L}^1$ of which the group SO(3) of isometries acts. Its orbits are spheres labelled by *radiuses* r, and it should be stressed that for a sphere $\mathcal{S}$ the value $r(\mathcal{S})$ is *not* the distance to the 'centre' (the centre may not exist at all, as is the case with the wormholes), but $\sqrt{A/(4\pi)}$, where A is the area of $\mathcal{S}$. Correspondingly, the quantity $r(\mathcal{S}_1) - r(\mathcal{S}_2)$ is the difference in size between $\mathcal{S}_1$ and $\mathcal{S}_2$ and *not the distance between them.*

Example 9 Consider the spacetime

$$ds^2 = -dt^2 + r\, dx^2 + r^2(d\vartheta^2 + \sin^2\vartheta\, d\varphi^2),$$
$$x \in \mathbb{R}, \qquad r(x) \leftrightharpoons \mathfrak{p} + \tfrac{1}{4}x^2, \qquad \mathfrak{p} \leftrightharpoons const.$$

Making r a new coordinate (at $x \neq 0$, of course), transform the metric to

$$ds^2 = -dt^2 + \frac{r}{r - \mathfrak{p}}\, dr^2 + r^2(d\vartheta^2 + \sin^2\vartheta\, d\varphi^2), \qquad \text{at } r \neq \mathfrak{p}.$$

Thus, the spacetime under consideration is a simplest Morris–Thorne wormhole with $\Phi = 0$, $b = \mathfrak{p}$. Remarkably, its scalar curvature R is zero. This means, in particular, that the classical massless scalar field [8], i.e. the field obeying the equation

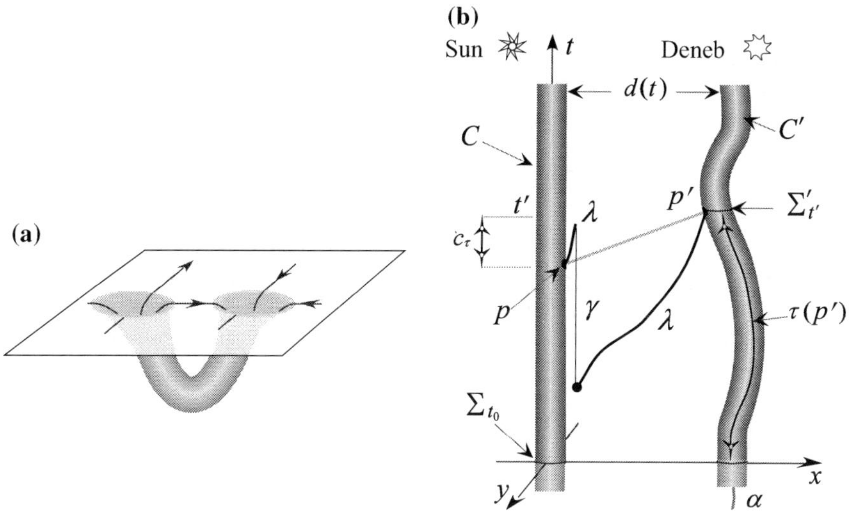

Fig. 3 Sections of an intra-universe wormhole. **a** The section $t = const$, $\varphi = const$ is merely a plane with a handle. This is the same surface as in Fig. 1b. **b** The section $z = const$. Actually, λ is a *continuous* curve traversing the wormhole and the grey curve pp' is a *closed* geodesic

$$(\Box - \xi R)\,\phi = 0, \qquad \xi = const, \tag{$*$}$$

can be constant in this spacetime: $\phi = \phi_0$. The stress–energy tensor for such a solution is $T_{ab} = \xi G_{ab}\phi_0^2$, see [14, (3.190)]. Thus, this wormhole being filled with the field $\phi = 1/\sqrt{8\pi\xi}$, solves the entire system consisting of $(*)$ and Einstein's equations [95].

Consider now another type of spacetime. It is constructed exactly as the inter-universe wormholes with the only deviation: the two balls that are to be removed are taken not in two different spaces, but in different regions of the *same* space, so that instead of the surface depicted in Fig. 2a one gets a 'handle' depicted in Figs. 1b and 3a. It is the evolution of this surface that constitutes a wormhole W_8. The wormholes of this kind are called *intra-universe* [172] and of all spaces constructed in this section so far, only they satisfy the definition of shortcut (if the throat is sufficiently short). Especially interesting are intra-universe wormholes with *variable* distance between the mouths. Such spacetimes, however, do not always admit a (3+1)-splitting, so we have to modify our description.

Remove from the Minkowski space two open cylinders, C and C', bounded at each t by the spheres, respectively,

$$\Sigma_t \leftrightharpoons \{p\colon \quad x^2(p) + y^2(p) + z^2(p) = c^2\}, \quad \text{and}$$
$$\Sigma'_t \leftrightharpoons \{p\colon \quad [x(p) - d(t)]^2 + y^2(p) + z^2(p) = c^2\},$$

where $d(t)$ is the x-coordinate of the point $\alpha(t)$ of some timelike curve α, see Fig. 3b (instead of spheres one could take, say, tori, pretzels, or other appropriate two-dimensional closed surfaces and instead of the Minkowski space—some other asymptotically flat spacetime). Now we could just glue together the boundaries of the thus obtained cylindrical holes. The result would be a wormhole whose mouths move with respect to each other, but the space would not be smooth. So, the last step would be 'smoothing the seam'. If, however, we are to describe a wormhole with infinitely short and narrow throat (that the motion of the mouths may leave the throat short is seen from Fig. 5a), we can simplify our task and *model* the wormhole by the 'holed' space $\mathbb{L}^4 - C - C'$ in which the following rule acts. An inextendible curve reaching a point $p \in \Sigma_t$ with velocity $\boldsymbol{v}$ is regarded as differentiable only if it is continued by a curve with the initial velocity $\boldsymbol{v}'$ emanating from $p' \in \Sigma'_{t'}$, where

$$\begin{aligned} &v_t = v'_{t'}, \quad v_\vartheta = v'_{\vartheta'}, \quad v_r = -v'_{r'}, \quad v_\varphi = -v'_{\varphi'}, \\ &\vartheta'(p') = \vartheta(p), \quad \varphi'(p') = -\varphi(p), \quad \tau(p') = \tau(p) + c_\tau. \end{aligned} \tag{5}$$

In these relations, r, ϑ, and φ are the standard polar coordinates in a neighbourhood of Σ_t, while r', ϑ' and φ' are the coordinates in a neighbourhood of $\Sigma'_{t'}$ induced by an isometry ψ that maps the former neighbourhood to the latter. c_τ is a constant and $\tau(p^{(\prime)})$ denotes the (Lorentzian) length of the portion of $C^{(\prime)}$ confined between the plane $t = t_0$ and the point $p^{(\prime)}$

$$\tau(p) \leftrightharpoons t(p), \qquad \tau(p') \leftrightharpoons \int_{t_0}^{t(p')} \sqrt{1 - \dot{d}^2}\, \mathrm{d}\breve{t}, \tag{6}$$

where the dot stands for $\mathrm{d}/\mathrm{d}\breve{t}$. In other words τ is the time between the moments $t = t_0$ and $p^{(\prime)} \in \Sigma^{(\prime)}_{t^{(\prime)}}$ as measured by an observer at rest with respect to the corresponding mouth (we do not have to specify *where* in $\Sigma^{(\prime)}$ the point $p^{(\prime)}$ lies, that is what we required the narrowness of the wormhole for).

The just given description of W_8 suffers an important ambiguity: before the boundary of C is glued to the boundary of C' the former can be shifted in time and/or rotated as in the two-dimensional case, see Remark 4. In terms or the rules (5) this is a consequence of the arbitrariness of c_τ and ψ. The latter is defined only up to an inversion ι and a rotation ϕ mapping $\Sigma'_{t'}$ to itself (see Fig. 4).

Thus even a simplest (static, spherically symmetric, flat outside a compact—in the space directions—set, and having $d = const$) intra-universe wormhole W_8 has many free parameters, including the distance d, the shift c_τ and the three angles defining the mutual orientation of the mouths. Note that the last four parameters are *global*. Varying the values of these parameters one obtains quite different wormholes—suffice it to say, that c_τ determines whether causality holds in W_8—but that difference cannot be established in any local experiments where the relevant regions are simply connected. The same is true for the difference between intra- and inter-universe wormholes (Fig. 3).

Fig. 4 An oriented spherically symmetric wormhole. For a curve traversing it, the exit point is obtained from the point of entry by a translation ψ—it sends a sphere around one mouth to that around the other—and a reflection with respect to the 'zeroth' meridian. Which meridian is zeroth is determined by the choice of two freely specifiable angles

Another important characteristic of a wormhole is its 'traversability'. The point is that in the course of its evolution a wormhole can break so quickly that no observer will have time to travel through the throat. To put differently, the regions on opposite sides of the throat will not be connected by causal curves. It is such wormholes that are called *non-traversable*.[1] The best-known specimen of a non-traversable wormhole is the (maximally extended) Schwarzschild space, see Sect. 2.1 in Chap. 9 for a brief consideration and [58], [135, Sect. 31.6] for detailed ones. The *traversable* wormholes are exemplified by the static space (4).

Of all the types of shortcuts the wormholes are studied the most, see [172] for a review as of 1994. This is because their significance is not restricted to the subject of this book: whether wormholes (can) exist is interesting by itself. We have got used to the idea that the universe topologically may not be just $\mathbb{R}^4$ when Planckian physics or, on the contrary, cosmology are discussed. However, when it comes to the macroscopic scale, multiple connectedness of spacetime is often regarded exotic and essentially unphysical. It seems important to (in)validate this—somewhat strange—bias. Also, wormholes, if they exist, offer an exciting possibility to rule out the concept of electric charge and thus to make a great step towards uniting gravitation and electromagnetism. A beautiful idea of 'charge without charge' was proposed by Einstein and Rosen in [42] and by Wheeler [180, 181]. Consider the electrostatic field with the force lines entering one mouth of a wormhole and coming out of the other. The field satisfies the charge-free Maxwell equations, but to a distant observer the mouths of the wormhole *look* like a pair of opposite charges; in particular, the flux through a sphere enclosing[2] one of them is non-zero. This inspires one to speculate that perhaps in nature there is no such 'substance' as charge at all, and the charged elementary particles are merely mouths of wormholes. Unfortunately, today we know

[1] Originally, the requirements were much more restrictive [136].

[2] Of course, it only *seems* that the sphere encloses the mouth: the real boundary of the throat is a *pair* of spheres, each around its own mouth.

too little about wormholes for a serious and detailed discussion of the 'charge without charge' concept.

Remark 10 Though the wormhole does resemble a pair of charges, the resemblance is not perfect.

(1) A charge generates the electric field, but it also moves in response to the external electric field. Why and how a wormhole would move in such a situation one can only speculate;
(2) Just because a wormhole is not the Minkowski space, the solutions of the Maxwell equations do not have the Coulomb form near a point charge, even when the charge rests somewhere far from the throat of W_8 in a flat region. Physically speaking, this means that, generally, in the presence of a wormhole a point charge experiences a force from the field generated by that very charge (a 'self-force'), see A.1.

As of today there is no observational evidence either for or against the existence of wormholes [168] (some—rather weak—restrictions were inferred from the idea that a wormhole mouth with negative mass $m \sim -0.1M_\odot$ must produce—by lensing quasars in an unusual way, see [32, 165] and the references there, gamma ray bursts of yet unobserved characteristics [168], or—also unobserved—double images of the mentioned quasars [165]). So, the problem of the existence of wormholes is purely theoretical at the moment. It can be split into two parts:

(1) Can a wormhole *appear*? Proposition 51(c) in Chap. 1 implies that new wormholes can form only at the cost of the global hyperbolicity. But this gives us only a necessary condition. Sufficient ones are hard to be found as we know too little about the evolution of non-globally hyperbolic spacetimes, see Sect. 3 in Chap. 2 for a discussion. On the other hand, it is conceivable that the universe came into being already with a wormhole in it. And again, today nothing can be said either for or against such a possibility.
(2) Can a traversable wormhole be supported against collapse by *realistic* matter? The first models of such wormholes [18, 44] were quite exotic and this was no accident: as is shown in [136] the Einstein equations require the matter in the throat of a static spherically symmetric wormhole *always* to be *exotic*, i.e. to violate the weak energy condition.

An argument due to Page, see Remark in Sect. F2 of [136], shows that the same must be true in much more general case too. Consider a very large sphere Ξ around one of the mouths and the beam formed by the inward null rays normal to Ξ and crossing it at a given moment. The expansion θ of that beam, see Sect. 1.2 in Chap. 2 for definition, is initially negative in all points $p \in \Xi$ (unless the space inflates *very* fast, see below). However *after* the beam traverses the throat, the expansion, clearly, changes its sign—the geodesics begin to diverge. Thus, there are two possibilities:

1. θ becomes infinite in a certain point $p_* \in \gamma_p$ of each of the geodesics traversing the wormhole. Correspondingly, any point of γ_p lying *beyond* p_* can be reached from

Ξ along a timelike curve, which is in fact a corollary to [76, Proposition 4.5.14], see the paragraph following Proposition 6 in Chap. 2;
2. there is a point in γ_p where θ *increases*.

The first possibility is definitely excluded in the case of a Morris–Thorne wormhole, where the radial geodesics for lack of a centre never meet and θ, consequently, remains locally bounded. In more complex wormholes one might expect that *some* γ_p do have points conjugated to Ξ. However, the spacetime must be quite pathological if *every* null geodesic normal to Ξ (of those that traverse the wormhole) on its way to the infinity is run down and left behind by some other geodesic of the same beam. On the other hand, the *second* possibility can be realized only at the cost of the Weak energy condition, see (4) in Chap. 2. It is easy to verify, for instance, that in the wormhole from Example 9 the WEC is violated in *every* point. This brings us to the following hypothesis:

Hypothesis 11 No sufficiently regular spacetime satisfying the weak energy condition contains a traversable wormhole.

Our consideration was not quite rigorous, of course, and we left it unspecified which spacetimes should be regarded as 'sufficiently regular'. Apparently, they must at least be globally hyperbolic: dropping this requirement enables one to find many traversable wormholes in which the WEC holds. These are, for example, the Kerr spacetime, the 'dihedral' wormholes [172], and the spaces like those considered in Sect. 5 in Chap. 4. However, global hyperbolicity alone is not sufficient for the hypothesis to be valid. For example, in the space found in [123], both global hyperbolicity and the WEC hold. Yet there is a traversable wormhole in it (inflating so fast that even the *ingoing* null radial geodesics have a positive expansion, cf. [117]). Friedman, Schleich and Witt formulated the theorem—called 'Topological censorship' [54]–asserting that for a traversable wormhole to violate the WEC it is sufficient to be globally hyperbolic and asymptotically flat (the latter is a *term*, its precise meaning can be found in [64]). The theorem, however, remains unproven [105].

One can speculate that some day a kind of matter will be discovered able to violate the WEC, but usually, it is thought that classical matter does obey it. Then from Page's arguments, it follows that a wormhole can be stabilized against instant collapse and thus made traversable only by quantum effects. This possibility is discussed in part II.

2.2 *Wormholes as a Means of Transport*

Consider a short traversable—static, say—wormhole with one of its entrances located near the Earth and the other—near Deneb. Either the onward flight to Deneb through the throat—its duration is $t(f) - t(s) \approx c_\tau$, see (5)—or the return trip—it takes $t(w') - t(f) \approx -c_\tau$—will be faster than it would be in the Minkowski space. So, the wormhole is a shortcut and quite an effective one, because *irrespective of d and c_τ* the round trip takes *no time* at all: $\overset{\leftrightarrow}{T} \approx c_\tau - c_\tau = 0$. What makes it possible in

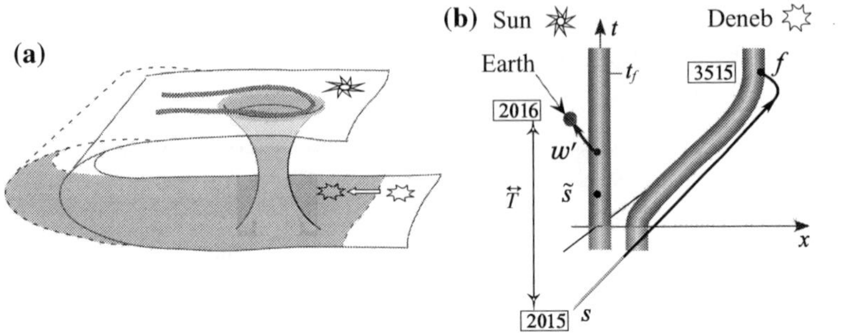

Fig. 5 **a** Two consecutive spacelike sections of a spacetime containing a wormhole one of whose mouths moves with respect to the other. The shape of the handle does not change, but the 'distance between the mouths' grows. **b** On its way back the spaceship plunges into the wormhole in 3515 and, according to the rule (5), emerges from it near the Earth in 2016, i.e. in $\overleftrightarrow{T} = \tau(f) \approx 1$ year after the start

this case to get around the 'light barrier', embodied in the inequality (1), is the fact that the *'real'* distance to the destination has nothing to do with the parameter d.

As a means of transport, such a shortcut has a serious drawback: it seems improbable that suitable wormholes can be found for *all* places of interest. So, let us discuss the possibility of *moving* a wormhole mouth (implying that it is easier to build—or to find—a wormhole with both mouths near the Earth and vanishing c_τ). An interstellar expedition, in this case, might look as follows. An astronaut starts from the Earth and heads to Deneb piloting a usual spaceship which tows one of the wormhole's mouths, see Fig. 5b. The mouth is pulled so gingerly that the throat remains short and our simplified description valid. The spaceship moves with a near-light speed and though by the Earth clock its trip takes $t(f) - t(s) \approx 1500$ yr, the *proper* time $\tau(f)$ of the flight is small, a year, say. So, when the traveller returns home through the wormhole it turns out that a mere

$$\overleftrightarrow{T} \risingdotseq t(w') - t(s) = \tau(w') = \tau(f) - c_\tau = \tau(f) \approx 1\,\text{yr} \tag{7}$$

has passed after the start. This solves the problem of the 'light barrier' though in quite a bizarre way: the expedition returns home 15 centuries before it reaches the destination.

Note that the trip described above is not a time journey. Roughly speaking, the 'time jump' is accompanied here by a 'space jump' and the latter is greater than the former. Indeed, even if the spaceship that plunges into the wormhole at the point f was launched from the Earth in $\tilde{s} \succ s$ (this is possible because the towed mouth moves with a *sub*luminal—though large—speed), it will return to the Earth in

$$\overleftrightarrow{T} = t(w') - t(\tilde{s}) > t(w') - \{t(f) - d[t(f)]\} =$$

$$t(s) + \int_{t(s)}^{t(f)} \sqrt{1 - \dot{d}^2}\, \mathrm{d}\check{t} - t(f) + d[t(f)] = \int_{t(s)}^{t(f)} \left(\sqrt{1 - \dot{d}^2} - 1 + \dot{d}\right) \mathrm{d}\check{t} =$$

$$= \int_{t(s)}^{t(f)} \sqrt{1 - \dot{d}}\left(\sqrt{1 + \dot{d}} - \sqrt{1 - \dot{d}}\right) \mathrm{d}\check{t}, \qquad (8)$$

where the inequality expresses the fact that to catch up with a photon emitted from the Earth the spaceship must leave the Earth *before* the photon. The next equality is obtained by substituting the expression (6) into the junction condition $t(w') = t(s) + \tau(f)$, see (5). In our scenario d only grows with time, so it follows from the bound (8) that $\overleftrightarrow{T}$ is positive and causality is not violated.

Remark 12 The difference $t(f) - t(w') = \int_{t(s)}^{t(f)} (1 - \sqrt{1 - \dot{d}^2})\, \mathrm{d}\check{t}$ is the time of flight through the wormhole (as measured by the Earth clock). We see that it is negative (this is, in effect, the twin paradox). If one used the naive formula $T = S/v$ one would have to conclude that the spaceship travels a *negative* distance. This effect does not have to be the result of moving a mouth. Any other method yielding $c_\tau \in (0, d)$ will do (for the homeward trip the best causality preserving c_τ is, clearly, $c_\tau = d - 0$).

The use of wormholes as a means of travel looks so simple merely because we have ignored all hard questions so far. One of them concerns the towing of the mouth. How *exactly* could this be done? The point is that the mouth is *not a body*, even though there is a remote resemblance. There is no use throwing rocks in it—the rocks will fly through the wormhole without stirring the mouths. It is equally useless to drag it with a net—a rope that clutches the mouth will take the form of a geodesic, see Fig. 5a, and pulling its ends further will result in breaking the rope, not in shifting the mouth. One could try to 'beckon' the mouth [137] by bringing to it a gravitating body, or an electric charge. But again, no reasons are seen to expect that the mouth, like a pointlike test particle, will move to or from the charge.

3 Warp Drives

Just as the wormhole can be obtained by the generalization to the four-dimensional case of the two-surface depicted in Fig. 1b, so one obtains another shortcut by multiplying by $\mathbb{L}^2$ the surface shown in Fig. 1a (or, to be more precise, the surface discussed in Example 2). And by slightly modifying this procedure one obtains what can be called 'warp drive'.

Example 13 (*The Alcubierre space*) Let Ω be the function defined by (2b). Then the spacetime M with topology $\mathbb{R}^4$ and metric

$$\mathrm{d}s^2 = -\mathrm{d}t^2 + \Omega^2(r)(\mathrm{d}x^2 + \mathrm{d}y^2 + \mathrm{d}z^2) \qquad r \leftrightharpoons \sqrt{x^2 + y^2 + z^2}, \qquad (9)$$

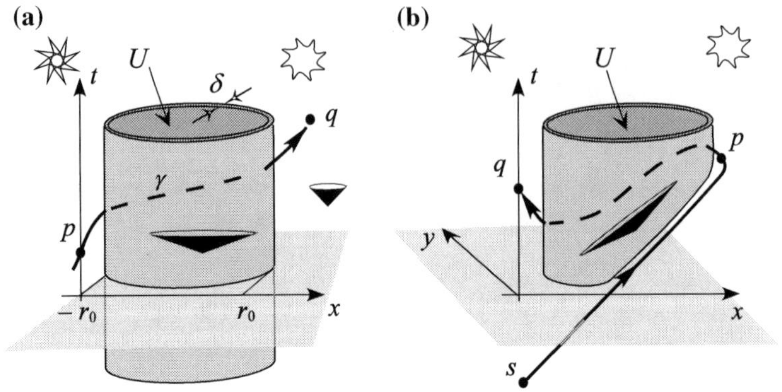

Fig. 6 **a** The Alcubierre space. The inside of U is flat, but the light cones there are $1/\Omega_0$ times more open (in these coordinates) than in the external space. So, even the very mildly sloping curve γ is actually timelike. Correspondingly, it will reach Deneb much sooner than in the case $\Omega \equiv 1$. **b** The Krasnikov tube. The light cones in U are tilted towards the Earth. Because of this $q \succ p$, even though $t(q) < t(p)$

is flat everywhere except the thin wall bounding the cylinder $U = \{r < (1-\delta)r_0\}$. But the light cones inside U are 'more open', then outside (this, of course, is a coordinate effect), and therefore the curves like γ from Fig. 6a are timelike, even though they would be spacelike be *the whole* cylinder $\{r < r_0\}$ flat. Obviously, M is a shortcut.[3]

Note that in using the Alcubierre space an advanced civilization[4] would come up against the same problem as with static wormholes: this type of warp drive does not enable one to quicken the *first* flight (unless, by a happy coincidence, one found an already existing shortcut just to the chosen destination) [93]. Indeed, assume that geometrically our Galaxy now and in the foreseen future is (almost) a region of the Minkowski space. Then in order to *create* the cylinder U between the Earth and Deneb one would have to distribute in a certain way some matter and wait until the metric governed by the Einstein equations will take the desired shape. In particular, one will have to wait until the metric changes appropriately in the vicinity of Deneb, which in the absence of superluminal—in the sense of Sect. 2 in Chap. 2—communication will take at least 1500 yr. By this time the flight—even with a superluminal speed—will not make sense anymore.

The problem can be solved as in the case of wormholes by making the 'distance' from the Earth to Deneb negative.

---

[3]The metric originally considered by Alcubierre [3] is slightly different from (9), which was proposed in [29] (in particular,the light cones in U are tilted forward), but the principle of operation is the same.

[4]To which in this context it is customary to refer since [136].

Example 14 (*The Krasnikov tube*) Let M be again $\mathbb{R}^4$, and let the metric be

$$g\colon \qquad ds^2 = (dx - dt)(k(r)dx + dt) + dy^2 + dz^2. \tag{10}$$

$k(r)$ stands for a monotone function such that

$$k\big|_{r>r_0} = 1, \qquad k\big|_{r<(1-\delta)r_0} = k_0, \qquad \text{where } -1 < k_0 = const < 0.$$

The space is curved only in a spherical layer of width δr_0 exactly as in the previous example. But there is a significant difference between the two situations: the return trip now *ends before it starts* [93] in perfect analogy to the trip through the wormhole discussed on p. 79. This effect is caused by the fact that in the region U, where $k = k_0$, the light cones are tilted so much that some *future*-directed null vectors are, nevertheless, directed towards decreasing values of t, see Fig. 6b. As a result, the timetable of the expedition is more like that with a portable wormhole mouth than that considered in Example 13: in 3515, immediately after the visit to Deneb is over (see Fig. 5b), the spaceship plunges into the curved region ('tube') and emerges from it near the Earth in 2016.

Remarkably, such a back-to-the-past journey—through solving, perhaps, the problem of prohibitively large interstellar distances—does not involve causality violation. M actually is globally hyperbolic.[5]

Proof Rewrite the metric as

$$g\colon \qquad ds^2 = -\omega_0^2 + \omega_1^2 + dy^2 + dz^2, \qquad \omega_0 \leftrightharpoons dt + \tfrac{k-1}{2}dx, \quad \omega_1 \leftrightharpoons \tfrac{k+1}{2}dx$$

and denote by $\boldsymbol{e}_{0,1}$ the vectors dual to 1-forms $\omega_{0,1}$

$$\boldsymbol{e}_0 = \partial_t, \qquad \boldsymbol{e}_1 = \tfrac{2}{k+1}(\tfrac{1-k}{2}\partial_t + \partial_x).$$

Consider an arbitrary inextendible future-directed timelike curve $\mu(l)$. Decompose its velocity in the orthonormal basis $\{\boldsymbol{e}_0, \boldsymbol{e}_1, \partial_x, \partial_y\}$:

$$\partial_l = A\partial_t + B\tfrac{2}{k+1}(\tfrac{1-k}{2}\partial_t + \partial_x) + C\partial_y + D\partial_z.$$

Because $\mu(l)$ is future-directed and timelike

$$A > 0, \quad \text{and} \quad A^2 > B^2 + C^2 + D^2. \tag{11}$$

Now introduce the auxiliary *Riemannian* metric g^R in M

$$g^R\colon \qquad ds^2 = \omega_0^2 + \omega_1^2 + dy^2 + dz^2$$

[5] Though a *pair* of tubes can combine into a time machine [47].

(it differs from g only in the sign at ω_0^2) and require that—yet unspecified—l be the natural parameter with respect to g^R, i.e. that $|l_1 - l_2|$ be the length of the segment of μ between $\mu(l_1)$ and $\mu(l_2)$. Then

$$A^2 + B^2 + C^2 + D^2 = 1.$$

Comparing this with inequalities (11) one gets

$$A > \sqrt{2}/2 > |B|.$$

Now let us define $F(t, x) = t + \frac{k_0 - 1}{2} x$ and find out how F changes along μ:

$$\begin{aligned} \mathrm{d}F/\mathrm{d}l = F,_i (\partial_l)^i = AF,_t + B\tfrac{2}{k+1}(F,_x + F,_t \tfrac{1-k}{2}) \quad &= A + B\tfrac{k_0-1}{k+1} + B\tfrac{1-k}{k+1} \geqslant \\ \geqslant \sqrt{2}/2(1 - \tfrac{k-k_0}{k+1}). \end{aligned}$$

Thus, F changes with the speed bounded away from zero. On the other hand, since (M, g^R) is, obviously, complete, l takes all values from minus to plus infinity. Hence, μ intersects—exactly once—the surface $F = 0$. So, this surface is a Cauchy surface of M, which, correspondingly, is globally hyperbolic by Proposition 50 in Chap. 1.

The existence of the points p and q appearing in the definition is evident and we conclude that M is a shortcut.

The purpose of the just considered examples was to show the possibility *in principle* of superluminal—in the above-specified sense—travel. Below, however, we shall also touch on some practical aspects and see that the large size of a shortcut can make its creation impossible by requiring too much 'exotic matter'. It should be noted therefore that for an interstellar trip one does not need a huge cylinder with r_0 of the order of light years. In fact, it would suffice to enclose the traveller in a small sphere—in the case of Example 13 it is called the *Alcubierre bubble*—with diameter $h \ll d$ (see Fig. 7). The transition from the cylinder to the bubble enables one to reduce the relevant quantity quite considerably: by the factor 10^{32}, when $d \sim 100$ lyr and $h \sim 100$ m.

Whether putting the shortcuts into a separate category of spacetimes is useful, only the future will tell, but one *drawback* in Definition 5 is already seen. The point is that the space far from a wormhole or an Alcubierre bubble is normally thought of as flat. But in the definition of a shortcut requires this flatness must be *exact*. Unfortunately, this, seemingly technical, restriction apparently excludes the spacetimes satisfying the WEC. Indeed, if the energy density is everywhere non-negative, and in some regions even positive, one would expect the mass of the shortcut to be also positive, while according to our definition it is zero.[6]

---

[6] This reasoning is not rigorous. As of today the conjecture that any shortcut violates the WEC is not proven (see [121, 140], though).

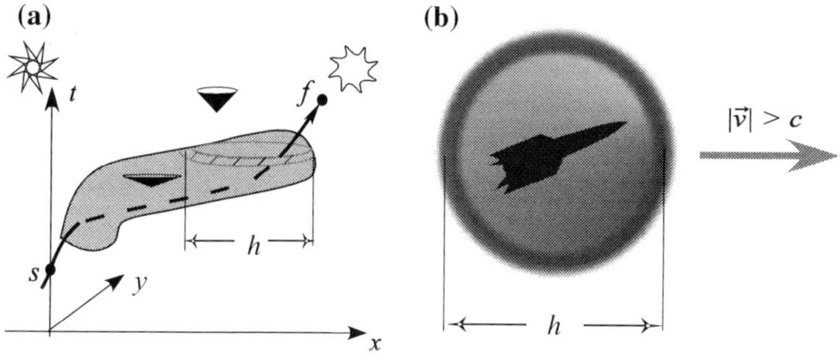

Fig. 7 The Alcubierre bubble. **a** The section $z = 0$. The curve from s to f is timelike, its little slope being merely a coordinate effect. At the same time, the little slope of the front of the grey piece of the spacetime means (since the front propagates in the region where the cones are narrow) that the speed of its propagation is greater than the speed of light. **b** The section $t = const$. The spaceship moves together with the thin spherical shell the speed of which is superluminal. In the absence of superluminal signals, this means that the 'bubble' is not a single moving sphere, but rather a sequence in time of independent motionless spheres, cf. Example 10 in Chap. 2. They could be created by some devices placed *in advance* along the travel path and programmed to come into operation at preassigned moments [93]

At first glance, this shortcoming might be overcome simply by changing $\mathbb{L}^4$ in the definition of a shortcut to an appropriate curved spacetime and the cylinder C to an appropriately defined world tube. A thus defined 'curved shortcut' may satisfy the WEC, as is shown in A.2. It is, however, hard to say how 'efficient' such a shortcut is. A positive mass is something opposite to the shortcut, something that gives rise to the signal delay [59] called the 'Shapiro effect'. So, a pessimist could suspect that all a shortcut satisfying the WEC can do is to compensate the said tiny delay.

Chapter 4
Time Machines

Everyone knows that dragons don't exist. But while this simplistic formulation may satisfy the layman, it does not suffice for the scientific mind. [There were] three distinct kinds of dragon: the mythical, the chimerical, and the purely hypothetical. They were all, one might say, nonexistent, but each nonexisted in an entirely different way.

The Third Sally, or The Dragons of Probability [112]

In this chapter we start with the introduction, in Sect. 1, of the concept of appearing time machine and give a simple example thereof. In the following section, a few less trivial examples are considered. The first of them is the Misner spacetime. It is one of the oldest and presumably the most important time machines: indeed, being just a flat cylinder it is in a sense the time machine in its pure form. And anyway, with its rich and counter-intuitive structure the Misner space deserves studying at least as a wonderful source of counter-examples [133]. Its simplest generalizations to the non-flat case are considered too. Then, in Sect. 3, we briefly discuss the process of evolution of a wormhole into a time machine, a widely known and popular process. Its importance lies in the fact that it is one of the most 'realistic' scenarios of how the universe might lose its global hyperbolicity.

All the listed time machines as well as that from Example 8 share one property: they have no "holes" like those in the Deutsch–Politzer space, see Example 73 in Chap. 1. In Sect. 4 we formalize that similarity by introducing the notions of compactly generated and compactly determined Cauchy horizons. All time machines with such horizons are shown to have some important common properties. However, none of those properties are compulsory for a *general* time machine. We show that by example in Sect. 5.

Among the time machines *not* considered in this book particularly interesting are the Clifton–Pohl torus [141, Example 7.16] and the flat four-dimensional time machines proposed by Gott [67] and Grant [68].

S. Krasnikov, *Back-in-Time and Faster-than-Light Travel in General Relativity*, Fundamental Theories of Physics 193, https://doi.org/10.1007/978-3-319-72754-7_4

1 The Time Machine

In any spacetime where the causality condition (not the *principle* of causality!) is violated an inhabitant and their younger/older self can meet or at least exchange signals, with all related consequences. In this sense any such space might be called a time machine. Historically, the Gödel, see [76, Sect. 5.7] and van Stockum [171] spacetimes were first to be called so, but there is a much simpler spacetime—the cylinder C_M from Example 42 in Chap. 1—that is just as good a model of time machine as those two. Indeed, the recognized merit of the latter is that they solve the Einstein equations with "realistic" sources and that they are free from singularities. But exactly the same is true, of course, in the case of C_M.

The spacetimes like these three, i.e. such that causality is violated in every point: $\overset{\looparrowright}{M} = M$, may be called *eternal* time machines. Can it be that our universe is among them? On the one hand, it seems premature to rule out this possibility. One still can imagine that the expansion of the universe will change to contraction [179] some day. It is tempting then to identify the initial and the final singularities and to obtain (after resolving somehow that singularity) an eternal time machine.[1] On the other hand, there is not a grain of evidence that this is really how the universe is arranged. Surprisingly, there is a chance that the uncertainty may be resolved experimentally. Indeed, causal loops which come to mind immediately (for example, world lines of particles resting with respect to the cosmological fluid) are *huge*. However, their existence automatically implies the existence also of arbitrarily short (in the Lorentz sense) closed causal curves. The latter are obtained from the former by approximating them with piecewise null broken lines. This suggests that cosmological scale causality violations may affect laboratory physics. Unfortunately, the role of an observer will be passive in any case: if the universe is not an eternal time machine *now*, it never will be.

More interesting are time machines in which the causality violating curves lie to the future of some regular region M^{r}. Let us require for definiteness that

$$M^{\mathrm{r}} \leftrightharpoons M - \mathrm{Cl}\big(I^{+}(\overset{\looparrowright}{M})\big) \qquad \text{is intrinsically globally hyperbolic,} \tag{1}$$

that is before the closed causal curves appeared the causal structure of the spacetime was "as nice as possible". Note for future reference that M^{r}

(a) is a past set;
(b) is causally convex by Proposition 37(*a*) in Chap. 1;
(c) is globally hyperbolic by Proposition 47 in Chap. 1.

Condition (1) is satisfied, for example by the DP space. Unless the existence of such time machines is somehow excluded either theoretically, or at least observationally, one may speculate that the part of the universe we live in is just M^{r}. Which means

---

[1] True, within such a scenario the arrow of time looks mysterious [120]. Then again, it looks so in *any* scenario.

that the universe is causal and predictable, cf. Sect. 3 in Chap. 2, *so far*, but one day it will lose both properties—a time machine will appear.

Definition 1 Suppose, the set $\overset{\looparrowright}{M}$ of an inextendible spacetime M is non-empty and satisfies (1). Then M, (or, sometimes, a connected component of $\overset{\looparrowright}{M}$) will be called a *time machine*, or an *appearing*[2] time machine.

Comment 2 As the example of Minkowski space populated by tachyons shows, the principle of causality may be violated even in a spacetime with most innocent geometry. Such a spacetime, however, may not be a time machine according to our definition.

Comment 3 We require connectedness because by Proposition 35(*b*) in Chap. 1 the set $\overset{\looparrowright}{M}$ is the union of disjoint and obviously connected subsets

$$\overset{\looparrowright}{M}_\alpha \rightleftharpoons J^+(p_\alpha) \cap J^-(p_\alpha), \qquad p_\alpha \in M, \qquad \alpha = 1 \dots k.$$

Correspondingly, *each* $\overset{\looparrowright}{M}_\alpha$ is called a time machine and we speak of k time machines in this M.

Corollary 4 *Any two points in a time machine lie on a common closed causal curve.*

Local physical laws inside the time machine are the same as outside. However, the pathological causal structure imposes non-trivial—and counter-intuitive, as in the case of "temporal paradoxes", see Chap. 6—constraints on possible solutions, cf. 3.3 in Chap. 2. Surprisingly little is known today about those constraints (see [52, 114], though). So, the study of the time machine reduces mostly to the study of M^{r}, i.e. of the moment, when the spacetime loses the global hyperbolicity (see also [88, 109, 132], where a number of facts concerning the geometry of Bd $\overset{\looparrowright}{M}$ are established).

Proposition 5 *The boundary of* M^{r} *is its future Cauchy horizon.*

$$\mathrm{Bd}\, M^{\mathrm{r}} = \mathcal{H}^+(M^{\mathrm{r}}).$$

Proof In M^{r} pick a spacelike Cauchy surface $\mathcal{S}$ (it exists there by Proposition 52 in Chap. 1). The proof will be divided into three steps.
1. First, let us prove that $\mathcal{D}(\mathcal{S})$ is an open set. To this end note that $\mathcal{S}$ being achronal in the spacetime $M^{\mathrm{r}} \subset M$ is also achronal as a subset of the whole M (indeed, a causal curve emanating from a point of $\mathcal{S}$ cannot return back to $\mathcal{S}$ without leaving M^{r}, but once it has left M^{r} it will never return—and hence never meet $\mathcal{S}$ again—because M^{r} is causally convex, see property (*b*) above). The openness of $\mathcal{D}(\mathcal{S})$ now follows from

[2] In contrast to eternal ones mentioned above.

Proposition 54 in Chap. 1.
2. Next, let us establish the equality

$$M^{\mathrm{r}} = \mathcal{D}(\mathcal{S}). \qquad (*)$$

The inclusion $M^{\mathrm{r}} \subset \mathcal{D}(\mathcal{S})$, as was mentioned just below Proposition 54 in Chap. 1, is obvious. Suppose, the converse inclusion does *not* hold, i.e. in $\mathcal{D}(\mathcal{S})$ there is a point p of $\mathrm{Cl}\big(I^+(\overset{\looparrowright}{M})\big)$. In such a case there must be also a point $p' \in \mathcal{D}(\mathcal{S}) \cap I^+(\overset{\looparrowright}{M})$, since $\mathcal{D}(\mathcal{S})$ is open. And this is impossible, because, moving from p' to the past along a timelike curve α (there is no loss of generality in assuming α to be future inextendible: not leaving the set $I^+(\overset{\looparrowright}{M})$, which is separated from $\mathcal{S}$, it cannot meet $\mathcal{S}$ to the future of p'), we could reach a point through which a closed causal curve ℓ passes. The curve consisting of α and ℓ (the latter being counted infinitely many times) is inextendible, but lies entirely in $I^+(\overset{\looparrowright}{M})$, see Proposition 23(*d*) in Chap. 1, and therefore will never meet $\mathcal{S}$, which contradicts the definition of $\mathcal{D}(\mathcal{S})$.
3. $\mathcal{D}(\mathcal{S})$, inasmuch as it coincides by $(*)$ with M^{r}, is a past set. Hence the second term in the right-hand side of (18) in Chap. 1 is disjoint with the first one and therefore can be dropped. This is our assertion. □

Proposition 6 *The surface $\mathcal{S}$ from Proposition 5 is closed in M.*

Proof Let λ be an inextendible timelike curve through a point $p \in \mathrm{Cl}_M\,\mathcal{S}$. The chronological future of that curve must contain points of a neighbourhood of p, and hence, some points of $\mathcal{S}$ and thus of M^{r}. But M^{r} is a past set, so λ itself also has points in M^{r}. It follows that λ (being timelike) meets $\mathcal{S}$ in a unique point p'. If $p' \neq p$, then $p \in \big(I^+(p') \cup I^-(p')\big)$ (since these points are connected with the timelike λ), and hence $\mathcal{S} \cap \big(I^+(p') \cup I^-(p')\big)$ is non-empty. This, however, is impossible, because $\mathcal{S}$ was shown to be achronal in M, see the proof of Proposition 5. So, $p' = p$, which means, in particular, that $p \in \mathcal{S}$. □

Combining Propositions 56 in Chap. 1, 5, 6 and Corollary 58 in Chap. 1 immediately gives

Corollary 7 *The boundary of a time machine is a closed achronal hypersurface generated by past inextendible null geodesics. If γ_1 and γ_2 are such generators, then $J^+(\gamma_1) \cap J^-(\gamma_2)$, when non-empty, is their common future end point.*

Of all spacetimes considered so far in this book only the DP space is a time machine. But its topology is quite non-trivial: it has non-contractible loops, "holes", etc., see p. 35. However, the impression that the time machine is an attribute of some too pathological spacetimes is deceptive. Let us check that it appears as a result of even a slight deformation of a most dull spacetime.

Example 8 Consider a Euclidean space $(\mathbb{R}^4, \boldsymbol{\varepsilon})$ with the cylindrical coordinates t, z, ρ, φ. It is covered by two regions U and K defined by the inequalities

$$U\colon\quad \chi^2 > \frac{1}{4}, \qquad K\colon\quad 0 \leqslant \chi^2 < \tfrac{1}{2}, \qquad \text{where}\quad \chi \leftrightharpoons \sqrt{t^2 + z^2 + (\rho - 1)^2}.$$

K is a solid torus which appears for a brief time around the circle

$$\ell\colon \qquad t = z = 0, \quad \rho = 1$$

and inflates to a maximum at $t = 0$, while U is the, slightly expanded, complement of K. Further, define a vector field $\boldsymbol{v}$ by the equality

$$\boldsymbol{v} = \epsilon_U \partial_t + \epsilon_K \partial_\varphi,$$

where $\{\epsilon_U, \epsilon_K\}$ is the partition of unity subordinate to the open cover $\{U, K\}$. Clearly, the spacetime $(\mathbb{R}^4, g)$ with

$$g_{ab} \leftrightharpoons \varepsilon_{ab} - 2v_a v_b$$

is the Minkowski space [cf. (12a) in Chap. 1] outside of K, that is in a region where $\epsilon_K = 0$ and $\epsilon_U = 1$. However, *inside* K there are closed causal curves[3]: the circle ℓ, for instance.

2 Misner-Type Time Machines

2.1 Misner Space

Consider a usual flat two-dimensional cone M. It is convenient to perceive it as being obtained from a plane (x_0, x_1) by cutting out the sector W bounded by the rays $\nu_{1,2}$

$$\nu_1\colon\ x_0 = ax_1, \quad \nu_2\colon\ x_0 = bx_1, \qquad \text{where}\quad x_0 < 0, \quad |a|, |b| > 1,$$

and gluing these rays together, see Sect. 6 in Chap. 1. It turns out that the properties of the cone depend crucially on the signature or the metric. In particular, if the metric of the initial plane is

$$\mathrm{d}s^2 = -\mathrm{d}x_0^2 + \mathrm{d}x_1^2 = -\mathrm{d}\alpha \mathrm{d}\beta \qquad \alpha \leftrightharpoons x_0 - x_1, \quad \beta \leftrightharpoons x_0 + x_1, \tag{2}$$

then the geometry of M—which is called the *Misner space* in this case—is quite bizarre. Its properties are the subject of the present subsection.

We begin with representing the Misner space as a *cylinder*. Consider a boost

$$\varpi\colon\quad (\alpha, \beta) \mapsto (\kappa\alpha, \kappa^{-1}\beta), \quad \kappa \leftrightharpoons \sqrt{\tfrac{(a+1)(b-1)}{(a-1)(b+1)}},$$

---

[3]This spacetime is, basically, a variant of the time machine proposed as an example, in [74].

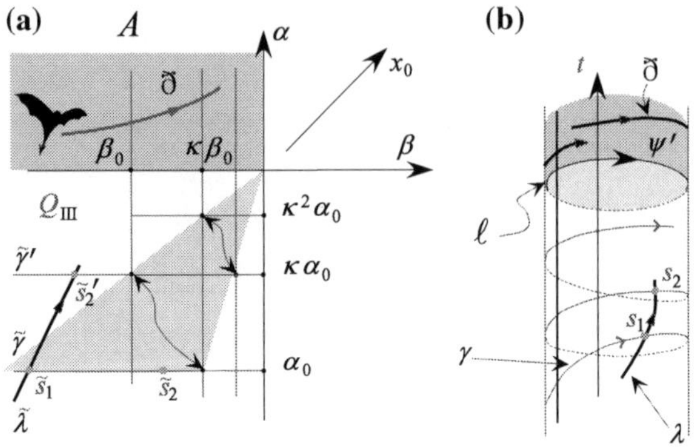

Fig. 1 **a** In the half-plane $\beta < 0$ the bounding lines of the vertical strips are identified thus giving rise to the cylinder, **b** In its upper half causality is violated, but its lower part ($t < 0$) is globally hyperbolic. It is the latter that is called the Misner space. In order to obtain it from Q, one takes W (the light gray sector) as the fundamental domain. The null geodesic γ meets the t-axis—which is the other generator of the null cone—infinitely many times.

mapping the quadrant $Q\colon\ \alpha, \beta < 0$ to itself, see Fig. 1, and ν_1 to ν_2. The parameter κ is taken to be less than unity (which involves no loss of generality, as will become obvious). Now M can be defined as the quotient of Q by the group of isometries generated by ϖ (cf. p. 37):

$$M \leftrightharpoons Q/G, \qquad G = \dots \varpi^{-1}, \mathrm{id}, \varpi, \varpi^2 \dots; \tag{3}$$

(that is how it is defined, for instance in [76]), i.e. as the space

$$\mathrm{d}s^2 = t^{-1}\mathrm{d}t^2 - t\mathrm{d}\psi^2 \qquad t < 0, \quad \psi = \psi + 2\ln\kappa, \tag{4}$$

where t and ψ are the coordinates in M induced by the projection $\pi\colon\ Q \to M$ and the rule

$$\forall p \in Q \qquad t(\pi(p)) = -\tfrac{1}{4}\alpha(p)\beta(p), \qquad \psi(\pi(p)) = \ln\frac{\alpha(p)}{\beta(p)} \quad (\mathrm{mod}\ 2\ln\kappa).$$

Written as in (4) the Misner space is associated rather with a cylinder, than with a cone. In support of this picture note also that the fundamental domain is defined non-uniquely. In the representation (3) we may take it to be not the sector W, but the strip bounded by the null geodesics $\beta = \beta_0$ and $\beta = \kappa\beta_0$, see Fig. 1a, that is, obtain M, by 'rolling up Q into a tube'. At first glance, the fact that M can be perceived both as a cylinder and as a cone seems paradoxical. The clue is in the fact that in the Lorentzian case a sequence of circles with circumferences tending to zero (the case in point are the circles $t = const$) can converge to a point (to the tip of a cone, say), *or* to a closed null curve.

The cylinder under discussion can be simplified further by choosing different 'time' or 'angle' coordinates:

$$\tau \leftrightharpoons -\ln|t|, \qquad \psi' \leftrightharpoons \psi - \ln|t|. \tag{5}$$

With the aid of these coordinates (4) can be cast in more convenient forms:

$$ds^2 = e^{-\tau}(-d\tau^2 + d\psi^2) \qquad \tau \in \mathbb{R}, \quad \psi = \psi + 2\ln\kappa, \tag{6a}$$

$$ds^2 = -2dt d\psi' - t d\psi'^2 \qquad t < 0, \quad \psi' = \psi' + 2\ln\kappa. \tag{6b}$$

The Misner space is causal and, moreover, globally hyperbolic (the circles $\tau =$ *const* being evident Cauchy surfaces). It is *extendible*, however, which, of course, comes as a complete surprise to one who thinks of this space as a cone. In particular, a (maximal, as we shall see) extension $M_A \leftrightharpoons A/\overset{G}{\sim}$ can be built by simply replacing everywhere above the quadrant Q to the left half-plane $A \leftrightharpoons \{p\colon\ \beta(p) < 0\}$. The transformation $\alpha \leftrightarrow \beta$ maps Q to itself. Hence by starting from the lower half-plane $B \leftrightharpoons \{p\colon\ \alpha(p) < 0\}$ instead of A we would arrive at a space M_B, which is isometric to M_A, but does not coincide with it (in particular, some inextendible in M_A curves are extendible in M_B and vice versa).

As α vanishes in some points of A, not all of the coordinates defined above (not, for example ψ and τ) can be extended to the whole M_A, but t and ψ' can. Thus, M_A is obtained merely by replacing in (6b) the condition $t < 0$ with $t \in \mathbb{R}$. The spacetime M_A, in contrast to M, is not causal—it is an appearing (if t grows in the future direction) time machine. In the coordinates ψ, τ [the latter is the coordinate, defined at $t > 0$ by relation (5)] the region $M_A - \overline{M}$ takes the form

$$ds^2 = -e^{-\tau}(-d\tau^2 + d\psi^2), \qquad \tau \in \mathbb{R}, \quad \psi = \psi + 2\ln\kappa,$$

which—up to the conformal factor $e^{-\tau}$ and the replacement of the period from 3 to $2\ln\kappa$—is the cylinder C_M from Example 42 in Chap. 1. Closed causal curves pass through any point of this region and the closed null geodesic $\ell\colon\ t = 0$ (it is the image of the ray $\alpha = 0$ under the map π) separates it from the causal region M, see Fig. 1b.

Now, let us consider the geodesic structure of the time machine M_A. Clearly (rigorously it can be proven by using [141, Lemma A9]), all geodesics here are the projections of geodesics lying in the half-plane A. And each of the latter either is parallel to the α-axis (and hence is complete), or terminates at it (of course, the end points belong neither to the geodesic, nor for that matter to A at all). Every geodesic γ_A of this second type is incomplete in A, and therefore its image $\gamma = \pi(\gamma_A)$ is incomplete in M_A, which means that M_A is *singular*. The geodesic γ_A is a straight line $\alpha = c\beta + d$ with $c \neq 0$ and, as a corollary, its image $\gamma(\beta)$ is given by a parameter equation

$$t = -\tfrac{1}{4}\beta(c\beta + d), \qquad \psi' = \ln 4 - \ln\beta^2.$$

Thus, as β tends to zero γ spirals infinitely winding itself onto the circle ℓ. Now, pick a pair of points in ℓ and let U_a, U_b be their neighbourhoods. Since M_A is Hausdorff, they can be chosen so that

$$\mathrm{Cl}_{M_A} U_a \text{ and } \mathrm{Cl}_{M_A} U_b \text{ are compact}, \qquad \mathrm{Cl}_{M_A} U_a \cap \mathrm{Cl}_{M_A} U_b = \varnothing. \tag{7}$$

If γ—which is partially imprisoned in either of the neighbourhoods—had an end point, that point would lie both in U_a and U_b at once, but this is impossible by (7). So, γ can be extended neither in M_A, nor in its extensions, if they exist (we shall see that they do not). Thus, the singularity in the Misner space is true, i.e. non-removable. On applying the same argument to the region $|t| \leqslant t_0$ instead of the whole M_A we come to an unexpected conclusion: singularities may be present even in a *flat* and *compact* subset of a spacetime.

Thus, there are three types of geodesic in M_A:

1. Null curves $\psi' = const$ (the generators of the cylinder). The affine parameter (it can be α, for example) of such a curve is unbounded, so it is complete;
2. Incomplete geodesics infinitely winding themselves onto the Cauchy horizon $t = 0$. Some of them are imprisoned in M (or in $M_A - M$), the others intersect the Cauchy horizon, reach the maximal (minimal) t, and approach the horizon from above (correspondingly, from below[4]);
3. The closed null geodesic ℓ. It is incomplete (because the ray $\alpha = 0,\ \beta < 0$ is). The reason is that (in)completeness is an attribute of a geodesic understood as a map $\mathbb{R} \to M_A$, not as a set of points of M_A, cf. Remark 7 in Chap. 1, and the *parameter* length of the circle becomes less (if we move to the future) with each return of the geodesic to its start. If, for example ℓ is parameterized by β, then, as is seen from Fig. 1a, in the future direction ℓ becomes $1/\kappa$ times shorter with each passage.

Above we have established that no maximal—in M_A—geodesic can be extended in any $\tilde{M} \supset M_A$, hence by Test 66 in Chap. 1 the spacetime M_A is *inextendible*.

2.2 (Anti-) de Sitter Time Machines

A time machine similar to Misner space, but with different global structure, can be obtained from a *non*-flat plane. In this subsection, we build a few such spacetimes, see also [57, 90, 116].

Let S be the strip

$$g\colon \quad \mathrm{d}s^2 = -\tfrac{8}{R}\sin^{-2}(u+v)\mathrm{d}u\mathrm{d}v, \qquad -\pi < u + v < 0. \tag{8}$$

[4]The geodesics of the latter kind have self-intersections, in M. This, however, does not violate causality, all such geodesics being spacelike.

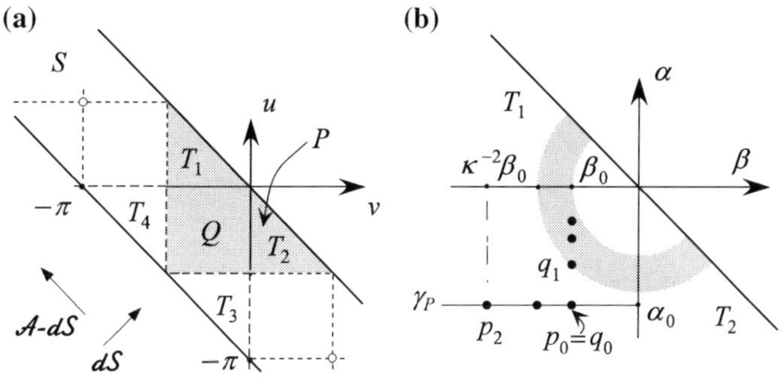

Fig. 2 **a** S is the two-dimensional (anti-, if the *upper* of the two arrows is timelike) de Sitter spacetime. It is conformally flat and the conformal factor diverges on the boundary of S. **b** P is the region $\alpha + \beta < 0$. The grey half-annulus is one of the possible fundamental domains. The horizontal and vertical rows of black dots depict, respectively, p_m and q_m

The free parameter is denoted by R because it is equal to the scalar curvature. If R is positive, the spacetime (S, g) is called the (two-dimensional) *de Sitter space*. This spacetime is highly symmetric and its global structure is as nice as that of Minkowski plane. Note, in particular, that it is non-singular (the straight lines bounding S are infinities, not singularities, see A.3 in Appendix) and globally hyperbolic (which is obvious from Fig. 2a).

If $R < 0$ the spacetime is called *anti-de Sitter space*. Sometimes the same name is given to the space—denote it $\tilde{S}$—which is locally isometric to ours, but differs from the latter topologically. $\tilde{S}$ is a hyperboloid embedded in an auxiliary flat space and is obtained from S by rolling the latter up into a cylinder, that is by quotienting it by the group the generator of which is a translation in the $(\partial_u - \partial_v)$-direction [that the translation is an isometry is seen from (8)]. The direction is timelike and it turns out that $\tilde{S}$ is an eternal time machine. We do not consider it further.

In the region $P \subset S$ defined by the inequality $-\frac{\pi}{2} < u, v < \frac{\pi}{2}$ (it is the shaded triangle in Fig. 2a) introduce the coordinates

$$\alpha \leftrightharpoons \mathrm{tg} u, \qquad \beta \leftrightharpoons \mathrm{tg} v.$$

In these coordinates, P is a half-plane bounded by the straight line $\alpha = -\beta$, see Fig. 2b. The metric in P is

$$\mathrm{d}s^2 = -\tfrac{8}{R}(\alpha + \beta)^{-2}\, \mathrm{d}\alpha \mathrm{d}\beta.$$

It is easy to see now that there is an isometry $\varpi_\kappa \colon P \to P$ which sends every point (α, β) to $(\kappa\alpha, \kappa\beta)$, where κ is a positive parameter taken—for the sake of definiteness—to be less than unity. The group G generated by ϖ_κ is properly discontinuous and acts freely, see conditions (27) in Chap. 1. This implies that $M_P \leftrightharpoons P/G$ is a spacetime and it is M_P that we shall call (anti-) de Sitter time machine. The projection $P \to M_P$ will be denoted by π: $M_P = \pi(P)$. As usual, M_P can be described also in terms of "cutting and pasting". To this end on the plane (α, β) consider a circle of radius β_0 with centre at $\alpha, \beta = 0$. Note that ϖ_κ merely "contracts" this circle, i.e. sends its every point to the point with the same polar angle and with radius $\kappa\beta_0$. Thus, M_P can be obtained by cutting a half-annulus bounded by the mentioned circles out of P and glueing together its outer and inner boundaries, see Fig. 2b.

To understand the structure of M_P first note that P can be split into five invariant, i.e. mapped by ϖ_κ to themselves, subsets. These are the half-axes $\alpha = 0$, $\beta = 0$ and the regions separated by them:

$$Q: \ \alpha, \beta < 0, \qquad T_1: \ \alpha > 0, \qquad T_2: \ \beta > 0$$

(a quadrant and two triangles). In each of these three the following coordinates are defined

$$\widetilde{\eta} \leftrightharpoons \ln|\alpha\beta|, \qquad \widetilde{\chi} \leftrightharpoons \ln|\alpha/\beta|.$$

Correspondingly, M_P consists of three regions, T_1/G, Q/G and T_2/G, separated by two closed null geodesics, ℓ_α and ℓ_β (which are the images of the half-axes $\alpha = 0$ and $\beta = 0$, respectively).

Coordinates in these regions are induced by the projection π (cf. the coordinates t and ψ in Misner space):

$$\forall\, p \in P \qquad \chi(\pi(p)) = \widetilde{\chi}(p), \qquad \eta(\pi(p)) = \widetilde{\eta}(p) \qquad (\mathrm{mod}\ 2\ln\kappa).$$

and the spacetimes take the form

$$Q/G: \qquad \mathrm{d}s^2 = \frac{2}{R}\mathrm{ch}^{-2}\frac{\chi}{2}\,[\mathrm{d}\chi^2 - \mathrm{d}\eta^2], \qquad \eta = \eta + 2\ln\kappa, \quad \chi \in \mathbb{R},$$

$$T_{1(2)}/G: \qquad \mathrm{d}s^2 = \frac{2}{R}\mathrm{sh}^{-2}\frac{\chi}{2}\,[\mathrm{d}\eta^2 - \mathrm{d}\chi^2], \qquad \eta = \eta + 2\ln\kappa, \quad \chi \lesseqgtr 0$$

[$\chi(p) \to \mp\infty$, when p approaches $\ell_{\alpha,\beta}$]. In which of these cylinders causality holds is determined by the sign of the term $\sim \mathrm{d}\eta^2$, i.e. by the sign of R. For example, in the de Sitter case (when $R > 0$) the regions $T_{1,2}/G$ are globally hyperbolic, while in Q/G there is a closed causal curve $\chi = const$ through every point. Thus, depending on how the future direction is chosen, M_P is either a time machine with two regions M^{r}, or an eternal time machine evolving in two causally unrelated causal regions. In the *anti*-de Sitter case the situation is reverse: there are two causality violating regions—T_1/G and T_2/G—separated by the globally hyperbolic Q/G.

Remark 9 The causal regions of the three time machines are similar:

$$\mathrm{d}s^2 = \Omega^2\big(-\mathrm{d}\tau^2 + \mathrm{d}\psi^2\big) \qquad \psi = \psi + 2\ln\kappa,$$

where

$$\Omega = e^{-\tau/2} \qquad \tau \in \mathbb{R}^1 \qquad \text{Misner}$$
$$\Omega = \frac{1}{\sqrt{R/2}}\operatorname{sh}^{-1}\tau/2 \qquad \tau < 0 \qquad \text{de Sitter}$$
$$\Omega = \frac{1}{\sqrt{-R/2}}\operatorname{ch}^{-1}\tau/2 \qquad \tau \in \mathbb{R}^1 \qquad \text{anti-de Sitter.}$$

As a point approaches the Cauchy horizon its τ-coordinate tends to ∞, $-\infty$, and $\pm\infty$ in the first, second and third cases, respectively.

M_P is singular. Indeed, each null geodesic in M_P is the image of the corresponding geodesic segment in (anti-)de Sitter space. This segment lies between the hypotenuse and one of the legs of P, see Fig. 2a, and hence is incomplete. As with Misner space this singularity is non-removable.

Proof Let $\gamma_P(\beta) \subset P$ be a null geodesic defined by the equation

$$\alpha(\beta) = \alpha_0 \neq 0$$

(note that the parameter β is not affine, so, its unboundedness from below does not imply the completeness—in the corresponding direction—of the geodesic). Pick a negative number β_0 and consider two sequences of points:

$$p_m \leftrightharpoons \gamma_P(\kappa^{-m}\beta_0), \quad \text{and } q_m: \quad \alpha(q_m) = \kappa^m\alpha_0, \quad \beta(q_m) = \beta_0 \qquad m \in \mathbb{N}, \qquad (9)$$

see Fig. 2b. For each m, q_m is in the orbit of the p_m [that is $q_m = \varpi^i_\kappa(p_m)$ at some i, specifically at $i = m$], When $m \to \infty$ the sequence $\{q_m\}$ converges to the point q_∞ with the coordinates $\alpha(q_\infty) = 0$, $\beta(q_\infty) = \beta_0$. Hence, its image $\pi(q_\infty)$ in each of its neighbourhoods contains points $\gamma(\beta) \leftrightharpoons \pi \circ \gamma_P(\beta)$ with unboundedly large β. So, if γ had the end point it would be $\pi(q_\infty)$. Now recall that β_0 was chosen arbitrarily, therefore the same must be true for any other point of ℓ_α. But a curve cannot have more than one future (past) end point, whence it follows that γ has no end points at all, it is maximal in all possible extensions of M_P. And for the same reason (with $\alpha \leftrightarrow \beta$) so are the vertical—in Fig. 2—geodesics. Thus, using again Test 66 in Chap. 1, we come to the conclusion that M_P is inextendible. □

In contrast to the Misner case, this time the isometry $\alpha \leftrightarrow \beta$ does not give rise to any new extensions. Consider, however, the transformation $I: S \to S$ sending

$$w \mapsto -\pi/2 - w, \qquad \text{where} \quad w \leftrightharpoons u, v$$

(this is a coordinate inversion with respect to the point $u = v = -\pi/4$). I maps the region P to $P' \leftrightharpoons T_3 \cup Q \cup T_4$, see Fig. 2a. Quotienting P' by the group G' generated by the isometry $\varpi'_\kappa \leftrightharpoons I \circ \varpi_\kappa \circ I^{-1}$ one gets a spacetime M'_P isometric to M_P. Now note that, first, $I(Q) = Q$ and, second, the orbits of G and G' in Q coincide [which is easy to see, if we express I in terms of the coordinates α, β:

$$I: \quad (\alpha, \beta) \mapsto (1/\alpha, 1/\beta)$$

and infer that $\varpi'_\kappa = \varpi_\kappa^{-1}$]. This means that $Q/G' = Q/G$ and, consequently, M'_P, which is an extension of the former, is also an extension of Q/G. As in the Misner case, it is impossible to extend Q/G both to T_1/G, *and* to T_4/G' (or to T_2/G and to T_3/G' at once); otherwise, as we know, there would be a sequence (the just considered sequence $\{p_n\} = \{q_n\}$, for example) with two different limits. It is possible, however, to obtain two more time machines of the same kind, by applying the whole procedure not to P or P', as above, but to the parallelogram cut in S by one of the following conditions: $-\pi/2 < u < 0$ or $-\pi/2 < v < 0$.

2.3 Some Global Effects

Both Minkowski and de Sitter spaces are static, but the surgery used in transformation them into the corresponding time machines violates this symmetry. As a consequence, the resulting time machines have an unusual property ('local staticity', see [57] and Sect. 1.1 in Chap. 2): each of their simply connected regions is static, but the whole spaces are not. What changes there with time are some of their *global* properties (in particular, causal loops appear only to the future of some surface). Such a non-standard feature has interesting consequences. In particular, conservation laws in Misner space show in an unexpected way.

Consider, for example a free falling observer $\lambda(\xi)$ who meets the same photon $\gamma(\zeta)$ twice: in $s_1 \leftrightharpoons \lambda(\xi_1) = \gamma(\zeta_1)$ (ζ and ξ are some affine parameters) and in $s_2 \leftrightharpoons \lambda(\xi_2) = \gamma(\zeta_2)$, see Fig. 1b. The energy ε of the photon as measured by the observer is

$$\varepsilon_i = c g_M(\boldsymbol{u}_i, \boldsymbol{v}_i), \qquad \boldsymbol{u}_i \leftrightharpoons \partial_\zeta(s_i), \quad \boldsymbol{v}_i \leftrightharpoons \partial_\xi(s_i), \quad \varepsilon_i \leftrightharpoons \varepsilon(s_i) \quad i = 1, 2$$

(c in this expression is a constant of appropriate dimension and g_M is the metric). Between the meetings both the photon and the observer move steadily in a flat space. So, one might naively expect that $\varepsilon_2 = \varepsilon_1$. This, however, is not the case: in fact, ε_2 is κ^{-1} times greater than ε_1.

Proof Let us, first, introduce a few useful objects. Pick a point $\tilde{s}_1 = \pi^{-1}(s_1)$ in A. Denote by $\tilde{\gamma}(\zeta)$ and $\tilde{\lambda}(\xi)$ the pre-images (under the projection π) of, respectively, $\gamma(\zeta)$ and $\lambda(\xi)$ that pass through $\tilde{s}_1$. For definiteness we take $\tilde{\gamma}$ to be a horizontal geodesic $\alpha = const$. By $\tilde{\boldsymbol{u}}_1$ and $\tilde{\boldsymbol{v}}_1$ we denote the vectors in $T_{\tilde{s}_1}$ tangent, respectively, to $\tilde{\gamma}(\zeta)$ and $\tilde{\lambda}(\xi)$. The isometry $d\pi$ sends them to $\boldsymbol{u}_1$ and $\boldsymbol{v}_1$, whence

$$\varepsilon_1 = c g_A(\tilde{\boldsymbol{u}}_1, \tilde{\boldsymbol{v}}_1) = -c \tilde{u}_1{}^\beta \, \tilde{v}_1{}^\alpha \tag{10}$$

[we have used (2) in the last equality].

The full pre-image $\pi^{-1}(s_2)$ comprises infinitely many points, we need two of them:

$$\tilde{s}_2 \leftrightharpoons \tilde{\gamma}(\zeta_2) \quad \text{and} \quad \tilde{s}_2' \leftrightharpoons \tilde{\lambda}(\xi_2) = \varpi(\tilde{s}_2),$$

see Fig. 1a. Finally, $\tilde{\boldsymbol{u}}_2$ denotes the velocity vector of the geodesic $\tilde{\gamma}$ in $\tilde{s}_2$.

The proof consists in comparing the expression (10) for ε_1 with its analogue for ε_2. In deriving the latter we shall need the vectors $\tilde{\boldsymbol{u}}_2', \tilde{\boldsymbol{v}}_2' \in T_{\tilde{s}_2'}$, which are the velocity vectors of, respectively, $\tilde{\gamma}' \leftrightharpoons \varpi(\tilde{\gamma})$ and $\tilde{\lambda}$.

It is obvious that in the spacetime (2) the components, in the coordinate basis, of the velocity vector of any geodesic remains constant. Hence,

$$\tilde{v}_2'^{\sigma} = \tilde{v}_1{}^{\sigma}, \quad \tilde{u}_2{}^{\sigma} = \tilde{u}_1{}^{\sigma}, \qquad \sigma \leftrightharpoons \alpha, \beta.$$

On the other hand, $\tilde{\gamma}$ and $\tilde{\gamma}'$ are pre-images of the same geodesic. So, if $\tilde{\gamma}(\zeta)$ satisfies the equation $\beta = c_1\zeta$, the equation for $\tilde{\gamma}'(\zeta)$ is $\kappa^{-1}\beta = c_1\zeta$. Hence $\tilde{u}_2'^{\beta} = \mathrm{d}\beta/\mathrm{d}\zeta = \kappa^{-1}\tilde{u}_2{}^{\beta}$. Combining this with the previous relations we get

$$\varepsilon_2 = c g_A(\tilde{\boldsymbol{u}}_2', \tilde{\boldsymbol{v}}_2') = -c\tilde{u}_2'^{\beta}\tilde{v}_2'^{\alpha} = -c\kappa^{-1}\tilde{u}_2{}^{\beta}\tilde{v}_1{}^{\alpha} = -c\kappa^{-1}\tilde{u}_1{}^{\beta}\tilde{v}_1{}^{\alpha} = \kappa^{-1}\varepsilon_1.$$

□

Thus, we have established that there are 'dangerous' null geodesics in Misner space—a free falling observer before reaching the Cauchy horizon meets such a geodesic infinitely many times. As measured in the observer's frame, the energy of a photon moving on the said geodesic exponentially grows with each encounter. The same geodesics are present in de Sitter time machines and, in fact, as will be shown below, see Sect. 4, in quite a broad class of time machines.

Consider now the wave equation $\Box f = 0$ in the (anti-) de Sitter time machine. To this end define on P the function $\tilde{f} \leftrightharpoons f \circ \pi^{-1}$. Since π is a—local—isometry, $\tilde{f}$ satisfies the corresponding equation

$$\partial_\alpha \partial_\beta \tilde{f} = 0. \tag{11}$$

Obviously, $\tilde{f}$ is a sum of two arbitrary functions—one of α and the other of β:

$$\tilde{f} = a(\alpha) + b(\beta).$$

It follows from the definition of $\tilde{f}$ that

$$a(\kappa\alpha) + b(\kappa\beta) = a(\alpha) + b(\beta).$$

By differentiating this equality we get

$$\kappa\partial_\alpha a(\kappa\alpha) = \partial_\alpha a(\alpha), \qquad \kappa\partial_\beta b(\kappa\beta) = \partial_\beta b(\beta).$$

So, if in some point (α_0, β_0) the derivative $f_{,\alpha}$ is non-zero, then $f_{,\alpha}(\kappa^m\alpha_0)$ diverges at $m \to \infty$, and consequently, f cannot be smooth in the point $(0, \beta_0)$. The same reasoning applies with $\alpha \leftrightarrow \beta$. Thus, in the class of smooth functions defined on the whole M_P *the wave equation has only constant solutions*, cf. [137].

3 Wormholes as Time Machines

Consider a wormhole so short and narrow that it may be described by the spacetime discussed in Sect. 2.2 in Chap. 3. This time, however, we shall choose the curve $\alpha(t)$ which determines the motion of the moving mouth in a special way. Specifically, we require that the distance $d(t)$ between the mouths would *decrease* starting from some time. Note that the 'time jump', i.e. the difference $t(f) - t(w')$, where f the passage of the spaceship through the right (moving) mouth and w' is its appearance from the left one, keeps growing, see Remark 12 in Chap. 3. (This is, in fact, a variety of the 'twin paradox', it comes into play because we affixed a clock to either of mouths and identify points with the same clock readings τ, not with the same t.) Thus, when $d(t)$ becomes sufficiently small, that is when the mouths come sufficiently close together—this happens to the future of some three-dimensional surface—$\overleftrightarrow{T}$ turns negative (the spaceship now returns home *before the departure*), which means that closed causal curves have appeared in the spacetime[5] [137]. In particular, a traveller traversing the throat from right to left now can meet their younger self (event o in Fig. 3a). Since the spacetime evolves from a globally hyperbolic initial region, it satisfies the definition of a time machine given in Sect. 1, the above mentioned -three-dimensional surface being its Cauchy horizon.

The importance of the time machines of this type lies in their apparent feasibility. They look exotic today, still, as we shall see in Chap. 9, it is not impossible that traversable wormholes do exist (or at least that they existed for some time in the early universe). Then it is natural to assume that they and/or their mouths move multidirectionally. So, the idea that the universe lost (or will lose one day) its global hyperbolicity via the just considered mechanism does not seem too far-fetched.

To develop it a bit further consider a directed to the right null geodesic γ emanating from a point between the mouths [in particular, at $y(0) = z(0) = 0$]. In some time it arrives at a point of Σ', or, which is the same, of Σ. In doing so it still lies in the plane $y = z = 0$ and still is directed to the right (if we have chosen an appropriate wormhole, cf. Sect. 2.1 in Chap. 3). Consequently, it will traverse the throat again (this is the event $p_2' = p_3$), and so on. Thus, γ will never leave the surface depicted in Fig. 3b. This surface is a two-dimensional time machine (its Cauchy horizon is the closed geodesic qq') and it has the following property in common with Misner and de Sitter spaces. The geodesic under consideration cannot meet qq', because the latter, being a Cauchy horizon is also null. So, γ will approach qq' plunging

[5] Likewise, they appear in a spacetime where two wormholes—both with a constant distance between the mouths—move with respect to each other.

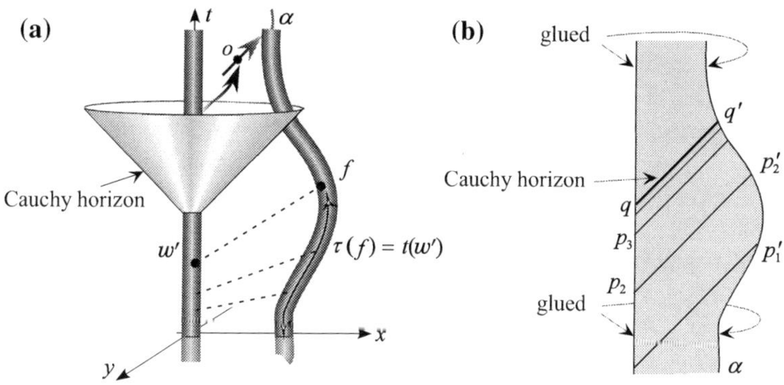

Fig. 3 A wormhole turning into a time machine. **a** The dashed lines connect identified points (so, each of the lines is, in fact a circle). The portion of the spacetime bounded from above by the Cauchy horizon is globally hyperbolic. **b** The section $y = z = 0$ of the wormhole. The thin tilted lines are a single geodesic γ

again and again into the right mouth and coming out from the left one. An observer with the world line lying between the mouths will meet a photon moving on γ, infinitely many times before they reach the horizon. Moreover, with each encounter the photon becomes more and more energetic, because q', as was already mentioned in discussing (8) in Chap. 3, may lie only in the *return* leg of the journey.

Thus, in (simplest) wormhole-based time machines—in their initial globally hyperbolic regions!—there are 'dangerous' null geodesics [137] similar to those present in the Misner space. A photon moving on such a geodesic returns infinitely many times in an arbitrarily small vicinity of a point in the Cauchy horizon, each time more and more blue-shifted. This brings up a few questions:

1. Do not the existence of such geodesics and the pathology in the behaviour of the solutions to the wave equation, see Sect. 2.3, indicate some instability of the relevant time machines[6]?
2. How typical is it for time machines to have this property?

The answer to the first question is not clear. If we wish to treat the problem classically (for the semiclassical analysis, see [115] and Chap. 7), we must speak not of a photon, but rather of a wave packet and, correspondingly, not of a null geodesic, but rather of a five-parameter set thereof. What would happen with such a bundle if it contains a 'dangerous' geodesic? On the one hand, the energy of a part of the bundle grows with each return. But on the other hand, the wormhole acts as a diverging lens, as is discussed on page 77. Therefore that part may become smaller and smaller [137]. So, which factor will prevail? In any case the whole effect can be easily neutralized by simply building a brick wall between the mouths. Such a wall would not prevent one from making a time trip (one can merely go round it), but would absorb the potentially dangerous photons, at the classical level, at least.

[6]This question is quite old, see, for example [64].

The second question is examined in the next section. We shall see that the appearance of dangerous geodesics inheres in quite a broad class of time machines and, in particular, in those originated from wormholes irrespective of their shapes, speeds, etc. (this might seem strange, because a small rotation of Σ_t before identifying it with $\Sigma'_{t'}$ would prevent γ from getting again in the throat [139]. The resolution is that in the modified wormhole some of *non-radial* geodesics would become dangerous). The existence of such a universal effect suggests the idea to estimate the abundance of wormholes from astronomical observations, the wormholes must manifest themselves as sources of well collimated high energy rays. One of the non-trivial properties of the rays would be that they look as a bundle of photons with the total energy $E = \sum_i \varepsilon_i$, where ε_i is the energy of the ith photon (thus, in a smoky room one would see a set of rays of different colours and brightness). However, a screen on the way of the bundle would absorb (if at all) only the photon of the highest energy.

4 Special Types of Cauchy Horizons

In time machines with 'perforations' (such as the DP space) the Cauchy horizon emerges, loosely speaking, from nowhere. It is intuitively clear that such spacetimes must differ strongly from those in which the horizon originates from the spacetime itself (as in the wormhole-based time machines). To capture this difference, a special class of time machines was introduced in [74].

Definition 10 A Cauchy horizon $\mathcal{H}^+$ is termed *compactly generated*, if it has a compact subset $\mathcal{K} \subset \mathcal{H}^+$ in which every generator of $\mathcal{H}^+$ is totally past imprisoned.

There are good mathematical reasons to single out in a special category the time machines with compactly generated Cauchy horizons (CGCHs): on the one hand, the category is rich enough to comprise many non-trivial specimens[7] and on the other hand, the defining condition is sufficiently restrictive to enable one to formulate and prove meaningful statements, cf. [26, 74, 84, 88]. However, from the physical point of view that condition is less transparent.[8] Indeed, as a distinctive feature of these time machines Hawking mentions that '...one might hope to predict events beyond the Cauchy horizon if it is compactly generated, because extra information will not come in from infinity or singularities'. This argument, however, is not quite convincing: *in any case* one cannot predict an event $e \notin M^{\mathrm{r}}$ without information that lacks in M^{r}: some of the causal curves, simply *by definition*, arrive at e without ever entering M^{r}. And any information brought in by such a curve is apparently extra, whether the curve originates from a singularity/infinity, or is just a loop of indefinite origin (though in

[7]These are, in particular, all time machines considered in this book so far, excluding the DP space.

[8]It is worth mentioning, therefore, that slightly different categories were considered too. In [88] the set which is required to be compact is the entire $\overset{\looparrowright}{M}$, while in [142] it is a subset $\mathcal{Q}$ of a Cauchy surface of M^{r} such that first CTCs appear on the boundary of $\mathcal{D}(\mathcal{Q})$.

the latter case unpredictability of that information may be restricted by some self-consistency requirement). The extra information is essential: the geometry of M^{r} taken alone does not predetermine even whether the horizon will prove compactly generated, see [97] or Theorem 2 in Chap. 5.

Still, in a sense the Misner space *is* more predictable than the DP space. We cannot predict *exactly what* information will come into the former, but we know at least that *some* information will: the appearance of the horizon is inevitable, as there is simply no extensions of M^{r} in which $\mathcal{D}(M^{\mathrm{r}})$ is larger than M^{r}

Definition 11 The Cauchy horizon of a time machine M is said to be *forced*, if M^{r} has no globally hyperbolic maximal extension.

The spacetime with a forced Cauchy horizon is perhaps the closest approach to the "artificial" time machine: as will be demonstrated in Chap. 5, whatever one does with the geometry of the world around them, one cannot compel the universe to give birth to a closed causal curve (unless there is a yet unknown non-local law of Nature; this could be, for example a law prohibiting 'holes', see Sect. 3.2 in Chap. 2). But by preparing a globally hyperbolic spacetime isometric to the region M^{r} of a time machine with forced Cauchy horizon, one constrains the universe to *choose* between remaining causal or being hole free.

What makes the compactly generated Cauchy horizons interesting from the physical point of view is just the fact that they *seem* to be forced. This, however, is yet to be established and that is why we shall consider also another, though quite similar, type of horizon.

Definition 12 A Cauchy horizon is *compactly determined*, if it has an open[9] subset $\mathcal{U}$ such that for some Cauchy surface $\mathcal{S}_0$ of M^{r} the set $\mathcal{L} \equiv \overline{J^-(\mathcal{U}) \cap J^+(\mathcal{S}_0)}$, is non-empty and compact.

As will be proven in the next subsection, all compactly determined Cauchy horizons (CDCHs) *are* forced.

To be compactly determined a horizon need not be compactly generated. An example is the spacetime

$$ds^2 = d^2y - 2dt d\psi - t d\psi^2 \qquad y, t \in \mathbb{R}^1, \quad \psi = \psi + 1,$$

which is the Misner time machine, see (6b), multiplied by the real axis. In contrast, the geometry of a CGCH which is not a CDCH is quite bizarre. Even if so pathological spacetimes exist (which is not clear) they definitely do not fit the idea of the 'laboratory-made' time machine, while this is the only kind of time machines we are interested in.

In this section, we consider both types of horizon (see also [26, 74, 84, 88]), but first we need to introduce a few useful objects. Pick a smooth future-directed timelike vector field $\boldsymbol{\tau}$ normalized by the equality $\tau^a \tau_a = -1$ [such fields always can be found, see Proposition 18 in Chap. 1]. Now the condition

---

[9] In the topology of the horizon, of course.

$$g(\boldsymbol{\varsigma}, \boldsymbol{\tau}) = -1, \qquad \text{where } \boldsymbol{\varsigma} \leftrightharpoons \partial_l, \tag{12}$$

fixes uniquely (up to an additive constant) the 'arc length parameter' l on any smooth causal curve $\beta(l)$. To see the geometric sense of l consider a tetrad in $\beta(l)$ whose zeroth unit vector is $\boldsymbol{\tau}$ and, if $\boldsymbol{\tau}$ and $\boldsymbol{\varsigma}$ are linearly independent, the first unit vector is chosen to be their linear combination. In such a basis $(\varsigma^0)^2 \geqslant (\varsigma^1)^2$ (since β is causal), $\varsigma^0 = 1$ [by (12)], and $\varsigma^i = 0$ at $i \neq 0, 1$. Thus, $1 \leqslant g^R(\boldsymbol{\varsigma}, \boldsymbol{\varsigma}) = (\varsigma^0)^2 + (\varsigma^1)^2 \leqslant 2$, where

$$g^R(\boldsymbol{x}, \boldsymbol{y}) \leftrightharpoons g(\boldsymbol{x}, \boldsymbol{y}) + 2g(\boldsymbol{x}, \boldsymbol{\tau})g(\boldsymbol{\tau}, \boldsymbol{y}) \tag{13}$$

is an auxiliary Riemannian metric in M. So, we can relate the difference $l_1 - l_2$ to $d_{l_1 l_2}$, which is the length (with respect to g^R) of the segment $\beta\big|_{|l_1 - l_2|}$

$$d_{l_1 l_2} < |l_1 - l_2| < \sqrt{2} d_{l_1 l_2}.$$

And the following evident statement shows why g^R is so important (we shall use it throughout this section).

Proposition 13 *Let $\beta(l)$ be a smooth causal curve which starts (ends) at a compact set and lies in it at all positive (respectively, negative) l. Then β is future (past) extendible, if and only if its length with respect to g^R is finite or, to put it differently, if and only if l is bounded.*

4.1 Imprisoned Geodesics and Compactly Determined Horizons

In this subsection, we study the geometry of the region immediately preceding a CDCH. One of its peculiarities is the fact that all past-directed null geodesics $\{\gamma_m \neq \gamma\}$ emanating from a point of a horizon generator γ must meet $\mathcal{S}_0$ in some $\{q_m\}$ lying within a compact spot $\mathcal{L} \cap \mathcal{S}_0$, see Fig. 4. This means that there is a subsequence of $\{\gamma_m\}$ which converges both to γ and to the geodesic through a limit point of $\{q_m\}$. This latter geodesic, as we shall see, is "dangerous". The property of the said subsequence to have two limit curves is too exotic for a globally hyperbolic space, which will enable us to prove that all CDCHs are forced.

We begin with proving that the past of any CDCH contains a geodesic infinitely winding itself onto the horizon (we do not assert at the moment that this geodesic is "dangerous", because its energetic properties are not established yet).

Proposition 14 *In any spacetime with compactly determined horizon there exists a future inextendible null geodesic α, totally imprisoned in $\mathcal{L} \cap M^{\mathrm{r}}$.*

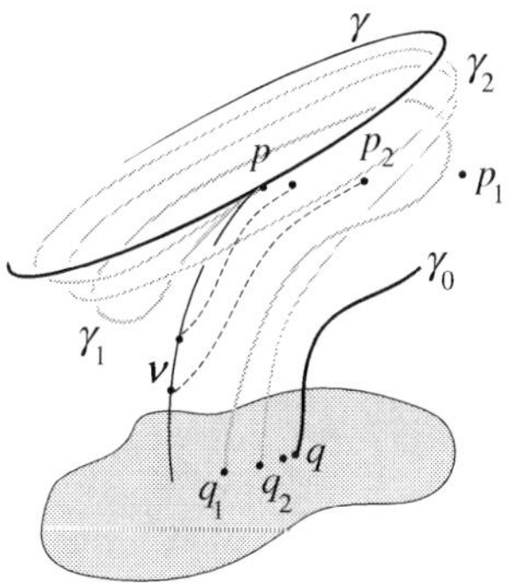

Fig. 4 The shadowed spot is the set $\mathcal{L} \cap \mathcal{S}_0$ [106]. The dashed lines are β_i

Remark 15 This fact does not contradict a Proposition 44 in Chap. 1, because the set $\mathcal{L} \cap M^{\mathrm{r}}$ is non-compact, while its closure $\mathcal{L}$, though compact, does not entirely lie in M^{r}.

Remark 16 In the two-dimensional case our proof would not work, but the proposition itself holds true, see the discussion at p. 99.

Proof Denote by H the set of the points that lie, together with all past-directed null geodesics through them,[10] in the horizon $\mathcal{H}^+$:

$$\mathsf{H} \leftrightharpoons \{x \in \mathcal{H}^+ : \ \big(J^-(x) - I^-(x)\big) \subset \mathcal{H}^+\}.$$

Note that

$$\mathsf{H}, \text{ when non-empty, consists of isolated points,} \tag{14}$$

because for any $x \in \mathsf{H}$ the set $J^-(x) - I^-(x)$ is a neighbourhood in $\mathcal{H}^+$ of x and this set does not contain any other points of H (which is immediate from Corollary 7).

Now pick a point p in $\mathcal{U} - \mathsf{H}$, where $\mathcal{U}$ is the set appearing in Definition 12. The velocities of the horizon generators parameterized by the "length" l form in T_p a—clearly closed and hence compact—proper subset of the sphere (12). So, there are future-directed null geodesics γ and $\{\gamma_m\}$, $m = 1, 2\ldots$ terminating at p and satisfying

$$\gamma \subset \mathcal{H}^+, \qquad \dot\gamma_m(p) \xrightarrow[m\to\infty]{} \dot\gamma(p), \qquad \forall m \quad (\gamma_m - p) \subset M^{\mathrm{r}}.$$

Every γ_m enters M^{r} and hence meets $\mathcal{S}_0$ in some point q_m (whence, in particular, $\gamma_m \neq \gamma$). From now on we understand γ_m to be geodesic *segments* from $\mathcal{S}_0$ to p (not *entire* geodesics) and measure the length parameter $l^{(m)}$ on each of them, from the corresponding q_m:

---

[10] That the union of such geodesics includes the set $J^-(x) - I^-(x)$ follows from Proposition 23(d) in Chap. 1. The reverse inclusion is, in fact, a corollary to Proposition 56 in Chap. 1.

$$l^{(m)} \in [0, l^{(m)\max}]: \qquad \gamma_m(0) = q_m, \quad \gamma_m(l^{(m)\max}) = p.$$

The superscript (m) will be dropped sometimes for simplicity of notation. All q_m lie in the compact (by Definition 12) set $\overline{J^-(\mathcal{K})} \cap \mathcal{S}_0 = \mathcal{L} \cap \mathcal{S}_0$, so passing to a subsequence if necessary, we state that

$$q_m \to q, \qquad \partial_l(q_m) \to \varsigma,$$

where q is a point of $\mathcal{L} \cap \mathcal{S}_0$, and ς is a null vector satisfying (12). We define $\alpha(l)$ to be the future inextendible geodesic fixed by the conditions

$$\alpha(0) = q, \qquad \partial_l(q) = \varsigma.$$

The proposition will be proved once we prove that α is, indeed, totally imprisoned in $\mathcal{L} \cap M^{\mathrm{r}}$.

At any $l < \overline{\lim}\, l^{(m)\max}$ the sequence $\gamma_m(l)$ contains, by the compactness of $\mathcal{L}$, a convergent subsequence $\{\gamma_j(l)\}$. And $\gamma_j(l) \to \alpha(l)$ since α is a solution of the geodesic equations (their solutions are known to depend continuously on the initial conditions). So, α cannot leave the set $\Gamma \leftrightharpoons \overline{\bigcup_j \gamma_j}$ as long as $l \leqslant \overline{\lim}\, l^{(j)\max}$, that is until it passes through $\alpha(\overline{\lim}\, l^{(j)\max}) = p$. But it *never* passes through p, because otherwise $\alpha = \gamma$ (these geodesics not only intersect in p, but also have in it equal velocities $\lim \dot{\gamma}_m$), which contradicts the fact that γ when followed into the past does not leave $\mathcal{H}^+$, see Corollary 7. Hence

$$\alpha \subset \Gamma \subset \left(\overline{J^-(p)} \cap J^+(q)\right) \subset \mathcal{L},$$

so it remains to prove that α does not leave M^{r} either.

To derive a contradiction assume that α does meet $\mathcal{H}^+$ in a point r. Then what will happen with α to the future of r? Three possibilities are conceivable: (a) α enters $M - \overline{M^{\mathrm{r}}}$; (b) it remains in $\mathcal{H}^+$; (c) it passes through some point $o \in M^{\mathrm{r}}$. But in the case (a) the extensions of some γ_j would also enter $M - \overline{M^{\mathrm{r}}}$. So, they would have to cross $\mathcal{H}^+$ in some points different from p. And that contradicts Corollary 7. The same corollary excludes (b), too: the extension of α to the future of r is a (part of a) horizon generator, it cannot leave the horizon in the past direction. Finally, (c) is also impossible, because otherwise one could arrive at r (which, being a point of $\mathcal{H}^+$, lies off M^{r}) moving in the past direction from $o \in M^{\mathrm{r}}$, which contradicts the fact that M^{r} is a past set. □

Proposition 17 *Any compactly determined Cauchy horizon is forced.*

Proof To obtain a contradiction, suppose that there is an isometry ϕ mapping M^{r} to a proper subset $\hat{M}$ of a globally hyperbolic inextendible spacetime M^e.

Conventions. (1) The field $\boldsymbol{\tau}$ used in (12) is defined on M^e and the corresponding field on M^{r} is obtained by pulling $\boldsymbol{\tau}$ back via ϕ. (2) For simplicity of notation, we shall write throughout the proof $\hat{A}$ for $\phi(A)$. (Note that the geodesic γ^e and the point p^e of it—they are defined below—both lie *off* $\hat{M}$. That is why they are not denoted by, respectively, $\hat{\gamma}$ and $\hat{p}$.)

1. Our proof starts with the observation that every neighbourhood U intersecting $\mathcal{H}^+$ contains a sequence of points $p_k \in M^{\mathrm{r}}$ and a point p such that

$$p_k \to p \in \mathcal{H}^+, \qquad \hat{p}_k \to p^e \in \mathrm{Bd}\,\hat{M}, \qquad (*)$$

otherwise by Test 68 in Chap. 1 we could extend the—inextendible by definition—spacetime M^e by glueing to it U via the isometry ϕ. As U is arbitrary, the set of points p satisfying $(*)$ is everywhere dense in $\mathcal{H}^+$. With (14) taken into consideration this means that there is a point in the just mentioned set that lies in $\mathcal{U} - \mathsf{H}$. From now on it is such a point that will be denoted by p, while q, q_m, γ and γ_m, $m = 1, 2\ldots$ are the same points and geodesic segments as in the proof of Proposition 14, with the only difference that γ_m are now understood to be half-open: $l \in [0, l^{(m)\max})$ and, correspondingly, $p \notin \gamma_m$.

Now consider a future-directed timelike curve $\nu(\upsilon)$ terminating at p. Our goal is to prove that the future end point of $\hat{\nu}$ is p^e. For this purpose we need yet another family of geodesics. Specifically, we pick a sequence of points $\nu(\upsilon_i)$ converging to p and by geodesics $\beta_i(\xi^{(i)})$ connect each of these points with a point p_k (k depends on i). In doing so the curves β_i, affine parameters $\xi^{(i)}$, and numbers $k(i)$ are chosen so—obviously, both requirements can always be satisfied—that beginning from some number

$$\beta_i \quad \text{are timelike and lie in a normal neighbourhood of } p; \qquad (15a)$$

$$g(z, \partial_{\xi^{(i)}}) = -1 \quad \text{at each } \upsilon_i, \qquad (15b)$$

where z is an arbitrary, but fixed from now on timelike vector field parallel transported along ν. Since M^{r} is a past set, condition (15a) guarantees the inclusion $\beta_i \subset M^{\mathrm{r}}$ and, consequently, the existence of $\hat{\beta}_i(\xi^{(i)})$ for all i. Denote by $L(\beta_i)$ and $L(\hat{\beta}_i)$ the affine lengths of β_i (i.e. $\xi^{(i)}[p_{k(i)}] - \xi^{(i)}[\nu(\upsilon_i)]$) and $\hat{\beta}_i$, respectively. Combining two obvious facts, $L(\hat{\beta}_i) = L(\beta_i)$ and $L(\beta_i) \to 0$, we conclude that

$$L(\hat{\beta}_i) \to 0 \qquad \text{at } i \to \infty. \qquad (**)$$

Now note that for any i_0 all $\hat{p}_j$ with sufficiently large j lie in the chronological future of $\hat{\nu}(\upsilon_{i_0})$. Therefore,

$$\forall \upsilon \qquad p^e \in \overline{I^+(\hat{\nu}(\upsilon))} = \overline{J^+(\hat{\nu}(\upsilon))} = J^+(\hat{\nu}(\upsilon)) \text{ and hence } \hat{\nu}(\upsilon) \in J^-(p^e),$$

where the last equality is due to the global hyperbolicity of M^e. Thus, $\hat{\nu}(\upsilon > \upsilon_0)$ lies in the compact set $J^-(p^e) \cap J^+(\hat{\nu}(\upsilon_0))$. But this means that the sequence $\hat{\nu}(\upsilon_i)$ (or some of its subsequences) converges and its limit coincides with p^e, since these two are connected with a geodesic (this is the limit curve of the family $\{\hat{\beta}_i\}$) of non-zero [as follows from (15b)] initial velocity and zero [as is seen from (∗∗)] length. The result will be the same if we replace $\{v_i\}$ by any other increasing sequence with the same limit.[11] Thus, p^e is, indeed, the future end point of $\hat{\nu}$.

Now note that the reasoning above remains valid if we replace simultaneously p^e by ν^e defined to be the future end point of $\hat{\nu}$ and $\{p_k\}$ by $\{\gamma_m(l_k)\}$ where m is fixed and $l_k \to l^{(m)\max}$. It follows that for all m, ν^e (which, as we already know, coincides with p^e) *is the future end point of* $\{\hat{\gamma}_m\}$.

2. Let us examine now the boundedness of the affine lengths of the geodesics $\{\gamma_m\}$ and $\{\hat{\gamma}_m\}$. For this purpose define the affine parameter $s^{(m)}$, or $\hat{s}^{(m)}$, respectively, on each of those geodesics by the conditions

$$s^{(m)}(p) = \hat{s}^{(m)}(p^e) = 0, \qquad g(\partial_v, \partial_{s^{(m)}})(p) = g(\partial_v, \partial_{\hat{s}^{(m)}})(p^e) = 1$$

(note that by the second condition geodesics are past-directed) and denote for brevity $L_m \leftrightharpoons s^{(m)}(q_m)$, $\hat{L}_m \leftrightharpoons s^{(m)}(\hat{q}_m)$. Next observe that the sequence $\{L_m\}$ is unbounded, because otherwise γ would pass through $q \in \mathcal{S}_0$ at some finite positive s, which is impossible, since γ cannot leave $\mathcal{H}^+$. But the *un*boundedness of the mentioned sequence combined with the obvious equality $L_m = \hat{L}_m$ implies the unboundedness of $\{\hat{L}_m\}$. And this means that irrespective of $\hat{s}$ the geodesic

$$\gamma^e(\hat{s}) \leftrightharpoons \lim_{\substack{m\to\infty \\ \hat{s}^{(m)}\to\hat{s}}} \hat{\gamma}_m(\hat{s}^{(m)})$$

will never reach $\phi(\mathcal{S}_0)$ and consequently, will never leave the compact set

$$J^-(p^e) \cap J^+\big[\phi(\mathcal{L} \cap \mathcal{S}_0)\big].$$

This, by Proposition 44 in Chap. 1, contradicts the strong causality of M^e. □

4.2 Deformation of Imprisoned Geodesics

Consider now a future-directed null geodesic $\gamma_0(l)\colon\ \mathbb{U} \to \mathcal{N}$, where l is still the arc length parameter, $\mathcal{N} \subset M$ is a compactum (for example, γ_0 and $\mathcal{N}$ may be, respectively, α and $\mathcal{L}$ from Proposition 14), and $\mathbb{U}$ is the half-axis $\mathbb{R}_+$ or $\mathbb{R}_-$. On γ_0 in addition to l we introduce an *affine* parameter s chosen so that the velocity $\boldsymbol{\eta} \leftrightharpoons \partial_s$ is future-directed and $s = 0$ at $l = 0$. Now γ_0 is characterized by the—clearly negative—function

$$h \leftrightharpoons \eta^a \tau_a,$$

which relates l to s:

[11] By the example of Misner space, one sees that in a non-globally hyperbolic M^e even this may not be the case.

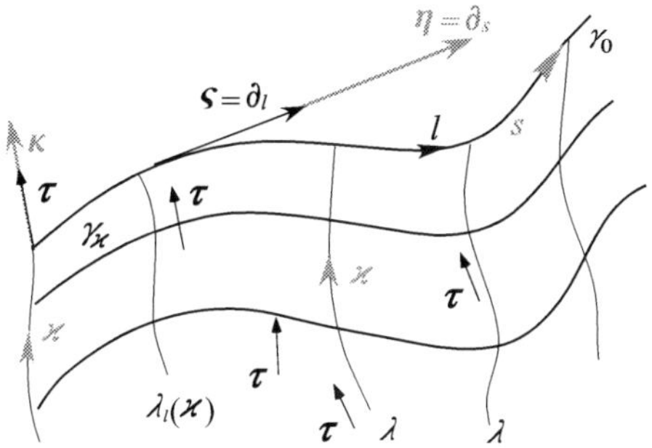

Fig. 5 The homotopy Λ. The curve γ_0 is null and λs are timelike

$$h = -\frac{\mathrm{d}l}{\mathrm{d}s}, \qquad s(l) = \int_l^0 \frac{\mathrm{d}\breve{l}}{h(\breve{l})} \tag{16}$$

(the first equality is obtained by multiplying both ends of the chain $\boldsymbol{\eta} = \partial_s = \frac{\mathrm{d}l}{\mathrm{d}s}\partial_l = \frac{\mathrm{d}l}{\mathrm{d}s}\boldsymbol{\varsigma}$ by $\boldsymbol{\tau}$) and one velocity to the other:

$$\boldsymbol{\eta} = -h\boldsymbol{\varsigma} \qquad \text{on } \gamma_0. \tag{17}$$

The function has a transparent physical meaning (unless γ_0 is a loop, this case will be considered separately):

$$h(l_1)/h(l_2) = \varepsilon(l_1)/\varepsilon(l_2), \tag{18}$$

where $\varepsilon(l)$ is the energy at a point $\gamma_0(l)$ of a photon moving on γ_0, as measured by an observer with velocity $\boldsymbol{\tau}[\gamma_0(l)]$, cf. Sect. 2.3.

Let us examine the curves obtained by moving every point of γ_0 some distance to the past along the integral curves of $\boldsymbol{\tau}$. Clearly, simply by choosing an appropriate rate of decreasing of the mentioned distance, the resulting curve—let us denote it γ_*—can be made also causal. At the same time γ_* cannot be imprisoned in any compact $\mathcal{O} \subset I^-(\mathcal{N})$, see Remark 15. A few important properties of CGCHs can be derived from these facts. This will be done in the next subsection, while now we are going to prove a lemma that will enable us to find the deformations in question explicitly.

Pick a positive (and sufficiently small, see below) constant $\varkappa_*$ and a smooth function f defined on $\mathbb{U}$ and obeying the inequalities $\underline{f} \leqslant f \leqslant \overline{f}$, where $\underline{f}$ and $\overline{f}$ are non-negative constants. Define a homotopy

$$\Lambda(l, \varkappa)\colon \quad \mathcal{G} \to M, \qquad \text{where } \mathcal{G} \leftrightharpoons \mathbb{U} \times [-\varkappa_*, 0],$$

see Fig. 5, by requiring that

(a) the first "horizontal" curve be γ_0:

$$\Lambda(l,0) = \gamma_0(l);$$

(b) each "vertical" curve $\lambda_c(\varkappa) \leftrightharpoons \Lambda(c,\varkappa)$ be (a part of) an integral curve of the field $\boldsymbol{\tau}$;
(c) the velocity $\boldsymbol{\kappa} = \partial_\varkappa$ in every point $p \leftrightharpoons \lambda_l(\varkappa)$ be equal to $f(l)\boldsymbol{\tau}(p)$.

In other words, Λ is constructed so that for any $\varkappa$ the "horizontal" line $\gamma_\varkappa(l) \leftrightharpoons \Lambda(l,\varkappa)$ is obtained from γ_0 by moving each point of the latter to the past—along the integral curves of $\boldsymbol{\tau}$—by $f(l)|\varkappa|$, the distance being measured by the natural parameter (loosely speaking f defines the shape of the deformation and $\varkappa$—its amplitude).
List. To summarize, we have introduced two types of curves:

1. Horizontal curves γ. One of them— the one whose deformation is under examination (it will be denoted, depending on the context, γ_0, α, or ℓ)—is parameterized by the "arc length parameter", which is the quantity l defined by (12). That curve is a null geodesic. In addition to l there is an *affine* parameter s on γ_0 and the corresponding velocity $\boldsymbol{\eta} \leftrightharpoons \partial_s$. All other horizontal curves are parameterized only by l, which in this case is defined by the requirement [see item (b) in the definition of Λ] that it is constant along λ's (note that when $\varkappa$ is non-zero, l need not be an arc length parameter on $\gamma_\varkappa$). The velocity vector corresponding to l is denoted by $\boldsymbol{\varsigma}$;
2. Vertical curves λ, which are the integral curves of the field $\boldsymbol{\tau}$. In addition to the natural parameter there also defined the parameter $\varkappa$ on λs, which differs from the former only by a constant (on each λ) factor, [see item (c) in the definition of Λ]. The velocity $\partial_\varkappa$ is denoted by $\boldsymbol{\kappa}$;

and a scalar function f, which is initially defined on $\mathbb{U}$, but which we shall extend to the entire $\mathcal{G}$ by the relation $f(p) \leftrightharpoons f[l(p)] \quad \forall p \in \mathcal{G}$.

The surface $\Lambda(\mathcal{G})$ may have self-intersections, so a technical remark is necessary. First, we consider $\boldsymbol{\eta}$ and h as functions of l, not of a point of M. And, second, the vectors $\boldsymbol{\varsigma}$ and $\boldsymbol{\kappa}$ are maps $\mathcal{G} \to T_M$. So, a derivative like $\varsigma^a{}_{;b}(p)$, $p \in \mathcal{G}$ is actually shorthand for $\left[\varsigma^a \circ \tilde{\Lambda}^{-1}\right]_{;b}\left(\Lambda(p)\right)$, where $\tilde{\Lambda}$ is the restriction of Λ to a neighbourhood $O_{\Lambda,p}$ of p in which Λ is injective.

Our way of defining l gives rise to a useful relation. To derive it pick a coordinate system $\{l, \varkappa, x_1, x_2\}$ in a convex neighbourhood of $\Lambda(p)$ so that within this neighbourhood $\Lambda(O_{\Lambda,p})$ is the surface $x_{1,2} = 0$ [this can be done, for example, by picking in T_p a pair of vectors $\boldsymbol{e}_1$ and $\boldsymbol{e}_2$, enlarging $\{\boldsymbol{\kappa}(p), \boldsymbol{\varsigma}(p)\}$ to a basis in T_p, and assigning the coordinates a, b, c, d to the end point of the geodesic that has unit (affine) length and emanates from the point q: $l(q) = a$, $\varkappa(q) = b$ with initial velocity $c\boldsymbol{f}_1 + d\boldsymbol{f}_2$, where $\boldsymbol{f}_1$ and $\boldsymbol{f}_2$ are the result of the parallel transfer of $\boldsymbol{e}_1$ and $\boldsymbol{e}_2$ from p to q along the geodesic]. Then the curves λ and γ are coordinate lines and hence, see Corollary 5 in Chap. 1,

$$\varsigma^a{}_{;\varkappa} - \kappa^a{}_{;l} \leftrightharpoons \varsigma^a{}_{;b}\kappa^b - \kappa^a{}_{;b}\varsigma^b = \varsigma^a{}_{,\varkappa} - \kappa^a{}_{,l} = 0, \tag{19a}$$

which yields, in particular,

$$\varsigma_{a;\varkappa} \leftrightharpoons \varsigma_{a;b}\kappa^b = \kappa_{a;b}\varsigma^b = (f\tau_a)_{;b}\varsigma^b = f'\tau_a + f\tau_{a;b}\varsigma^b. \tag{19b}$$

The existence—when $\varkappa_*$ is sufficiently small—of the homotopy Λ is guaranteed, even for a singular M, by the compactness of $\mathcal{N}$ (recall that $\mathcal{N} \supset \gamma$). Indeed, for any point $p \in M$ let $t_{\rm m}(p)$, be the length, with respect to the Riemannian metric g^R, of the shortest inextendible curve emanating from p. For example, in the DP space $t_{\rm m}(p)$ is the distance from p to the nearest "hole" and for the time machine from Example 8 it is infinite. Clearly, in the general case $t_{\rm m}(p)$ is also either infinite, or continuous and positive in $\mathcal{N}$. Hence, $\underline{t} \leftrightharpoons \inf_{\mathcal{N}} t_{\rm m} > 0$ due to the compactness of $\mathcal{N}$. So, from any point it is possible to travel any (natural) distance $T < \underline{t}$ along the integral curve λ of the field τ. Thus any number less than $\underline{t}/(2\overline{f})$ can be taken to be $\varkappa_*$.

As noted above, to properly use the fact that a causal curve is imprisoned in a compactum, that is to derive a contradiction from that fact, we need the compactum to lie in $M^{\rm r}$, not just in the causal past of γ_0, cf. Remark 15. We shall now formulate a convenient sufficient condition. Let

$$F(p,t): \qquad \mathcal{N} \times [\overline{f}\varkappa, \underline{f}\varkappa] \to M$$

be a map that displaces every point $p \in \mathcal{N}$ a parameter distance t to the future along the corresponding λ. Any $\gamma_\varkappa$ with $\varkappa \in [-\varkappa_*, 0]$ lies in $\mathcal{O} \leftrightharpoons F\big(\mathcal{N} \times [\overline{f}\varkappa, \underline{f}\varkappa]\big)$. Being the image of a compact set under a continuous map, $\mathcal{O}$ is compact. Also it lies, by construction, in $J^-(\mathcal{N})$. Generally, however, it may happen, that $\mathcal{O} \not\subset I^-(\mathcal{N})$. Such a possibility is excluded, if F moves to the past *every* point of $\mathcal{N}$. Thus we have established the existence of a compact set $\mathcal{O}$ such that

$$\begin{array}{ll} \forall \varkappa \in [-\varkappa_*, 0] & \gamma_\varkappa \subset \mathcal{O}, \\ \text{If } \underline{f}, \varkappa \neq 0, & \text{then } \mathcal{O} \subset I^-(\mathcal{N}). \end{array} \tag{20}$$

Lemma 18 *If f'/f is bounded and for some positive constant c_1*

$$h'/h < -f'/f - c_1 f, \qquad \forall l \in \mathbb{U}, \tag{21}$$

then there is $\varkappa_0$ such that the curve $\gamma_{\varkappa_0}$ is timelike and inextendible in the same direction as γ_0.

Before proceeding to the proof proper (it will be similar to the proof of [76, Lemma 8.5.5]) we have to establish the boundedness—at sufficiently small $\varkappa_*$—of a number of relevant quantities in $\Lambda(\mathcal{G})$. First, note that f and f' are bounded, by definition and by hypothesis, respectively. And so are τ^a, $\tau^a{}_{;b}$ and $\tau^a{}_{;bc}$, since they are smooth in the entire[12] compact set $\mathcal{O}$, which, as follows from (20), contains $\Lambda(\mathcal{G})$. Next, consider the components ς^a. At $\varkappa = 0$ the boundedness of ς^a follows from the fact that the length of $\boldsymbol{\varsigma}$ in the Riemannian metric (13) is constant on γ_0:

$$g^R(\boldsymbol{\varsigma}, \boldsymbol{\varsigma}) = g(\boldsymbol{\varsigma}, \boldsymbol{\varsigma}) + 2[g(\boldsymbol{\varsigma}, \boldsymbol{\tau})]^2 = 2.$$

Next, the function $\varsigma^a(l, \varkappa)$ is bounded also on the entire $\mathcal{G}$, being the solution of the differential equation (19b) with bounded coefficients and with bounded, as we just have established, initial value $\varsigma^a(l, 0)$. Finally, the chain

$$h' = (\eta^a \tau_a)_{;b}\varsigma^b\big|_{\gamma_0} = \eta^a \tau_{a;b}\varsigma^b = -h\varsigma^a \tau_{a;b}\varsigma^b, \tag{22}$$

in which the second equality follows from the fact that γ_0 is a geodesic, and the last one—from (17), proves that *irrespective of the validity of* (21)

$$h'/h \quad \text{is bounded on } \gamma_0. \tag{23}$$

Proof of the lemma. Consider the function $y \leftrightharpoons \varsigma^a\varsigma_a$. It plays the role of an indicator: a curve $\gamma_\varkappa$ is timelike at some l_0 when and only when $y(l_0, \varkappa)$ is negative. Write down the following chain of equations valid in every point of γ_0

$$\begin{aligned} \tfrac{1}{2}y_{,\varkappa} = \varsigma_{a;\varkappa}\varsigma^a = \kappa_{a;l}\varsigma^a = (f\tau_a)_{;l}\varsigma^a\big|_{\gamma_0} &= -f' - f\tau_a\varsigma^a{}_{;l} = -f' + f\tau_a(h^{-1}\eta^a)_{;l} \\ &= -f' + fh(h^{-1})_{,l} + f\tau_a h^{-1}\eta^a{}_{;l} = -f' - f\ln'|h|, \end{aligned} \tag{24}$$

which is derived by using, in turn, Eq. (19a) (in the second equality), the normalizing condition (12) defining l (in the fourth one), the relation (17) (in the next equality), and, finally, the fact that $\boldsymbol{\eta}$ satisfies the geodesic equation $\eta^a{}_{;l} = -\frac{1}{h}\eta^a{}_{;b}\eta^b = 0$ on γ_0. Combining the resulting equation with the hypothesis (21) we get

$$y_{,\varkappa}(l, 0) > 2c_1 f^2. \tag{25}$$

On the other hand, $y = 0$ on γ_0. So, we conclude that for any l there is a (negative, of course) $\varkappa_{**}$ such that

$$y(l, \varkappa) < 0, \qquad \forall \varkappa \in (\varkappa_{**}(l), 0) \tag{26}$$

(the value of $\varkappa_{**}$ depends on l). Thus, $\gamma_\varkappa$ for the relevant $\varkappa$ is timelike at l. We, however, are looking for an *inextensible* timelike curve or, equivalently, a $\varkappa_{**}$ such that the inequality (26) holds for all l *simultaneously*. So, let us write down one chain more:

[12] For the sake of simplicity, in discussing the boundedness of tensor components we shall assume that the whole $\mathcal{O}$ is covered by a single coordinate system. The generalization to the case when such a system does not exist is straightforward, because $\mathcal{O}$ always can be covered by a *finite* number of compact sets $\mathcal{O}_m$ each of which *is* covered by a single chart.

$$\forall\varkappa \quad \tfrac{1}{2}y,_{\varkappa\varkappa} = [(f\tau_a)_{:l}\varsigma^a],_{\varkappa} = (f\tau_a)_{:l\varkappa}\varsigma^a + (f\tau_a)_{:l}\varsigma^a{}_{:\varkappa}$$
$$= f'\tau_{a:\varkappa}\varsigma^a + f\tau_{a:l\varkappa}\varsigma^a + (f'\tau_a + f\tau_{a:l})(f'\tau^a + f\tau^a{}_{:b}\varsigma^b)$$

[the last parenthesized factor is $\varsigma^a{}_{:\varkappa}$ transformed with the use of (19b)]. Substitute the formulas

$$\tau_{a:l} = \tau_{a:b}\varsigma^b, \qquad \tau_{a:\varkappa} = \tau_{a:b}\kappa^b = f\tau_{a:b}\tau^b,$$
$$\tau_{a:l\varkappa} = \tau_{a:bc}\varsigma^b f\tau^c + \tau_{a:b}\varsigma^b{}_{:\varkappa} = f\tau_{a:bc}\tau^c\varsigma^b + \tau_{a:b}(f'\tau^b + f\tau^b{}_{:c}\varsigma^c),$$

in its rightmost part to obtain

$$y,_{\varkappa\varkappa}(l,\varkappa) = b_1 f'^2 + b_2 f' f + b_3 f^2,$$

where b_k are some bounded (as follows from the established above boundedness of ς^a and the derivatives of τ^a) functions. Dividing this relation by the inequality (25) one gets (recall that by hypothesis f'/f is bounded)

$$|y,_{\varkappa\varkappa}(l,\varkappa)/y,_{\varkappa}(l,0)| < c_2, \qquad \forall\varkappa\in[\varkappa_{**},0], \tag{27}$$

where, as usual, c_2 is a constant. And since, as we already know, $y(l,0) = 0$ and $y,_{\varkappa}(l,0) > 0$, the bound (27) means that for some negative $\varkappa_0$

$$y(l,\varkappa_0) < 0, \qquad \forall l\in\mathbb{U},\ \varkappa\in[\varkappa_0,0),$$

and hence $\gamma_{\varkappa_0}$ is *timelike*.

Similar reasoning can be applied to the quantity $\omega \leftrightharpoons \varsigma^a\tau_a$. Namely, $\omega = -1$ at $\varkappa = 0$, while its derivative in the κ-direction

$$\omega_{:\varkappa} = \varsigma_{a:\varkappa}\tau^a + \varsigma^a\tau_{a:\varkappa} = f'\tau_a\tau^a + f\tau_{a:b}\varsigma^b\tau^a + f\varsigma^a\tau_{a:b}\tau^b$$

is bounded. Hence, $\omega(l,\varkappa_0)$ at a sufficiently small $\varkappa_0$ is greater (in absolute value) than $\frac{1}{2}$. Consequently, the length of the corresponding $\gamma_{\varkappa_0}$ in the Riemannian metric (13) is infinite and $\gamma_{\varkappa_0}$ is *inextendible*, see Proposition 13. □

4.3 Some Properties of Time Machines with CG(D)CH

The WEC Violation

As we have seen above, exotic spacetimes such as shortcuts and wormholes need exotic matter for their existence. The same is true for time machines, too. To prove this claim we begin with an important technical result (essentially, this is Lemma 8.5.5 of [76] proven more rigorously [107]. One more proof can be found in [130]).

Proposition 19 *If a Cauchy horizon $\mathcal{H}^+$ is compactly generated, then its generators are past complete.*

Proof Assume, a generator γ_0 is past *in*complete. Then its affine parameter s is bounded from below and the integral (16) converges at $l \to -\infty$. This enables us to define on the semi-axis $(-\infty, 0]$ a smooth positive function

$$f(l) \leftrightharpoons \frac{1}{h}\left[-\int_l^0 \frac{\mathrm{d}\breve{l}}{h(\breve{l})} + 2\int_{-\infty}^0 \frac{\mathrm{d}\breve{l}}{h(\breve{l})}\right]^{-1}. \tag{28}$$

For the thus defined f

$$f'/f + h'/h = -f \tag{29}$$

and hence condition (21) is satisfied.

The fact, by itself, that the integral in (16) converges does not mean, of course, that the integrand is bounded. It is quite imaginable that as $l \to -\infty$ the integrand takes increasingly large values in increasingly narrow intervals. However, such a behaviour would contradict (23) and we conclude that $1/h$ is bounded after all. It follows then that f [by (28)] and, consequently, f'/f [by (29) and (23)] are bounded too. Thus, all conditions of Lemma 18 (with $\mathcal{K}$ as $\mathcal{N}$) are satisfied and there must be a timelike past inextendible curve $\gamma_{\varkappa 0}$ in $\mathcal{O}$, see (20). This curve is totally imprisoned in a compact subset of the globally hyperbolic spacetime M^{r}, namely, in the intersection of the compactum $\mathcal{O}$ with $J^-[\gamma_{\varkappa 0}(0)]$, which is closed by Proposition 48 in Chap. 1 (note that the case in point is $\gamma_{\varkappa 0}$, not γ_0, whose causal past can be non-closed). And such an imprisonment is forbidden by Proposition 44 in Chap. 1. □

Corollary 20 *If in some point of a CGCH $\mathcal{H}^+$ the expansion θ is positive,*[13] *then somewhere on the horizon the Weak energy condition does not hold.*

This corollary—in [74] it is proven by combining Proposition 19 with Corollary 8 and Proposition 6 in Chap. 2—is really important, because the condition $\theta > 0$ to all appearance must hold in (some, at least) points of the horizon of *any* time machine that is of a laboratory rather than of a cosmological scale. Indeed, in such a spacetime one expects the horizon to have (asymptotically, at least) the shape of an expanding cone.

To make that argument more (though not quite) rigorous we introduce, following [74], a special class of spacetimes: it consists of time machines whose Cauchy horizons are compactly generated, while the Cauchy surfaces $\mathcal{S}$ of the initial globally hyperbolic regions are non-compact (that is the region from which the time machine originates in such a universe is "smaller than the whole universe"). Let $\gamma(s)$ be an affinely parameterized generator of the Cauchy horizon of such a time machine. Let, further, μ_t be the map that translates every[14] point of the horizon to the past along

[13] As before, we consider geodesic generators of $\mathcal{H}^+$ to be *future* directed.

[14] Strictly speaking, μ_t is not defined in points in which a few generators meet, so one has to show that such points form a set of measure zero.

the corresponding γ by the parameter distance t

$$\mu_t: \quad \gamma(s) \mapsto \gamma(s-t).$$

At $t = 0$ the 3-volume $\mathcal{V}$ [with respect to the Riemannian metric g^R, see (13)] of the three-dimensional region $\mathcal{K}$ is transformed under the action of μ_t by the following law

$$\frac{\mathrm{d}}{\mathrm{d}t}\mathcal{V}(\mu_t(\mathcal{K})) = \frac{\mathrm{d}}{\mathrm{d}t}\int_{\mathcal{K}} \mathrm{d}\mathcal{V} = -\int_{\mathcal{K}} \theta \, \mathrm{d}\mathcal{V},$$

where the last (by no means obvious) equality is, essentially, [76, Eq. (8.4)]. On the other hand, $\mathcal{H}^+$ is non-compact (the integral curves of the field $\boldsymbol{\tau}$ provide the—possibly non-smooth—embedding of $\mathcal{H}^+$ in $\mathcal{S}$, that is $\mathcal{S}$ is an extension of $\mathcal{H}^+$, which would be impossible, be the latter compact, see Corollary 70 in Chap. 1). Hence, $\mathcal{K}$ is only a *part* of $\mathcal{H}^+$ and $\mu_t(\mathcal{K})$ is only a part of $\mathcal{K}$

$$\mathcal{V}(\mu_t(\mathcal{K})) < \mathcal{V}(\mathcal{K}), \qquad \forall t.$$

Comparing this with the previous formula, we conclude that somewhere θ must, indeed, turn positive, which implies that the WEC is violated in *all* time machines with CGCH (and non-compact Cauchy surfaces of M^{r}). It is apparently this result that gave rise to the widespread idea that violations of causality and of the WEC are inseparably connected.

The just reproduced reasoning contains a lacuna [25]: the horizon is implied to be sufficiently smooth [for example, second derivatives enter Eq. (3) in Chap. 2]. However, generally, horizons are not that smooth, see [25, 109]. It is important, therefore, that a proof was found [131] that does not lean on the smoothness assumptions.

(Almost) Closed Generators

The Cauchy horizon bounds a region containing closed causal curves. It would be natural to expect that it too contains some closed null geodesics, at least when it is compactly generated and its generators are not "open" by singularities. However, the inevitable presence of closed null curves in CGCHs is not proved in the general case, an example due to Carter, see [76, Fig. 39] might give a clue to why.

Proposition 21 [74] *Every CGCH has a generator that lies entirely in $\mathcal{K}$ and that is both future and past*[15] *inextendible.*

Proof Let $\lambda(l)$ be a future-directed horizon generator lying in $\mathcal{K}$ at all $l \leqslant 0$. Since λ is past inextendible, the parameter l is unbounded from below, see Proposition 13. Hence, we can define an infinite sequence of points $p_m \leftrightharpoons \lambda(-m)$, $m = 1, 2 \ldots$. Denote by ς_m the velocities of λ in the points p_m. Due to the compactness of both $\mathcal{K}$ and the sphere (12), there is a subsequence (p_k, ς_k) such that

$$p_k \to p \in \mathcal{K} \quad \text{and} \quad \varsigma_k \to \varsigma,$$

[15]The "past" part is trivial, because such are *all* generators by Corollary 7.

where ς is a null vector satisfying (12). Consider a maximal geodesic $\mu(l)$ emanating from $p \leftrightharpoons \mu(0)$ with the initial velocity ς. By Corollary 7, $\mathcal{H}^+$ is closed, so μ is one of its generators.

Suppose, μ is *not* a sought-for curve. But by construction it is a null geodesic inextendible in both directions. So, the only possibility is that $\mu(d) \notin \mathcal{K}$ at some d, which by the closedness of $\mathcal{K}$ implies, in particular, that around $\mu(d)$ there exists a coordinate ball B disjoint with $\mathcal{K}$. This means that every geodesic emanating from a point sufficiently close to p with initial velocity sufficiently close to ς also meets B and thus leaves $\mathcal{K}$. In particular, this is true for the segment of λ bounded by $\lambda(-i)$ and $\lambda(d-i)$ with sufficiently large i. A contradiction. □

Unfortunately, the proposition does not tell whether the endless generator is closed. Let us dwell on the horizons (not necessarily compactly generated or determined) that do have closed generators, an example is the horizon of the Misner time machine, which is generated by the circle $t = 0$, see Sect. 2.1.

Let $\ell \subset \mathcal{H}^+$ be a closed null geodesic. Then the *function* $\ell(l)$ is periodic:

$$\ell(l)\colon \quad \mathbb{R} \to \mathcal{H}^+, \qquad \ell(l) = \ell(l + \tilde{l}), \quad \tilde{l} \text{ is a positive constant.}$$

On the other hand, ℓ does not need to be periodic as function of an affine parameter s. Correspondingly, $h(l)$ also may not be periodic. It is, however, a solution of the differential equation (22), whose coefficients are periodic. Therefore, in the general case

$$h(l) = a^{l/\tilde{l}}\chi(l), \qquad \text{where } \chi(l) \text{ is periodic}, \quad a = const > 0. \tag{30}$$

As we are going to prove, a cannot be less than unity. Whether it can be equal to unity is unknown, while geodesics with $a > 1$, such as the Misner horizon, see p. 91, have two curious properties. First, they cannot be the world lines of photons (it would be impossible to assign a definite energy or momentum to such a photon) and, second, the geodesic ℓ is *incomplete*. Thus, the following proposition implies that time machines with CGCH are "almost for sure" singular.

Proposition 22 [74] *$a \geqslant 1$ with inequality only if ℓ is future incomplete.*

Proof Pick a constant c and on the segment $[0, \tilde{l}]$ (this is a period of ℓ) define the function

$$f(l) \leftrightharpoons \frac{1}{h(l)}\left[c^2 \int_l^{\tilde{l}} h(\breve{l})\,\mathrm{d}\breve{l} - 1\right].$$

Clearly, it is smooth and positive. Besides, it solves the equation

$$f' + f\,\ln' h = -c^2.$$

Now denote by $\widetilde{\Lambda}$ the restriction

$$\widetilde{\Lambda} \leftrightharpoons \Lambda\big|_{[0,\tilde{l}]\times[-\varkappa_*,0]}$$

of the homotopy Λ (which is defined on p. 107). By replacing Λ with $\widetilde{\Lambda}$ and γ_0 with ℓ everywhere between the definition of Λ and inequality (26) (we do not have to estimate $y,_{\varkappa\varkappa}$, because now we deal not with the semi-axis $\mathbb{U}$, but merely with its compact subset $[0, \widetilde{l}]$), we get the proof of the fact that the curve $\ell_{\varkappa_0} \subset M^{\rm r}$, obtained by translating each point of ℓ the (natural parameter) distance $f\varkappa_0$ along the integral lines of $\boldsymbol{\tau}$, is timelike and future directed.

However, if $a < 1$, then at sufficiently small c^2

$$f(\widetilde{l}) = -1/h(\widetilde{l}) = -\frac{1}{ah(0)} > \frac{1}{h(0)}\left[c^2\int_0^{\widetilde{l}} h(\breve{l})\,\mathrm{d}\breve{l} - 1\right] = f(0),$$

that is the future end point $\ell_{\varkappa_0}(\widetilde{l})$ of $\ell_{\varkappa_0} \subset M^{\rm r}$ chronologically precedes the past end point $\ell_{\varkappa_0}(0)$ (the latter is obtained by translating to the past the same point $\ell(0) = \ell(\widetilde{l})$ along the same line as the former, but by a smaller distance), which is impossible, $M^{\rm r}$ being causal. Thus, $a \geqslant 1$.

If $a > 1$, then the integral

$$s(l) = -\int_0^l \frac{\mathrm{d}\breve{l}}{h(\breve{l})} = -\int_0^l \frac{a^{-\breve{l}/\widetilde{l}}\,\mathrm{d}\breve{l}}{\chi(\breve{l})}$$

[we have consecutively used (16) and (30)] converges at $l \to \infty$, which proves the second assertion of the proposition. □

"Dangerous" Geodesics [106]

It was established above, see Proposition 14, that a CDCH forms *after* the appearance of a geodesic $\alpha(l)$ that is totally future imprisoned by a compact set. How pathological is the existence of such a geodesic? To answer this question we shall consider the energy properties of the photon travelling on α. Specifically, we split α into segments of unit (Riemannian) length and find—employing (18)—the maximal values of the photon's energy in each of these segments taking the energy of the photon in the point $\alpha(l_0)$ to be unit (the energy thus becomes a function of two variables—l and l_0). Then we sum those maximal values up and see that an appropriate choice of the initial point $\alpha(l_0)$ makes the sum *arbitrarily large*. We shall now prove this fact before discussing its physical meaning.

Proposition 23 *Let $\alpha(l)$ be the geodesic from Proposition 14. Then for an arbitrarily big constant E there is a positive number l_0 such that*

$$\frac{1}{|h(l_0)|}\sum_{k=0}^{\infty} h_k \geqslant E, \qquad \textit{where}\ \ h_k \leftrightharpoons \max_{l\in[l_0+k,\,l_0+k+1]} |h(l)|.$$

Proof Suppose, the assertion is false. Then the series in the inequality converges for any choice of l_0. Hence, first, h tends to zero at $l \to \infty$ and, second, the positive (since h is negative) function

$$f(l) \leftrightharpoons \frac{1}{h(l)} \int_l^\infty h(\breve{l})\, d\breve{l} < E, \qquad l > 0$$

is defined. It is bounded by E and its numerator tends to zero, which enables us to use the Cauchy formula [192, Sect. 120] and to establish that

$$f(l) = -\frac{h(l_*)}{h'(l_*)} \qquad \text{at some } l_* > l.$$

By (23) this means that f is bounded away from zero

$$0 < c_3 < f. \qquad (\star)$$

It is easy to check that $h'/h + f'/f = -1/f$ and hence condition (21) is fulfilled (as f is positive). The same equality being combined with the boundedness of h'/h, see (23), and $1/f$, see ($\star$), implies the boundedness of f'/f. Thus the conditions of Lemma 18 are satisfied, whence there must exist a future inextendible timelike curve which, as follows from (20), is totally imprisoned in a compact subset of $I^-(\alpha) \subset M^{\mathrm{r}}$. But this contradicts Proposition 44 in Chap. 1 because M^{r} is strongly causal. □

It is this result that suggests, see the discussion in the end of Sect. 3, that compactly determined Cauchy horizons are unstable and do not actually form. What appears instead are just bright flashes. Indeed, consider a ball $B \subset \mathcal{L}$ such that $\alpha \cap B$ consists of the infinite number of segments α_i. An observer in $p \in B$ will perceive α as a *bundle* of photons: within the "laboratory" B their world lines are α_i. If the 4-velocity of the observer is $\boldsymbol{\tau}(p)$, he interprets $E \leftrightharpoons \sum_i h(q_i)$, where $q_i \in \alpha_i$, as the "total energy" of the bundle (B is assumed to be small enough to justify neglecting the possible variation of h along α_i). In these terms, the just proven proposition says that for arbitrarily large E, there is a null geodesic in $\mathcal{L}$ such that a photon moving on it with the unit initial energy will cross the laboratory B, as a bundle of rays with the total energy larger than E.

5 A Pathology-Free Time Machine

The time machines considered so far possess a number of "pathological" features: they contain exotic matter, dangerous geodesics and singularities. This raises the question of whether these properties are associated with *all* time machines. As an answer, we build in this section a specific time machine (for a more general consideration see Appendix A.3) free from all just listed 'flaws' [92].

In the two-dimensional DP space M_{DP}, see Example 73 in Chap. 1, pick a smooth positive function $w(x_0, x_1)$ which is equal to 1 at $|x_0|, |x_1| > 1.5$ and tends to zero at $x_{0,1} \to \pm 1$. Also we require first two derivatives of w to be bounded (for instance, w can be the function from Example A.7 in Appendix). Then the desired time machine

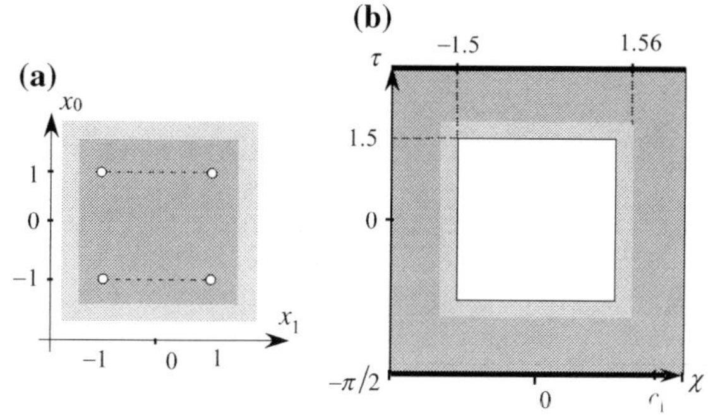

Fig. 6 **a** The spacetime M_w. $w = 1$ exterior to the dark grey region. The white circles depict missing points, w tends to zero as they are approached. **b** The space F. The dark grey region is a portion of the Friedmann universe. The light grey 'frames' are isometric. The 'integration of the time machine into the Friedmann universe' is achieved by cutting away the white regions and glueing the light grey frames together

is the spacetime (M_w, g) with $M_w = M_{DP} \times S^2$ and

$$g: \qquad ds^2 = w^{-2}(-dx_0^2 + dx_1^2) + r_*^2(dx_2^2 + \sin^2 x_2 dx_3^2),$$

where $x_2 \in [0, \pi]$, $x_3 \in [0, 2\pi)$ are angle coordinates and the constant r_* obeys the following inequalities

$$0 < r_* < \inf_{M_1} \left| w(w,_{x_1x_1} - w,_{x_0x_0}) + w,^2_{x_0} - w,^2_{x_1} \right|^{-1/2}. \tag{31}$$

(as is explained in Example 73 in Chap. 1, $x_{0,1}$ coordinatize only a *part* of the factor M_w, specifically, a plane with two cuts). The resulting spacetime is free from all pathologies mentioned above, in particular, it has no singularities, see Corollary A.8 in Appendix. This fact is not too surprising by itself: we have intentionally chosen the conformal transformation of the initially singular spacetime so as to send the singularities to infinity. It is remarkable, however, that we managed to do this without violating the WEC. Indeed, it follows from (A.39) in Appendix that the inequality (31) guarantees the fulfilment of this condition[16] throughout M_w.

What may provoke some objections is the $\mathbb{R}^2 \times S^2$ topology of the ambient space. So, let us show that the time machine under discussion can be integrated into the Friedmann universe, see Fig. 6. Let F be a space with topology $\mathbb{R}^1 \times S^3$ and metric

$$ds^2 = a^2(\tau)[-d\tau^2 + d\chi^2 + r^2(\chi)(d\vartheta^2 + \sin^2\vartheta d\varphi^2)], \qquad \chi \in [-\pi/2, \pi/2].$$

[16]And even the dominant energy condition. On the other hand, the strong energy condition never holds in spacetimes like M_w [124].

Here ϑ and φ are the standard angle coordinates, while r and a are smooth positive convex functions satisfying the following conditions:

$$a(x) = 1, \quad r(x) = r_* \quad \text{at } |x| < 1.55, \qquad r(x) = \cos x \quad \text{at } |x| > 1.56.$$

As is easy to see, the space F beyond the cylinder $|\tau|, |\chi| \leqslant 1.56$ is merely the closed Friedmann universe. Our requirements on a make the initial and final singularities inevitable, but these are the usual cosmological singularities unrelated to the time machine. Further, the direct calculations (they can be considerably lightened by the use of the formulas from [135, 14.16]) show that the WEC holds everywhere in F.

Now we are going to build a new space by cutting a part (comprising $\overset{\looparrowright}{M}_w$) out of M_w and using it to replace a region in F. To this end we, first, note that the simple change of notation

$$x_0 \to \tau, \quad x_1 \to \chi, \quad x_2 \to \vartheta, \quad x_3 \to \varphi$$

isometrically maps the "frame" $\{1.5 < |x_0|, |x_1| < 1.55\} \subset M_w$ to the corresponding region of F, see Fig. 6 (we denote this isometry by σ). The desired spacetime M_F is obtained by 1) removing the set $\{1.55 \leqslant |x_0|, |x_1|\}$ from M_w, 2) removing the set $\{|\tau|, |\chi| \leqslant 1.5\}$ from F, and 3) glueing the two thus obtained regions by σ:

$$M_F \leftrightharpoons \{q \in M_w \colon\ |x_0(q)|, |x_1(q)| < 1,55\} \cup_\sigma \{p \in F \colon\ |\tau(p)|, |\chi(p)| > 1,5\}.$$

The result is a universe that contains a time machine, but does not violate the WEC and is free from dangerous geodesics and non-cosmological singularities. There is a region in this universe the boundary of which is a 2-sphere existing for a finite time interval (so, the region may be interpreted as a laboratory) and the exterior of which is just the Friedmann universe minus a compactum.

Chapter 5
A No-Go Theorem for the Artificial Time Machine

1 The Theorem and Its Interpretation

So far we have divided time machines into 'eternal' and 'appearing', see Sect. 1 in Chap. 4. This classification, however, is too coarse. Suppose, one wishes to travel to the past. A possible strategy would be just to look for an already existing closed timelike curve, or to wait passively until such a curve appears (as is exemplified by the Deutsch–Politzer space, such waiting may not be hopeless however innocent the spacetime looks at the moment). The alternative would be to *create* such a curve. Appropriately distributing matter and thus, according to the Einstein equations, acting upon the metric, an advanced civilization could try to *force* the spacetime to evolve into a time machine. In fact, it was the conjecture that, given a suitable opportunity, the civilization can succeed [137], that initiated intensive studies of time machines and wormholes.

The difference between *finding* a closed causal curve (which appeared owing to circumstances beyond our control) and *manufacturing* it, i.e. between a 'natural' time machine and an 'artificial' one, is crucial to our discussion in this chapter. As we shall see, this difference is so large that while the spacetimes of the former kind seem (by now) to be feasible, the artificial time machines are impossible within classical relativity. Even more, the last assertion remains valid if general relativity is complemented by any local condition, see Sect. 1 in Chap. 2.

The difference between the two types of time machine has to do with intentions and will of its creators, so it is hard, if possible, to describe it in purely geometrical terms. Our goal, however, is only to formulate a *necessary* condition for considering a time machine artificial and *this* can be done without turning to so vague notions. We proceed from the idea that for the existence of a time machine (i.e. a region $\overset{\looparrowright}{M}$ of a c-space M, the role of the hypothetical condition c in this definition will be discussed in a moment) to be attributable to the activity in a region $U \subset M$, *at least* the following two conditions must hold:

S. Krasnikov, *Back-in-Time and Faster-than-Light Travel in General Relativity*, Fundamental Theories of Physics 193, https://doi.org/10.1007/978-3-319-72754-7_5

($\mathscr{C}_1$) $\overset{\looparrowright}{M}$ lies entirely in the future of U;

($\mathscr{C}_2$) A non-empty $\overset{\looparrowright}{M}$ satisfying condition ($\mathscr{C}_1$), exists in *all* C-maximal extensions of the C-space U, cf. the definition of the forced Cauchy horizon.

If such U does not exist, we regard the time machine as natural, spontaneous, rather than artificial.

Remark 1 Along with relativity, we include in our consideration theories in which some additional local condition C is imposed on the universe (put differently, the universe is described not by a mere spacetime, but by a C-space) and this fact complicates the analysis. However, the *meaning* of C turns out to be irrelevant, it is important only that C stands for the *same* condition throughout this chapter. In particular, C may be empty, so the reader not interested in too much generality can consecutively ignore the symbol C- up to Remark 26.

Condition ($\mathscr{C}_1$) is self-evident, but ($\mathscr{C}_2$) needs some comment. Consider, for example the case when M is the DP space (see Example 73 in Chap. 1) and U is its subset $x_0 < -2$. Note that, as already was mentioned in Sect. 3.1 in Chap. 2, *exactly the same* U could have different extensions: Minkowski space, for instance or $\mathbb{R}^4$ endowed with the metric

$$g = (1 + f^2)\eta, \qquad \text{where } f\big|_{x_0<-2} = 0, \quad f \not\equiv 0, \quad \eta \text{ is the flat metric}, \qquad (*)$$

Some of these extensions can be excluded by imposing appropriate local conditions (that is why we speak of C-extensions, rather than of mere extensions). For example, by requiring that the WEC hold in all points of M, we exclude the extension ($*$). However, generally, it is impossible to exclude in this manner *all* extensions but one [96]: if *some* extension is found, then, as a rule, infinitely many others also can be built (for instance, in the way described in Example 33 in Chap. 2). So, a hypothetical advanced civilization acting in U *in principle* cannot take credit for the geometry of $J^+(U)$. This fact, however, does not necessarily make efforts for creating a time machine pointless:

(1) The situation may change with progress in science. It is conceivable that one day an improved version of relativity will be able to predict the development of any initial data *unambiguously*, see Sect. 3.2 in Chap. 2. Or, speculating further, one can imagine that the progress in quantum gravity will allow one to assign *probabilities* to different developments/extensions of the same U. In both the cases, we shall have to revise the present concept of time machine creation.

(2) We may aim at causality violation, *in general*, not at a particular time machine. In such a case, we must consider as a successful result *any* development of the initial data if the future of U in that development contains a closed causal curve. Correspondingly, any non-causal C-extension of the C-space U deserves to be called an *artificial time machine* unless there is a C-maximal C-extension of U in which

causality does hold. Condition ($\mathscr{C}_2$) is just the requirement that there be no such causal C-extensions.

In just the same way, we could classify the non-globally hyperbolic spacetimes with initial globally hyperbolic regions (this type of spacetime is broader than time machines, because a non-globally hyperbolic space need not be non-causal). The example of the DP space shows that loss of global hyperbolicity not always can be related to something that happened in the past. On the other hand, there are spacetimes—an example is Misner space—in which the initial geometry makes it inevitable that the global hyperbolicity will be lost: M^{r} has *no* globally hyperbolic extensions, cf. discussion of the forced horizons in Sect. 4 in Chap. 4.

In this chapter, we prove a theorem stating that causality—in contrast to global hyperbolicity—*cannot* be violated forcedly either in classical relativity or in any of its generalizations obtained by postulating an additional local law. We emphasize that the theorem does not forbid causality violations, it only guarantees that they never *have to* appear. Imagine, for example that a certain $\mathscr{B}$ gets a wormhole and starts to move its mouths in a vigorous and precise manner [10]. At the same time another participant, $\mathscr{A}$, only broods and swallows pills. What the theorem tells us is that the results of their efforts will be exactly the same: in either case a time machine may appear and may not. So, if a time machine appears after all, we shall have no reasons to link it to $\mathscr{A}$ or $\mathscr{B}$ and thus to think of it as artificial.

Theorem 2 ([97]) *For any local condition* C, *every* C*-extendible space* U *has a* C*-maximal extension* $M^{\max}$ *such that all closed causal curves in* $M^{\max}$*—if they exist there—are confined to the chronological past of* U.

The assertion of the theorem is quite obvious in the case of DP-like spaces. It might seem, however, that the situation is different for time machines with CDCHs. For example, in the Misner space an observer sees that as time goes on causal cones 'open' more and more (in coordinates t, ψ'), see Fig. 1b in Chap. 4. So, one may expect the *inevitable*, at first glance, causality violation at $t = 0$, when $\partial_{\psi'}$ becomes horizontal. However these expectations are exactly as groundless as in the DP case: by the theorem, among maximal extensions of the Misner space there are causality respecting ones. At the next stage, the observer may wonder how the geometry of their spacetime is affected by additional (i.e. not stemming from general relativity) restrictions. What, for example, will happen, if the local condition holds, that the spacetime describing their universe is everywhere flat? The answer given by the theorem is that among flat extensions of Misner space there *must* be such that (i) they have no flat extensions and (ii) they adhere to the causality condition. Indeed, an extension of this kind is obtained by, first, cutting the cylinder M_A, see Fig. 1b in Chap. 4, along the semi-axis $\psi' = 0,\ t \geqslant 0$ and the Minkowski plane along the semi-axis $\beta = 0,\ \alpha \geqslant 0$, see Fig. 1a in Chap. 4, and, second, glueing the left bank of either cut to the right bank of the other, see Fig. 3 in Chap. 6.

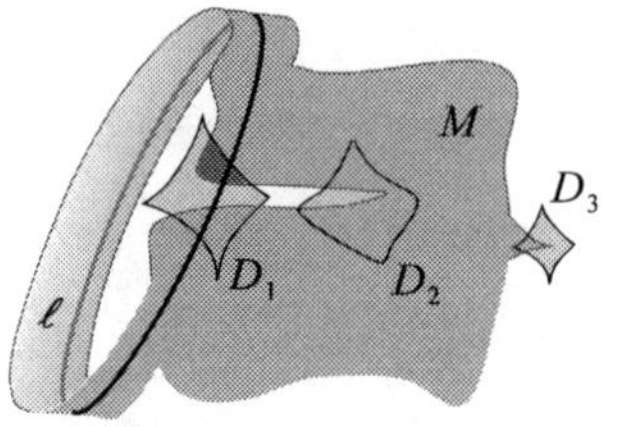

Fig. 1 D_i are variously situated perfectly simple subsets of M^e. R is the union of M and one of them. The intersections $M \cap D_1$ and $M \cap D_2$ are not causally convex in the corresponding D_i. Besides, in the latter case the intersection is connected, whence it follows that M is not convexly C-extendible. At the same time, in the former case one can 'unstick' the upper connected component of that intersection from M, and hence from R (so, the darkest region in the figure should be understood as a part of M seen through D_1). $M \cap D_1$ in this case is causally convex and $M_\triangle \leftrightharpoons M \cup D_1$ has no new, absent in $\overset{\looparrowright}{M}$, causal loops

2 Outline of the Proof

Let C be a local condition and M be a corresponding C-space. The proof will consist of two (quite unequal) parts: first, we shall prove Proposition 24 saying that for any M of a certain type there exists a C-maximal spacetime $M^{\max} \supset M$ such that $\overset{\looparrowright}{M}{}^{\max} = \overset{\looparrowright}{M}$ (that is $M^{\max}$ has no new—compared to M—causal loops). It will remain only to prove that any U has a C-extension M which (1) is of the type mentioned above and (2) satisfies the causality condition in all points outside of $I^-_M(U)$. This program will be realized as follows.

I. Assume, M has a C-extension M^e (otherwise, M itself would be the sought-for $M^{\max}$). Assume, further, that M is *convexly* C-*extendible*, that is (see Sect. 3) the following holds:

> For any choice of M^e and its perfectly simple subset D, each connected component of $D \cap M$ is causally convex in D (we emphasize that it need not be causally convex in M or in M^e).

Let us check that the convex C-extendibility of M guarantees the existence of a C-extension $M_\triangle \supset M$ (perhaps, not C-maximal yet) which satisfies the condition $\overset{\looparrowright}{M}_\triangle = \overset{\looparrowright}{M}$, i.e., which has no *new* closed causal curves. To this end, extend M to some $R \leftrightharpoons M \cup D$, where $D \subset M^e$ is perfectly simple, see Fig. 1. Consider both possible dispositions of D:

$D \cap M$ is connected. $M_\triangle \leftrightharpoons R$ is just the sought-for extension. Indeed, a new (i.e. not lying entirely in M) causal loop ℓ would have to pass through a point $p \in (D - M)$. But D, being normal, does not contain closed causal curves. So, ℓ would have to leave D—and, therefore, to get into $D \cap M$—both to the future and to the past of p. Thus, the existence of ℓ would imply the existence of p such that

$$p \notin M, \qquad J_D^{\pm}(p) \cap M \neq 0,$$

contrary to the assumed causal convexity of $D \cap M$ in D.

$D \cap M$ is non-*connected.* Unsticking from M all but one connected components of $D \cap M$ reduces this case to the previous one.

Thus, all convexly C-extendible spacetimes can be extended without violating the condition C and at the same time without giving birth to new causal loops. This, however, does not bring us much nearer to the completion of the first step of the proof: it may well happen that $M_\triangle$ is still C-extendible, but—in contrast to M—not *convexly* C-extendible, in which case we would not be able to repeat the procedure and thus to build $(M_\triangle)_\triangle$, $[(M_\triangle)_\triangle]_\triangle$, etc., as one might wish. In order to solve this problem we build in Sect. 4 one more extension of the (convexly C-extendible) space M. This new extension, denoted by $(M)_{\mathbb{M}}$, or simply by $M_{\mathbb{M}}$, is constructed from $M_\triangle$ by cutting and pasting and have the following properties (it is the proof of the last of them that makes up the most tiresome part of this chapter):

($\mathscr{P}_1$) $M_{\mathbb{M}}$ is a C-space [by Test 3(b) in Chap. 2 this follows from the fact that $M_{\mathbb{M}}$ is locally isometric to $M_\triangle$, which in its turn was a part of M^e];

($\mathscr{P}_2$) $M_{\mathbb{M}}$, as well as $M_\triangle$, have no causal loops apart from those lying entirely in M;

($\mathscr{P}_3$) If $M_{\mathbb{M}}$ is C-extendible, then—in contrast to $M_\triangle$—it *must* be convexly C-extendible.

The class of the spacetimes possessing the same properties will be denoted V, so the properties ($\mathscr{P}_{1,2,3}$) of $M_{\mathbb{M}}$ can be expressed as the mere membership $M_{\mathbb{M}} \in$ V.

Now, assuming M is an extension of U that satisfies the causality condition everywhere outside $I^-_M(U)$ (the existence of such M is a non-trivial fact, requiring a separate discussion, see below), we need only to find a C-maximal space in V. To this end we check whether M, denoted also by V_0 in this case, has a C-extension. If it does we define V_1 to be its C-extension belonging to V (it can be $M_{\mathbb{M}}$, for example). If V_1 is C-extendible, too, we denote by V_2 *its* extension lying in V [for example, $V_2 \leftrightharpoons (V_1)_{\mathbb{M}}$], etc. Eventually, we either build at some step a C-maximal V_k, and its existence will prove the theorem, or obtain an infinite chain of spaces

$$\{V_k\},\ k = 0, 1, \ldots, \qquad \forall k \quad V_k \in \mathsf{V},$$

each of which is an extension of the previous one. For such a chain it is easy to build a spacetime $V^{\cup} \in$ V in which *all* the spaces are imbedded at once. Hence, by the Zorn lemma in V there is a maximal[1] element V^*. Such an element has no extension in V. So, in particular, it cannot be extended to $(V^*)_{\mathbb{M}}$, which means that V^* is C-maximal. This proves the theorem, because V^* satisfies all the requirements of $M^{\max}$.

Thus, to prove the theorem it remains to verify that there exists an appropriate M for a general U, i.e. to show that any C-extendible spacetime U has a convexly

[1] With respect to some order relation $\preccurlyeq$ that we shall introduce in V. Up to some technical details $A \preccurlyeq B$ means $A \subset B$.

C-extendible (or C-inextendible) C-extension M which contains no causal loops outside $I^-_M(U)$. This is done as follows. Consider the set W of all possible C-spacetimes of the form $V = I^-_V(U)$. Clearly,

$$\overset{\looparrowright}{V} \subset I^-_V(U) \qquad \forall\, V \in \mathsf{W}.$$

So, all we need is to find a convexly C-extendible (or C-maximal) element in W and to declare it to be M. For this purpose we shall show (by the Zorn lemma again), that in W there exists a maximal element V^{m}. A past-directed causal curve cannot leave V^{m}, even if the latter is a part of a larger C-space (otherwise V^{m} would not be maximal). It follows that V^{m} is convexly C-extendible or C-maximal and hence can be chosen to be M.

3 Convexly C-Extendible Sets

We begin with introducing the notion that plays the central role in our proof. The convex C-extendibility is an analogue of causal convexity, but, in contrast to the latter, is an intrinsic characteristic of spacetime (cf. p. 21) even though it defines largely, as we shall see, the properties of the *boundary* of the spacetime in its extensions.

Let a C-extension M^e of a spacetime M have the form $D \cup M$, where D is perfectly simple. Let further $D_\curlyvee$ be a connected component of $D \cap M$.

Definition 3 M is *convexly* C*-extendible*, if each thus defined $D_\curlyvee$ is a causally convex subset of D.

Generally, neither convex, nor intrinsically globally hyperbolic sets, even if they are C-extendible, have to be convexly C-extendible, as is exemplified by, respectively, a rectangle in the Minkowski plane, and the set built in Example 55. However,

Proposition 4 *If an intrinsically globally hyperbolic spacetime M is both convex and* C*-extendible, then it is convexly* C*-extendible.*

Proof Consider a timelike curve $\gamma \subset D$, connecting a pair of points

$$p \prec p', \qquad p, p' \in D_\curlyvee$$

(the condition that M is C-extendible guarantees the existence of D and $D_\curlyvee$ in question. This exhausts its role). We must show that the entire γ lies in $D_\curlyvee$ or, equivalently, that the non-empty and obviously open (in the topology of γ) set $\gamma \cap D_\curlyvee$ is closed. Put differently, we only need to prove that if a convergent sequence s_k $k = 1, 2 \dots$ lies in $\gamma \cap D_\curlyvee$, then so does $s \leftrightharpoons \lim s_k$.

We begin with the observation that each s_k belongs to the set $\leqslant p, p' \geqslant_{D_\curlyvee}$. Indeed, by Proposition 13 in Chap. 1, $D_\curlyvee$ is convex. Hence, there is a geodesic $\lambda_{ps_k} \subset D_\curlyvee$ from p to s_k. This geodesic is causal and future directed [because the very existence of γ implies the membership $s_k \in I^+_D(p)$, thus the said geodesic connects p with a point of its chronological future in the—convex—set D], so $s_k \in J^+_{D_Y}(p)$. The membership $s_k \in J^-_{D_Y}(p')$ is proved in the same way.

From the intrinsic global hyperbolicity of M and D it follows that, respectively, $\leqslant p, p' \geqslant_D$ and $\leqslant p, p' \geqslant_M$ are compact. And s_k, as we just have established, belongs to both these sets. Consequently, $s \in \leqslant p, p' \geqslant_D \cap \leqslant p, p' \geqslant_M$ let alone

$$s \in \gamma \cap D \cap M$$

(the membership $s \in \gamma$ follows merely from the closedness of γ, which, in fact, is even compact, being a continuous image of the compactum [0, 1], see [186, Proposition III.3.9]). All connected components of $D \cap M$ are open. So, the one containing s, also contains some of s_k. Hence, it is just the component denoted by D_Y. □

Corollary 5 *If a perfectly simple set is* C*-extendible, then it is also convexly* C*-extendible.*

A convexly C-extendible space certainly must be C-extendible, but, as is shown by the example of $M_{\text{ꭙ}}$, it does not have to be convex, i.e. the converse to the just proven proposition is, generally, false. One of the reasons is that convex C-extendibility is, loosely speaking, a characteristic of the 'subsurface' (i.e. lying near the boundary) region of a spacetimes, not of their 'depths'. This, in particular, implies a special structure of the boundary of a convexly C-extendible region: 'mostly' it is achronal (though not *always* for the reasons obvious from inspection of D_3 in Fig. 1).

Proposition 6 *Let M_1 be a* C*-extension of a convexly* C*-extendible spacetime M and U be a neighbourhood of some point of $\mathrm{Bd}_{M_1} M$. Then there exists a perfectly simple set $P \subset U$ and a connected component P_Y of $P \cap M$ such that $\mathrm{Bd}_P P_Y$ is a non-empty closed imbedded achronal $(n-1)$-dimensional C^{1-} submanifold of P.*

Proof Assume U is perfectly simple (by Proposition 65 in Chap. 1 this does not lead to loss in generality) and denote by U_M a connected component of $U \cap M$. Then the convex C-extendibility of M implies (by definition) the causal convexity of U_M in U, i.e. the equality

$$U_{\updownarrow} = \varnothing, \qquad U_{\updownarrow} \leftrightharpoons \left(I^+_U(U_M) \cap I^-_U(U_M)\right) - U_M.$$

In the spacetime U_M, there may be inextendible timelike curves that are extendible as curves in U. The sets of all future and past end points of such curves are denoted by $\mathcal{B}^+$ and $\mathcal{B}^-$, respectively. Clearly both sets lie in $\mathrm{Bd}_U U_M$, but may not exhaust it, see Fig. 2a.

Pick a point $q \in \mathcal{B}^+$ (by Corollary 66 in Chap. 1, $\mathcal{B}^+$ and $\mathcal{B}^-$ cannot both be empty, so if $\mathcal{B}^+ = \varnothing$, then from now until the end of the proof one simply must replace 'past' $\leftrightarrow$ 'future' and $+ \leftrightarrow -$). $I^+_U(U_M)$ contains q and hence also some perfectly simple neighbourhood $P \ni q$. This neighbourhood is disjoint with $\mathcal{B}^-$ [because the latter is a subset of $I^-_U(U_M) - U_M$ and hence its intersection with P

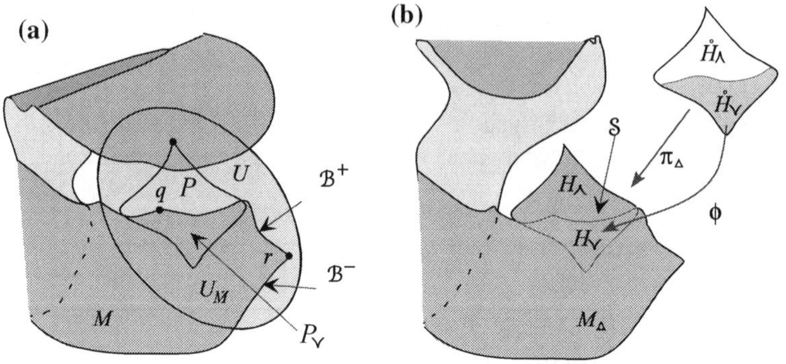

Fig. 2 **a** Of the two curves meeting in r, the upper one is $\mathcal{B}^+$ and the lower one is $\mathcal{B}^-$. r belongs to neither. **b** $M_\triangle$ is obtained from M by, first, pasting $\mathring{H}$ to the latter (the result is R) and, second, unsticking all but one connected component of the intersection of the initial space with the region pasted in (after which that region is renamed to H in order to avoid confusion with $P \subset M_1$ or $\mathring{H}$)

would have to lie in the—empty—set $U_\updownarrow$]. And this means that timelike curves in the space $P \cap U_M$ cannot have past end points in P.

Let $P_\curlyvee \subset U_M$ be a connected component of the set $P \cap M$. Then, if a past-directed timelike curve lying entirely in P leaves $P_\curlyvee$, it must also leave M, and, consequently (due to our choice of $P_\curlyvee$) $P \cap U_M$, too. But the last possibility is excluded, as we just have proven. So, $P_\curlyvee$ is a past set in P and our assertion follows from Proposition 31 in Chap. 1. □

The sets P and $P_\curlyvee$, defined in the course of proving the proposition, and numerous sets isometric to them will be used extensively in what follows. Therefore, it is worth mentioning that the main difference between P and U is that the former is sufficiently small to be contained in $I^+_U(U_M)$. It is this containment that enables P to avoid the presence in it of sharp turns of $\mathrm{Bd}_P M$, similar to the point r in Fig. 2a.

4 Construction of $M_{\wedge\wedge}$

In this section, for an arbitrary convexly c-extendible spacetime M, we build a c-extension $M_{\wedge\wedge}$ of a special type: it is also convexly c-extendible (as will be proven later) and has no closed causal curves besides those lying in M. $M_{\wedge\wedge}$ is built in a few steps. First, we glue to M a perfectly simple region H and thus obtain a c-extension denoted $M_\triangle$ (see Fig. 2b). Then one more copy of H is glued to the 'upper' (i.e. lying outside M) part of $M_\triangle$ (to ensure the Hausdorffness of the resulting extension $M_\diamond$—it is depicted in Fig. 3a—we delete the corresponding three-dimensional submanifold). Finally, an even smaller perfectly simple set G is glued to $M_\diamond$, see Fig. 4. The meaning of these manipulations is this. The spacetime M is convexly c-extendible and so is H. However, the boundaries of these two constituents in a common extension may intersect in such a manner (cf. the part of BdM covered by the set D_2 in Fig. 1) that the result of glueing H to M is *not* convexly c-extendible. To get around this problem, we remove the unwanted intersections and extend M to a spacetime $M_{\triangle\!\!\triangle}$ (to preserve the property c we do this by glueing to M sets isometric to some subsets of M) such that the removed points are missing in *any* extension of $M_{\triangle\!\!\triangle}$ (cf. Sect. 6.2 in Chap. 1.).

Remark on notation 7 In building $M_{\triangle\!\!\triangle}$ and analyzing its structure, we shall need a great many notations. In order not to get confused (see also Remark 11), I denote different sets of common origin by the same letters with different diacritics. The diacritics are chosen so as to refer to the distinguishing feature of the corresponding set. For example, $M_{\diamond}$ is the union of M and a perfectly simple set, which is diamond shaped in the simplest case, and $M_{\triangle}$ differs from M by the 'upper part' of that diamond.

4.1 The Spacetime $M_{\triangle}$

Proposition 8 *Each convexly* c*-extendible spacetime* M *has a* c*-extension* $M_{\triangle}$ *such that*

(*a*) $M_{\triangle} = M \cup H$, *where* H *is perfectly simple, and* $H_{\curlyvee} \leftrightharpoons M \cap H$ *is connected;*
(*b*) M *is a past or future set in* $M_{\triangle}$*;*
(*c*) *The boundary* $\mathcal{S} \leftrightharpoons \mathrm{Bd}_{M_{\triangle}} M$ *is a closed imbedded connected achronal* $(n-1)$*-dimensional* C^{1-} *hypersurface in* $M_{\triangle}$.

Let M_1, U, $P_{\curlyvee}$ etc., be the same as in Proposition 6. We are going to build $M_{\triangle}$ by cutting the region $R \leftrightharpoons M \cup P$ out of M_1 and then unsticking from R all connected components of P but $P_{\curlyvee}$, see Fig. 2. In doing so, we shall also introduce for future use a few objects more.

Proof Let $\mathring{M}$ and $\mathring{H}$ be the spaces isometric to, respectively, M and P:

$$\phi_M\colon\ M \to \mathring{M}, \qquad \phi_P\colon\ P \to \mathring{H}, \qquad \phi_M, \phi_P \text{ are isometries.}$$

Then, see Sect. 6 in Chap. 1, the just defined spacetime R can be represented as a result of glueing $\mathring{H}$ to $\mathring{M}$:

$$R = \mathring{M} \cup_{\phi_{MP}} \mathring{H}, \qquad \phi_{MP} \leftrightharpoons \phi_M \circ \phi_P{}^{-1}\colon\ \ \mathring{H} \to \mathring{M}.$$

Denote by ϕ the restriction of ϕ_{MP} to the *connected* component $\mathring{H}_{\curlyvee} = \phi_P(P_{\curlyvee})$ of its domain [which is the set $\phi_P(P \cap M) \subset \mathring{H}$]. Glueing $\mathring{H}$ and $\mathring{M}$ by ϕ, we obtain the sought-for space $M_{\triangle}$:

$$M_{\triangle} = \mathring{M} \cup_{\phi} \mathring{H}.$$

Now, considering that we shall not need R any longer, let us slightly change our definitions. Specifically, we shall redefine

$$H \leftrightharpoons \pi_\triangle(\mathring{H}), \qquad H_\curlyvee \leftrightharpoons \pi_\triangle(\mathring{H}_\curlyvee), \qquad M \leftrightharpoons \pi_\triangle(\mathring{M}),$$

where $\pi_\triangle$ is the canonical projection of $\mathring{M} \cup \mathring{H}_\curlyvee$ to $M_\triangle$, while so far they have been defined to be corresponding subsets of R. The new definition is consistent with the intuitive picture of $M_\triangle$ as R, in which P is replaced with H (i.e. is glued to M in a different way). As clear from Assertion 71 in Chap. 1, $M_\triangle$ is an extension of M (since so is R). At the same time, $M_\triangle$ being locally isometric to R belongs to C. This proves Proposition 8(a).

For the sake of definiteness assume, as in the proof of Proposition 6, that $\mathcal{B}^+ \neq \varnothing$. Consider then a past-directed timelike curve $\mu \subset M_\triangle$ starting in M. The only way for this curve to leave M would be to get, first, to $H_\curlyvee$ and then to $H - H_\curlyvee$. But, in such a case the curve would intersect the boundary of $H_\curlyvee$ in H and the intersection would lie in $\mathcal{B}^-$ contrary to the fact that $\mathcal{B}^-$ is disjoint with H (see the proof of Proposition 6). Thus, μ does *not* leave M and hence the latter is a past set in $M_\triangle$. Item 8(b) is proved.

Except for the connectedness, all properties of $\mathcal{S}$ listed in 8(c) follow from Proposition 31 in Chap. 1, when 8(b) is taken into consideration. But as shown in, for example the proof or [76, Proposition 6.3.1], there is a neighbourhood W of every point of $\mathcal{S}$ continuously projected to $\mathcal{S} \cap W$. Require that P (defined just as a sufficiently small perfectly simple neighbourhood of a point q) would lie entirely in W. Then the connectedness of $\mathcal{S}$ follows from that of P. □

Convention 9 From now on, for the sake of definiteness, we take M to be the *past* set in $M_\triangle$. We are entitled to adopt this convention, since the proof of the theorem is based only on Proposition 24 (which does not depend on the choice of the time orientation) and formula (⊛), p. 145, which does not exploit any properties of $H_\curlyvee$ (it contains only M).

Remark 10 As proven on p. 122, $M_\triangle$ contains no closed causal curves, besides those confined to M. So, be $M_\triangle$, generally, convexly C-extendible (which is not the case) we could proceed immediately to Proposition 24.

4.2 *The Spacetime* $M_\Diamond$

In this subsection, we build one more C-extension of M (note that it will not be an extension of $M_\triangle$) denoted $M_\Diamond$. Similarly to how $M_\triangle$ was built, we, first, represent an auxiliary space ($M_\triangle - \Sigma$ in this case, notation is specified below) as a result of glueing together two spaces ($\mathring{H}_{\multimap}$ and $\grave{M}_{\rtimes}$) and then unstick 'redundant' components of their intersection.

Let G be a perfectly simple set such that it contains a point of the surface $\mathcal{S}$ but not the *entire* $\mathcal{S}$. Further, let $\mathrm{Cl}_H G$ be compact. G splits $\mathcal{S}$ into three non-empty disjoint

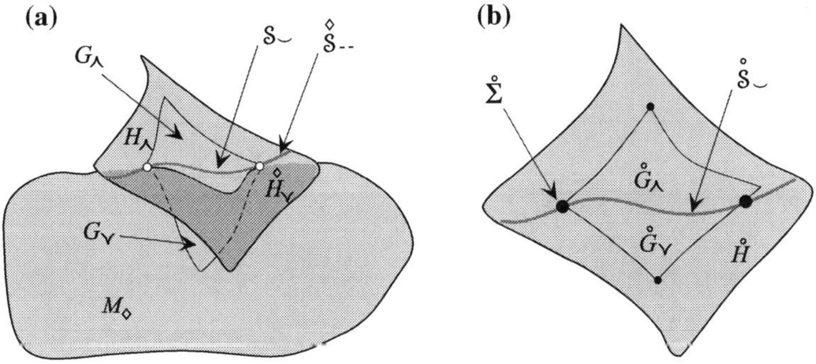

Fig. 3 $M_{\Diamond}$ is obtained by glueing the upper part ($\mathring{H}_{\wedge}$) of the space $\mathring{H}_{\circ\!-\!\circ} = \mathring{H} - \mathrm{Cl}\mathring{\mathcal{S}}_{\smile}$ [see the right picture] to $M_{\succ\!\prec} \rightleftharpoons M_{\triangle} - \mathrm{Cl}\mathcal{S}_{-}$ [see the left picture]. The dark region is the part of M seen through $H_{\circ\!-\!\circ}$. The white circles depict missing points

parts, which we denote as follows:

$$\mathcal{S}_{\smile} \rightleftharpoons \mathcal{S} \cap G, \qquad \mathcal{S}_{-} \rightleftharpoons \mathcal{S} - \mathrm{Cl}_H G, \qquad \Sigma \rightleftharpoons \mathcal{S} - \mathcal{S}_{\smile} - \mathcal{S}_{-}$$

(that $\Sigma \neq \varnothing$ follows from the connectedness of $\mathcal{S}$). We also separate out two regions in H:

$$H_{\vee} \quad \text{and} \quad H_{\wedge} \rightleftharpoons H - \mathrm{Cl} H_{\vee},$$

and two regions in G:

$$G_{\vee} \rightleftharpoons G \cap H_{\vee} \qquad \text{and} \qquad G_{\wedge} \rightleftharpoons G \cap H_{\wedge},$$

see Fig. 3a. Finally, we shall need a space $\grave{M}_{\triangle}$ isometric to $M_{\triangle}$ and notations for two isometries:

$$\psi_M\colon\ M_{\triangle} \to \grave{M}_{\triangle} \qquad \text{and} \quad \psi_H\colon\ H \to \mathring{H}.$$

To simplify the notation, we adopt the convention that $\grave{A}$ and $\mathring{A}$ stand for, respectively, $\psi_M(A)$ and $\psi_H(A)$. In particular,

$$\mathring{H}_{\wedge} \rightleftharpoons \psi_H(H_{\wedge}), \quad \mathring{G}_{\vee} \rightleftharpoons \psi_H(G_{\vee}), \quad \grave{\Sigma} \rightleftharpoons \psi_M(\Sigma), \quad \text{etc.}$$

Separate notations are introduced for $\mathring{H}$ cut along $\mathring{\mathcal{S}}_{\smile}$, and $M_{\triangle}$ cut along $\mathcal{S}_{-}$:

$$\mathring{H}_{\circ\!-\!\circ} \rightleftharpoons \mathring{H} - \mathrm{Cl}\mathring{\mathcal{S}}_{\smile}, \qquad M_{\succ\!\prec} = M_{\triangle} - \mathrm{Cl}\mathcal{S}_{-}.$$

Now, we are in a position to build $M_{\Diamond}$. To this end, represent $M_{\triangle} - \Sigma$ as the result of gluing $\mathring{H}_{\circ\!-\!\circ}$ to $\grave{M}_{\succ\!\prec}$:

$$M_{\triangle} - \Sigma = \mathring{H}_{\circ\!-\!\circ} \cup_{\psi_{MH}} \grave{M}_{\succ\!\prec}, \qquad \text{where } \psi_{MH} \rightleftharpoons \psi_M \circ \psi_H{}^{-1}.$$

Next, note that $\mathring{H}_{\circ\!\!-\!\!\circ}$ is stuck to $\grave{M}_{\rtimes}$ in two disjoint regions: $\mathring{H}_{\curlyvee}$ and $\mathring{H}_{\curlywedge}$. So, we can build a new spacetime by 'unsticking' one of them (viz. $\mathring{H}_{\curlyvee}$). Put formally, we define

$$M_{\Diamond} \leftrightharpoons \mathring{H}_{\circ\!\!-\!\!\circ} \cup_{\psi} \grave{M}_{\rtimes}, \qquad \text{where } \psi \leftrightharpoons \psi_{MH}\big|_{\mathring{H}_{\curlywedge}}.$$

As is easily seen, $M_{\Diamond}$—it is portrayed in Fig. 3a—is, indeed, an extension of M, but not of $M_{\triangle}$. So, strictly speaking, we should distinguish subsets of $M_{\triangle}$, such as $H_{\curlyvee}$, $\mathcal{S}_{-}$, etc. from the corresponding subsets of $M_{\Diamond}$, i.e. $\pi_{\Diamond}(\mathring{H}_{\curlyvee})$, $\pi_{\Diamond}(\grave{\mathcal{S}}_{-})$ etc., where $\pi_{\Diamond}$ is the canonical projection

$$\pi_{\Diamond}: \qquad (\mathring{H}_{\circ\!\!-\!\!\circ} \cup \grave{M}_{\rtimes}) \to M_{\Diamond}.$$

However, our notation needs to be simplified, so we shall write (with minor detriment to rigor)

$$G_{\curlyvee} = \pi_{\Diamond}(\grave{G}_{\curlyvee}), \qquad \mathcal{S}_{\smile} = \pi_{\Diamond}(\grave{\mathcal{S}}_{\smile}), \qquad \text{etc.,} \tag{1}$$

that is identify, when convenient, $\pi_{\Diamond}(\grave{M}_{\rtimes})$ and its subsets with $M_{\rtimes}$ and its corresponding subsets. For the same reason, we equate

$$H_{\circ\!\!-\!\!\circ} = \pi_{\Diamond}(\mathring{H}_{\circ\!\!-\!\!\circ}).$$

At the same time, the sets under discussion should not be confused with their copies lying in the 'outward' part of $M_{\Diamond}$ [which is the set $\pi_{\Diamond}(\mathring{H}_{\curlyvee})$, i.e. the lower part of the rhomb $\pi_{\Diamond}(\mathring{H}_{\circ\!\!-\!\!\circ})$ in Fig. 3a]. So, we shall mark the latter by the symbol $\Diamond$ over the relevant letter:

$$\overset{\Diamond}{G}_{\curlyvee} \leftrightharpoons \pi_{\Diamond}(\mathring{G}_{\curlyvee}), \qquad \overset{\Diamond}{H}_{\curlyvee} \leftrightharpoons \pi_{\Diamond}(\mathring{H}_{\curlyvee}), \qquad \text{etc.}$$

4.3 The Spacetime $M_{\curlywedge\!\curlywedge}$

Now, we are going to build a special c-extension of $M_{\Diamond}$: formally it is obtained by glueing a copy of G to $M_{\Diamond}$, or, to be precise, to its part $\overset{\Diamond}{G}_{\curlyvee}$, see Fig. 4. As a first step, through the use of yet another isometry, ς, we define a copy of G:

$$\tilde{G} \leftrightharpoons \varsigma(\mathring{G}),$$

notice that $\varsigma_{\curlyvee} \leftrightharpoons \varsigma \circ \pi_{\Diamond}^{-1}\big|_{\overset{\Diamond}{G}_{\curlyvee}}$ maps its 'lower' (i.e. isometric to $G_{\curlyvee}$) part to $M_{\Diamond}$, and form the spacetime in question by the following glueing:

$$M_{\curlywedge\!\curlywedge} \leftrightharpoons M_{\Diamond} \cup_{\varsigma_{\curlyvee}} \tilde{G}.$$

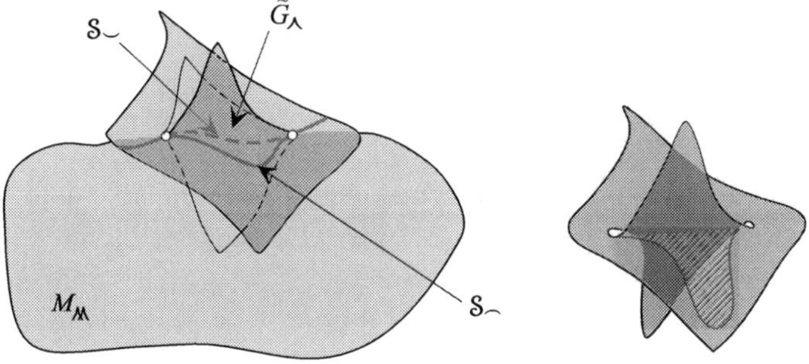

Fig. 4 $M_{\lambda\!\lambda}$—it is portrayed in the left picture—can be represented as the result of pasting the space shown at the right to M (not to $M_\Diamond$!). The hatched region, in the process, transforms into $\overset{\diamond}{G}_\vee$

Again, from now on we (1) identify the spacetime $\tilde{G}$ and its projection to $M_{\lambda\!\lambda}$ and (2) abbreviate $\widetilde{\mathbb{S}}_\smile$ to $\mathbb{S}_\frown$. It is easily seen that $M_{\lambda\!\lambda}$ is an extension of $M_\Diamond$ and, moreover, it is a c-extension of M, since $M_{\lambda\!\lambda}$ is locally isometric to the corresponding part of M_1.

Remark on notation 11 The structure of $M_{\lambda\!\lambda}$ is, actually, not that complex. However, we have to separate out a plethora of subsets and to consider a few auxiliary spacetimes. To sort them out, note that we have only three 'basic' sets: the perfectly simple spaces $\mathring{G}$ and $\mathring{H}$ and the achronal surface $\mathring{\mathbb{S}}$, see Fig. 3b. $\mathring{\mathbb{S}}$ splits $\mathring{H}$ into two parts: $\mathring{H}_\vee$ and $\mathring{H}_\wedge$. Likewise, $\mathring{\mathbb{S}}_\smile$, that is the part of $\mathring{\mathbb{S}}$, which lies in $\mathring{G}$ splits it into $\mathring{G}_\vee$ and $\mathring{G}_\wedge$. The cut along $\mathring{\mathbb{S}}_\smile$ transforms $\mathring{H}$ into $\mathring{H}_{\circ\!-\!\circ}$. Further, we have a set of isometries mapping the above listed sets to $M_{\succ\!\prec}$, $M_\Diamond$ and $\tilde{G} \subset M_{\lambda\!\lambda}$, or to the auxiliary space $\grave{M}_\triangle$. Different images of the same spaces are distinguished by diacritics showing in which of the spacetimes a set lies. Thus, for any set A it is tacitly understood that

$$A \subset M_{\succ\!\prec}, \qquad \mathring{A} \subset \mathring{H}, \qquad \overset{\diamond}{A} \subset M_\Diamond - M_{\succ\!\prec}, \qquad \tilde{A} \subset \tilde{G}, \qquad \grave{A} \subset \grave{M}_\triangle$$

(this notation system admits of aliases, e. g. $\mathring{\mathbb{S}}_\frown \equiv \mathring{\mathbb{S}}_\smile$). There are two exceptions to this rule: $H_{\circ\!-\!\circ}$ and $\mathbb{S}_\frown$ lack diacritics, though neither lies (entirely) in $M_{\succ\!\prec}$.

Thus, $M_{\lambda\!\lambda}$, for example can be represented as the union of (overlapping) regions

$$M_{\succ\!\prec}, \quad H_{\circ\!-\!\circ}, \quad \tilde{G},$$

or of disjoint sets

$$M, \quad \mathbb{S}_\smile, \quad H_\wedge, \quad \mathbb{S}_-, \quad \overset{\diamond}{H}_\vee, \quad \mathbb{S}_\frown, \quad \tilde{G}_\wedge.$$

The union of $H_\wedge$, $\mathbb{S}_-$, and $\overset{\diamond}{H}_\vee$ makes up $H_{\circ\!-\!\circ}$.

Proposition 12 *In the spacetime $M_{\wedge\!\!\wedge}$ the region M is a past set and $\tilde{G}_{\wedge}$ is a future set.*

Proof The first assertion is evident from Figs. 3 and 4. Formally, it can be derived from the fact that M is a past set in $M_{\vartriangle}$, see Proposition 8(b), and hence, by Test 28(b) in Chap. 1, in $M_{\rtimes}$ too. Our claim now follows from Test 28(c) in Chap. 1 with $W \leftrightharpoons M$, $A \leftrightharpoons M_{\rtimes}$, and $B \leftrightharpoons H_{\circ\!\!-\!\!\circ} \cup \tilde{G}$.

Now assume that $\tilde{G}_{\wedge}$ is *not* a future set. Then, it must contain a future-directed timelike curve μ terminating at $\mathrm{Bd}_{M_{\wedge\!\!\wedge}} \tilde{G}_{\wedge} = \mathrm{Bd}_{\tilde{G}} \tilde{G}_{\wedge}$ (the equality stems from the fact that by construction $M_{\diamond}$ has no point that would contain in each of its neighbourhoods a point of $\tilde{G}_{\wedge}$, whence the boundary thereof lies in $\tilde{G}$). The isometry ς^{-1}, see the beginning of this subsection, maps μ to a timelike future-directed curve lying in $\mathring{G}_{\wedge} \subset \mathring{H}_{\wedge}$ and terminating at $\mathrm{Bd}_{\mathring{G}} \mathring{G}_{\wedge} \subset \mathrm{Bd}_{\mathring{H}} \mathring{H}_{\wedge}$. But such curves do not exist, since $\mathring{H}_{\wedge}$ is a future set in $\mathring{H}$, as follows from Proposition 8(b) by Tests 28(a) and 28 (b) of Chap. 1 with $W \leftrightharpoons H_{\wedge}$, $A \leftrightharpoons M_{\vartriangle}$, and $\mathcal{P} \leftrightharpoons \overline{M - H}$. □

5 The Structure of $M_{\wedge\!\!\wedge}$

Our goal in this section is to prove that a C-extendible $M_{\wedge\!\!\wedge}$ is also convexly C-extendible. To this end, we divide $M_{\wedge\!\!\wedge}$ into three regions, M, $\tilde{G}$, and $H_{\circ\!\!-\!\!\circ}$, of which the first two are convexly C-extendible and—up to a cut—so is the third, and reduce the properties of $M_{\wedge\!\!\wedge}$ to properties of these regions. In doing so, we utilize the facts that in extensions of $M_{\wedge\!\!\wedge}$ there are nonhomotopic curves with the same ends (homotopy is meant to be fixed end point), and that a common neighbourhood of such curves cannot be perfectly simple. That these curves are nonhomotopic is due to a singularity, which is present in $M_{\wedge\!\!\wedge}$ in spite of the fact that $M_{\wedge\!\!\wedge}$ consists of a few *non*singular regions.[2] Thus, our immediate subject is homotopic properties of curves intersecting $\mathcal{S}_{\smile}$ and $\mathcal{S}_{\frown}$.

5.1 *The Surfaces $\mathcal{S}_{\smile}$ and $\mathcal{S}_{\frown}$*

Our way of constructing $M_{\wedge\!\!\wedge}$ (by unsticking relevant regions) implies the existence of a projection $\pi_{\wedge\!\!\wedge}$, which maps $M_{\wedge\!\!\wedge}$ to $M_{\vartriangle}$ locally isometric. In particular, it maps timelike curves to timelike, and each of the surfaces $\mathcal{S}_{\smile}$, $\mathcal{S}_{\frown}$ to the surface that we also denoted by $\mathcal{S}_{\smile}$ [even though it lies in $M_{\vartriangle}$, cf. (1)]. Thus, Proposition 8(c) yields

Corollary 13 *$\mathcal{S}_{\smile}$ and $\mathcal{S}_{\frown}$ are connected closed imbedded C^{1-} hypersurfaces in $M_{\wedge\!\!\wedge}$. They both and even their union are achronal in $M_{\wedge\!\!\wedge}$.*

---

[2]The nature of this singularity is the same as in the DP space, or, say, in the twofold covering of the punctured Minkowski plane.

Instead of appealing to Proposition 8(c) we could use the following reasoning. The definitions of $\mathcal{S}_\smile$ and $M_{\rtimes\ltimes}$ imply that $\mathcal{S}_\smile = \mathrm{Bd}_{M_{\rtimes\ltimes}} M$. On the other hand, $M_{\text{ж}}$ was obtained from $M_{\rtimes\ltimes}$ by glueing some open sets to regions disjoint with M. Therefore,

$$\mathcal{S}_\smile = \mathrm{Bd}_{M_{\text{ж}}} M. \tag{2}$$

Likewise, $\mathcal{S}_\frown = \mathrm{Bd}_{\tilde{G}} \tilde{G}_\wedge$ and, again, $M_{\text{ж}}$ is obtained from $\tilde{G}$ by glueing some open sets to regions disjoint with $\tilde{G}_\wedge$. Whence,

$$\mathcal{S}_\frown = \mathrm{Bd}_{M_{\text{ж}}} \tilde{G}_\wedge.$$

Now, almost all the listed properties of $\mathcal{S}_\smile$ and $\mathcal{S}_\frown$ follow from Propositions 12 and 31 in Chap. 1.

Our next goal is to show that those surfaces cannot be extended: roughly speaking, Σ, which was deleted in the process of building $M_{\text{ж}}$, cannot be 'glued back' into spacetime. In this sense, Σ generates the above-mentioned singularity.

Proposition 14 *In any extension M^e of the spacetime $M_{\text{ж}}$, the sets $\mathcal{S}_\frown$ and $\mathcal{S}_\smile$ are closed.*

Proof We prove the proposition for $\mathcal{S}_\smile$, the '$\mathcal{S}_\frown$' case being much the same. The idea of the proof is to show that the existence of a non-empty edge $\mathrm{Bd}_{M^e}\mathcal{S}_\smile - \mathcal{S}_\smile$ would lead to contradictory properties of a certain past-directed geodesic γ. This geodesic would have to enter both $H_\curlyvee$ (being limit to a family of geodesics that descend passing to the right of the left white circle in Fig. 3a) and $\overset{\diamond}{H}_\curlyvee$ (as a limit of geodesics descending to the left of the circle), but these two sets are disjoint.

Thus, suppose that $\{p_m\}$ is a sequence of points such that contrary to our claim

$$p_m \in \mathcal{S}_\smile, \qquad \lim_{m\to\infty} p_m = z \in (M^e - \mathcal{S}_\smile). \tag{$*$}$$

Denote by $\grave{p}_m$ the images of p_m in $\grave{M}_\vartriangle$, that is $\grave{p}_m \leftrightharpoons \psi_M(p_m)$. Now recall that, by construction, $\mathrm{Cl}_H G$ is compact and, consequently, so is $\mathrm{Cl}_{\grave{M}_\vartriangle} \grave{G}$. Being a closed subset of the latter $\mathrm{Cl}[\psi_M(\mathcal{S}_\smile)]$ is compact too. This entails the existence of a point $\grave{p}$ such that

$$\lim_{k\to\infty} \grave{p}_k = \grave{p} \in \grave{\Sigma},$$

where $\{\grave{p}_k\}$ is a subsequence of $\{\grave{p}_m\}$ and $\grave{\Sigma}$ stands in place of the whole set $\mathrm{Cl}[\psi_M(\mathcal{S}_\smile)]$, because $\grave{p}$ by the assumption ($*$) is not in $\psi_M(\mathcal{S}_\smile)$.

Now pick a point $\grave{o}$ in $I^+_{\grave{H}}(\grave{p})$ and consider a sequence of geodesics $\grave{\gamma}_k(\tau) = \exp_{\grave{o}}(\tau\grave{\boldsymbol{v}}_k)$ connecting $\grave{o}$ to $\grave{p}_k$ (they must exist, since $\grave{H}$, which contains both $\grave{o}$ and $\grave{p}_k$, is convex). The parameter τ on each of the $\grave{\gamma}_k$ is chosen so that

$$\grave{\gamma}_k(1) = \exp_{\grave{o}}(\grave{\boldsymbol{v}}_k) = \grave{p}_k.$$

Beginning from some k_0, all these geodesics are timelike and, therefore, do not meet $\grave{\mathcal{S}}_-$ [otherwise they would have more than one—including $\grave{p}_k$—common points with the *achronal*, see Proposition 8(*c*), surface $\psi_M(\mathcal{S})$]. So, each $\grave{\gamma}_k$ lies in the domain of the projection $\pi_\diamond$ and is projected to a geodesic $\gamma_k \subset H$ which starts at $o \leftrightharpoons \pi_\diamond(\grave{o}) \subset M_{\curlywedge\!\curlywedge}$, has the initial velocity $\boldsymbol{v}_k = \mathrm{d}\pi_\diamond(\grave{\boldsymbol{v}}_k)$, and passes through p_k at $\tau = 1$.

Since all $\grave{p}_k$, as well as $\grave{p}$, are in the convex set $\grave{H}$, where exp is a diffeomorphism, we have

$$\lim_{k\to\infty} \grave{\boldsymbol{v}}_k = \grave{\boldsymbol{v}}, \qquad \text{where } \grave{\boldsymbol{v}} \leftrightharpoons \exp_{\grave{o}}^{-1}(\grave{p}),$$

and, by the smoothness of $\pi_\diamond$,

$$\lim_{k\to\infty} \boldsymbol{v}_k = \boldsymbol{v}, \qquad \text{where } \boldsymbol{v} \leftrightharpoons \mathrm{d}\pi_\diamond(\grave{\boldsymbol{v}}).$$

Consider now the geodesics

$$\grave{\gamma}(\tau) \leftrightharpoons \exp_{\grave{o}}(\tau\grave{\boldsymbol{v}}) \quad \text{and} \quad \gamma(\tau) \leftrightharpoons \exp_o(\tau\boldsymbol{v}).$$

Let ϵ be a positive number, so small that, first, both of these geodesics can be extended to the values $\tau = 1 + \epsilon$ [that they can be extended to some $\tau > 1$ follows from the very existence of $z = \gamma(1)$ and $\grave{p} = \grave{\gamma}(1)$] and, second, $\grave{\gamma}(1+\epsilon)$ lies still in $\grave{H}_{\curlyvee}$. Then, by continuity,

$$\begin{aligned}\gamma(1+\epsilon) = \lim_{k\to\infty} \gamma_k(1+\epsilon) &= \lim_{k\to\infty} \pi_\diamond[\grave{\gamma}_k(1+\epsilon)] \\ &= \pi_\diamond[\lim_{k\to\infty} \grave{\gamma}_k(1+\epsilon)] = \pi_\diamond[\grave{\gamma}(1+\epsilon)] \in H_{\curlyvee}. \quad (3)\end{aligned}$$

However, since $\grave{p} \in \grave{\Sigma}$, we can find a sequence of points

$$\grave{q}_k \in \grave{\mathcal{S}}_-, \qquad \lim_{k\to\infty} \grave{q}_k = \grave{p},$$

and, repeating the reasoning above with $\grave{p}_k$ replaced by $\grave{q}_k$, obtain $\gamma(1+\epsilon) \in \overset{\diamond}{H}_{\curlyvee}$. This contradicts (3). □

Thus, we have established that the spacetime $M_{\curlywedge\!\curlywedge}$ contains incomplete geodesics that cannot be extended in any extension of $M_{\curlywedge\!\curlywedge}$.

Corollary 15 *$M_{\curlywedge\!\curlywedge}$ is singular.*

Now, let us find out how timelike curves intersect $\mathcal{S}_\frown$ and $\mathcal{S}_\smile$. Unfortunately, these surfaces, though achronal, are not necessarily spacelike (cf. p. 17). So, we have to *prove* some facts that would be standard be $\mathcal{S}_\frown$ and $\mathcal{S}_\smile$ at least C^1. Specifically, we shall show that a continuous deformation does not change the number of intersections of a timelike curve with $\mathcal{S}_\frown$ and $\mathcal{S}_\smile$ as long as the curve remains timelike and its end

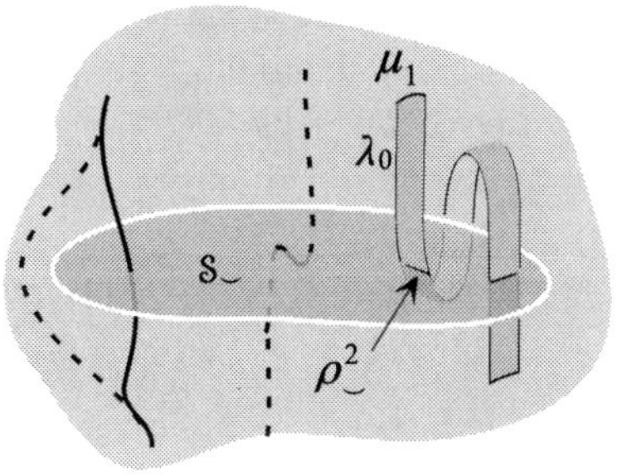

Fig. 5 The solid curve cannot be continuously deformed—keeping it timelike and its end points fixed—into a dashed one. The grey strip to the right is $\Lambda(\mathcal{Q})$

points do not cross the relevant surface. The curve can neither leave the surface (i.e. the solid curve in Fig. 5a cannot deform into the leftmost dashed one, being restricted by the singularity), nor bend so as to give rise to new intersections (being timelike the curve intersects the achronal surface $\mathcal{S}_\smile$ 'transversally').

Definition 16 Let $\lambda(\tau)$ be a timelike curve in an extension M^e of the spacetime $M_{\Join}$. A $\frown$*-root* (or a $\smile$*-root*) of λ, is a value τ_i of the parameter τ such that $\lambda(\tau_i)$ is in $\mathcal{S}_\frown$ (respectively, in $\mathcal{S}_\smile$). The number of $\frown$- and $\smile$-roots of λ is denoted $i_\frown[\lambda]$ and $i_\smile[\lambda]$, respectively.

Obviously, for any future-directed curve λ *lying in* $M_{\Join}$ and having end points a and b the following holds:

$$i_\smile[\lambda] = \begin{cases} 1 \text{ if } a \in \mathrm{Cl}_{M_{\Join}} M,\ b \notin M; \\ 0 \text{ otherwise} \end{cases} \tag{4a}$$

(the first line follows from the equality $\mathcal{S}_\smile = \mathrm{Bd}_{M_{\Join}} M$, see (2), which leads in this case to the restriction $i_\smile[\lambda] \geqslant 1$, and Corollary 13, which implies that $i_\smile[\lambda] \leqslant 1$; the second line—from the fact that on the strength of Proposition 27 in Chap. 1 the part of λ lying to the past of $\lambda \cap \mathcal{S}_\smile$ must be contained in M, while the remainder, on the contrary, must lie in $M_{\Join} - M$). Likewise,

$$i_\frown[\lambda] = \begin{cases} 1 \text{ if } b \in \mathrm{Cl}_{M_{\Join}} \tilde{G}_\wedge,\ a \notin \tilde{G}_\wedge; \\ 0 \text{ otherwise.} \end{cases} \tag{4b}$$

Denote by $\mathcal{Q}$ the square $[0, 1] \times [0, 1]$ in the plane (τ, ξ) and consider a homotopy

$$\Lambda(\xi, \tau)\colon\ \mathcal{Q} \to M^e, \tag{5}$$

such that for every fixed ξ_0 the curve $\lambda_{\xi_0}(\tau) \leftrightharpoons \Lambda(\xi_0, \tau)$ is timelike and future-directed. The 'horizontal' curves will be denoted $\mu_{\tau_0}(\xi) \leftrightharpoons \Lambda(\xi, \tau_0)$. We begin with establishing some properties of the set $\mathcal{R}$ of all points $(\tau_\alpha, \xi_\alpha) \in \mathcal{Q}$ such that τ_α is a root of λ_{ξ_α}. Clearly, $\Lambda(\mathcal{R})$, being an intersection of two closed sets, $\Lambda(\mathcal{R}) = \Lambda(\mathcal{Q}) \cap (\mathcal{S}_\frown \cup \mathcal{S}_\smile)$, is closed too [both in M^e and, as a consequence, in $\Lambda(\mathcal{Q})$]. Hence, by continuity, $\mathcal{R}$ is closed in $\mathcal{Q}$.

Notation 17 By analogy with $\pm$, we introduce the sign $\circ$. An expression or an assertion, containing this sign, is understood as a *pair* of expressions (respectively, assertions), linked by 'and': in one of them all $\circ$ must be replaced by $\smile$, and in the other—by $\frown$.

Proposition 18 *Suppose, curves* μ_0 *and* μ_1 *intersect neither* $\mathcal{S}_\frown$, *nor* $\mathcal{S}_\smile$. *Then,* $\mathcal{R}$ *is a finite number of disjoint continuous curves* $\nu_\circ^m$ $m = 1, \dots, m_\circ < \infty$ *from* $\xi = 0$ *to* $\xi = 1$. *They are graphs of continuous functions*

$$\tau = r_\circ^m(\xi), \qquad 0 < r_\circ^m < 1.$$

A curve (not necessarily causal) $\psi(s) \subset \Lambda(\mathcal{Q})$ *with the end points at the curves* $\rho_\circ^{m_1} \leftrightharpoons \Lambda(\nu_\circ^{m_1})$ *and* $\rho_\circ^{m_2} \leftrightharpoons \Lambda(\nu_\circ^{m_2})$ *leaves* $M_{\mathbb{X}}$ *if* $m_1 < m_2$.

Proof Consider a point $p_* = (\tau_*, \xi_*) \in \mathcal{Q}$ and assume that τ_* is a, $\frown$-, for definiteness, root of the curve λ_{ξ_*}. Then, by definition, $\lambda_{\xi_*}(\tau_*) \in \mathcal{S}_\frown = \mathrm{Bd}_{M_{\mathbb{X}}} \tilde{G}_\wedge$. But λ_{ξ_*} is future-directed, while $\tilde{G}_\wedge$ is a future set in $M_{\mathbb{X}}$. So, for any sufficiently small $\delta \neq 0$ (so small, that $\lambda\big|_{|\tau-\tau_*|\leqslant\delta}$ is still in $M_{\mathbb{X}}$, and $0 < \tau_* \pm \delta < 1$)

$$\lambda_{\xi_*}\big|_{\tau_*-\delta<\tau\leqslant\tau_*} \subset M_{\mathbb{X}} - \mathrm{Cl}_{M_{\mathbb{X}}} \tilde{G}_\wedge, \qquad \lambda_{\xi_*}\big|_{\tau_*<\tau\leqslant\tau_*+\delta} \subset \tilde{G}_\wedge,$$

see Proposition 27 in Chap. 1. Hence, by the continuity of Λ, there is $\epsilon = \epsilon(\delta) \neq 0$ such that at $|\xi - \xi_*| \leqslant \epsilon$ every segment $\lambda_\xi\big|_{|\tau-\tau_*|\leqslant\delta}$ also starts out of $\tilde{G}_\wedge$, traverses $M_{\mathbb{X}}$ and ends in $\tilde{G}_\wedge$. It follows then from (4) that in the rectangle

$$\square_{\delta\epsilon} \leftrightharpoons \{\tau, \xi : \ \tau, \xi \in [0, 1], \ |\tau - \tau_*| < \delta, \ |\xi - \xi_*| < \epsilon\}$$

for every ξ there is exactly one root.[3] It is now easily seen that

(a) $\mathcal{R} \cap \square_{\delta\epsilon}$ is the graph of some continuous (as we could choose an arbitrarily small δ) function $\tau = r_\frown^m(\xi)$ with the domain $|\xi - \xi_*| \leqslant \epsilon$;

b) the number of roots of λ_{ξ_*} is finite. Indeed, otherwise they would have a limit point in some $\tau_\infty \in [0, 1]$, and τ_∞, due to the closedness of $\mathcal{R}$ would be a root, which would contradict either the hypothesis of the proposition (if $\tau_\infty = 0, 1$), or the existence of $\square_{\delta\epsilon}$ (if $\tau_\infty \neq 0, 1$);

c) $\nu_\circ^m$ with different m are disjoint.

Now, let us establish that the domain of each $r_\frown^m$ is the whole interval $[0, 1]$ or, equivalently, that the domain is closed (that it is open we already know—every ξ in it has a neighbourhood also contained there). Consider to this end, a number τ_{**} that is *not* a root of λ_{ξ_*}. The point $p_{**} \leftrightharpoons \lambda_{\xi_*}(\tau_{**})$ is not in $\mathcal{S}_\circ$. But the latter is closed (by Proposition 14), and, consequently, some neighbourhood of p_{**} is also disjoint with $\mathcal{S}_\circ$. So, there is a rectangle $\square_{\delta'\epsilon'}$ around p_{**} that contains no points of $\mathcal{R}$. This means that the complement of the domain of $r_\frown^m$ is open.

---

[3] This is the reason why we require that λ be timelike, not just causal.

To analyse the last assertion, consider the segment of ψ between $\rho_\smile^{m_1}$ and $\rho_\smile^{m_1+1}$ (the case of $\rho_\frown$ is, naturally, perfectly analogous). For definiteness, λ are taken to be future directed, and the indices of roots to grow with τ. Then on one end of ψ—specifically, near $\rho_\smile^{m_1}$—there is a point belonging to $M_{\curlywedge\curlywedge} - M$, and on the other end—to M. The assertion now follows from (2). □

Corollary 19 Under the conditions of Proposition 18

$$i_\circ[\lambda_{\xi_1}] = i_\circ[\lambda_{\xi_2}] \qquad \forall \xi_1, \xi_2 \in [0, 1].$$

In particular, the homotopy under discussion cannot relate the solid curve in Fig. 5a with either of the dashed.

Thus, $i_\frown$ and $i_\smile$ are analogous of intersection indices, the timelikeness of the relevant curves playing the role of their transversality to the surfaces $\mathcal{S}_\circ$. The numbers $i_\circ$ do not change at a fixed ends homotopy leaving the curves timelike. But in a *convex* neighbourhood *all* timelike curves with common end points are related by a homotopy of that kind.[4] So, if $O \subset M^e$ is a convex neighbourhood, then to every pair of points $p, q \in O$ connected by a timelike curve $\lambda_{pq} \subset O$ a pair of numbers $i_\circ[pq]$ can be assigned:

$$i_\circ[pq] \rightleftharpoons i_\circ[\lambda_{pq}], \qquad \forall p, q\colon\, p \in I^\pm_O(q).$$

5.2 *Convex* c*-Extendibility of* $M_{\curlywedge\curlywedge}$

Let D be a perfectly simple subset of some extension M^e of $M_{\curlywedge\curlywedge}$. Let, further, $D_{M_{\curlywedge\curlywedge}}$ be a connected component of $D \cap M_{\curlywedge\curlywedge}$. To prove the convex c-extendibility of $M_{\curlywedge\curlywedge}$, we must establish the causal convexity of $D_{M_{\curlywedge\curlywedge}}$ in D, i.e. to show that any timelike curve $\varkappa \subset D$ with the ends in $D_{M_{\curlywedge\curlywedge}}$ lies there entirely. But we know (from the convex c-extendibility of the relevant sets) that this holds for curves that start and end in components of the intersections $D \cap A$, where $A \rightleftharpoons M, \tilde{G}$ or H (the symbol H here should not be understood too literally: strictly speaking, $M_{\curlywedge\curlywedge}$ contains $H_{\circ\!-\!\circ}$ rather than H). So, the main idea of our proof is to show that the above-mentioned curve $\varkappa$ is in this class, i.e. both its ends lie in the same component of $D \cap A$, (at least, when $\varkappa$ does not cross $\mathcal{S}_\circ$, that is when the difference between $H_{\circ\!-\!\circ}$ and H is immaterial). For this purpose, we show in Proposition 20 that $\varkappa$ ends in the same A, where it starts. At the same time, Proposition 21 says that $D_{M_{\curlywedge\curlywedge}} \cap M$ and $D_{M_{\curlywedge\curlywedge}} \cap \tilde{G}_\wedge$ are connected (this does not follow automatically from the connectedness of $D_{M_{\curlywedge\curlywedge}}$: the intersection

[4]Indeed, any of them is homotopic to the geodesic connecting its ends: it suffices to define λ_{ξ_*} as the geodesic segment from $\lambda_0(0)$ to $\lambda_0(\tau = \xi_*)$ joined with the segment $\lambda_0|_{\xi_* < \tau \leqslant 1}$.

of two connected sets need not be connected, of course). The combination of these two propositions proves (except when $A = H$) that $\varkappa$ returns to the same component of $D \cap A$ where it begins. It remains to prove—this is done in Proposition 23 (with the use of the convex c-extendibility of H)—that in the last case, i.e. when the curve starts and ends in $D_{M_{\mathbb{M}}} \cap H_{\circ\!-\!\circ}$—the curve does not leave $D_{M_{\mathbb{M}}}$ either.

Proposition 20 *If points $a, b \in D_{M_{\mathbb{M}}}$ are connected by a timelike curve $\varkappa \subset D$ that intersects neither $\mathcal{S}_{\frown}$, nor $\mathcal{S}_{\smile}$, then they both lie in one of the three sets: M, $H_{\circ\!-\!\circ}$ or $\tilde{G}_{\wedge}$.*

Proof Connect a and b by a curve φ that—in contrast, perhaps, to $\varkappa$—lies wholly in $D_{M_{\mathbb{M}}}$ (the existence of such φ is guaranteed by the very definition of $D_{M_{\mathbb{M}}}$). To prove the proposition, we plan to demonstrate that φ can be chosen so that, if it intersects $\mathcal{S}_{\frown}$ or $\mathcal{S}_{\smile}$, then later it intersects the same surface in the opposite direction. But φ lies in $M_{\mathbb{M}}$, where it is impossible to get into one of the sets A from another, without an uncompensated intersection of $\mathcal{S}_{\frown}$ or $\mathcal{S}_{\smile}$.

We already know how *timelike* curves intersect $\mathcal{S}_{\circ}$. So, we begin with replacing an *arbitrary* φ by a piecewise timelike broken line. To this end, we cover φ by a finite number of perfectly simple sets $F_k \subset D_{M_{\mathbb{M}}}$, $k = 1, \ldots, K$, numbered according to the rule

$$\varphi \subset \bigcup_k F_k, \qquad F_{k-1} \cap F_k \neq \varnothing, \qquad a \in F_1, \quad b \in F_K,$$

and pick $K + 1$ points f_k, so that:

$$f_1 = a, \qquad f_{K+1} = b, \qquad f_k \in F_{k-1} \cap F_k, \quad k \neq 1, K+1,$$

see Fig. 6. Each pair f_k, f_{k+1} lies in a common perfectly simple set F_k, whence there is a point

$$p_k \in I^+_{F_k}(f_k) \cap I^+_{F_k}(f_{k+1}), \quad p_k \notin \mathcal{S}_{\circ}.$$

Connecting at each k the point p_k with f_k and f_{k+1} by, respectively, future- and past-directed timelike curves σ_k and η_k, where $\sigma_k, \eta_k \subset F_k$, we obtain a continuous broken line from a to b. Since this broken line lies in $M_{\mathbb{M}}$, it can be used instead of φ in the sense that the proposition will be proven once we show that the line in question does not intersect $\mathcal{S}_{\circ}$ at all, or intersect it an equal number of times in either direction.

Let $x \in D$ be such a point that

$$f_k, p_k \in I^+_D(x) \quad \forall k$$

(x exists, since D is perfectly simple). The obvious equality

$$i_{\circ}[x, f_k] + i_{\circ}[\sigma_k] = i_{\circ}[x, p_k] = i_{\circ}[x, f_{k+1}] + i_{\circ}[\eta_k]$$

gives

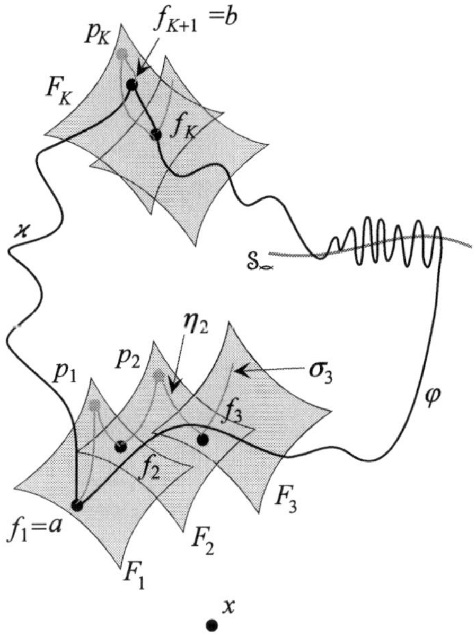

Fig. 6 $\varkappa \subset D$ may leave $M_{\text{ж}}$, even if its end points are in $D_{M_{\text{ж}}}$

$$i_\circ[x, f_{k+1}] - i_\circ[x, f_k] = i_\circ[\sigma_k] - i_\circ[\eta_k],$$

whence, by induction,

$$i_\circ[f_1, f_{K+1}] = i_\circ[x, f_{K+1}] - i_\circ[x, f_1] = \sum_{k=1}^{K} i_\circ[\sigma_k] - \sum_{k=1}^{K} i_\circ[\eta_k].$$

On the other hand, $i_\circ[f_1, f_{K+1}] = i_\circ[\varkappa] = 0$; the last equality being due to the fact that, by hypothesis, $\varkappa$ does not intersect $\mathcal{S}_\circ$. Thus,

$$\sum_{k=1}^{K} i_\circ[\sigma_k] = \sum_{k=1}^{K} i_\circ[\eta_k],$$

which means, see (4), that the broken line under consideration intersects $\mathcal{S}_\circ$ in the *outward* direction exactly as many times as in the *inward* direction. □

Proposition 21 *The sets $D_{M_{\text{ж}}} \cap M$ and $D_{M_{\text{ж}}} \cap \tilde{G}_\wedge$ are connected.*

Proof We give the proof only for $D_{M_{\text{ж}}} \cap M$. The connectedness of $D_{M_{\text{ж}}} \cap \tilde{G}_\wedge$ is proved similarly.

Let a, b be a pair of arbitrary points in $D_{M_{\text{ж}}} \cap M$. To establish the proposition, it suffices to find a curve $\varphi \subset D$ connecting a with b and lying entirely in M (such a curve lies automatically—as follows from its very existence—in a connected component of $D \cap M$, and hence in $D_{M_{\wedge\wedge}}$ and, furthermore, in $D_{M_{\wedge\wedge}} \cap M$). To this end, pick a curve

$$\varphi'(\xi)\colon\ [0, 1] \to D_{M_{\wedge\wedge}}$$

from a to b and a pair of points x, y such that

$$\varphi' \subset \langle x, y\rangle_D$$

[φ' exists, because, by definition, $D_{M_{\wedge\wedge}}$ is connected, and x, y—because D is perfectly simple, see Remark 62(c) in Chap. 1]. We shall build the desired φ by 'projecting' $\varphi' - M$ to $\mathcal{S}_\smile$. Specifically, suppose, φ' leaves M, p and q being, respectively, the first and the last points of $\varphi' - M$. Consider the homotopy (5) determined by the condition that λ_ξ at each ξ is a curve in D from x to y through $\varphi'(\xi)$ [λ_ξ can be taken, for example to be the broken line consisting of two geodesic segments in D: one is from x to $\varphi'(\xi)$ and the other—from $\varphi'(\xi)$ to y]. Since φ' does not leave $M_{\wedge\wedge}$, both points, p and q belong, by Proposition 18, to the same curve [it is $\Lambda(\rho^c_\smile)$ for some $c \in \mathbb{Z}$] lying in $\mathcal{S}_\smile \cap D$. The broken line consisting of the segment of φ' from a to p, the segment of $\Lambda(\rho^c_\smile)$ from p to q, and, finally, the segment of φ' from q to b, is contained in $\mathrm{Cl}_{M_{\wedge\wedge}} M$. By a small deformation (shifting a little all inner points to the past) it is transformed into the sought-for curve φ. □

Our plan, as explained in the beginning of Sect. 5.2, is to derive the relevant property of $\varkappa$ (its inability to reenter a once left component of $D \cap M_{\wedge\wedge}$) from the same property of curves having both ends in $A \leftrightharpoons M, \tilde{G}$. What could spoil this plan are $\varkappa$ that have both their end points in $D_{M_{\wedge\wedge}} \cap A$, but in different connected components of $D \cap A$. It is the existence of such $\varkappa$ that is excluded by the just proved proposition.

Corollary 22 *$D_{M_{\wedge\wedge}} \cap M$ and $D_{M_{\wedge\wedge}} \cap \tilde{G}_\wedge$ are connected components (not necessarily unique, $D \cap M_{\wedge\wedge}$ may not be connected) of spaces $D \cap M$ and $D \cap \tilde{G}_\wedge$, respectively.*

Proposition 23 *Any convexly* C*-extendible spacetime M has an extension that (1) is* C*-maximal or convexly* C*-extendible and (2) has the causality violating set coinciding with $\overset{\looparrowright}{M}$.*

Proof To prove the claim, we shall show that the mentioned extension can be taken to be $M_{\wedge\wedge}$.

(1) We begin by focusing on the first condition. To obtain a contradiction, suppose that $M_{\wedge\wedge}$ does *not* satisfy it, i.e. that $M_{\wedge\wedge}$ has a C-extension such that in a perfectly simple subset D of the latter there is a future-directed timelike curve $\varkappa(s)$ starting at $a = \varkappa(0)$, ending at $b = \varkappa(1)$, and leaving $M_{\wedge\wedge}$ somewhere in between:

$$a, b \in D_{M_{\wedge\wedge}}, \quad \varkappa \not\subset M_{\wedge\wedge} \qquad (\star)$$

($\varkappa \subset D$, hence $\varkappa \not\subset M_{\wedge\wedge}$ is equivalent to $\varkappa \not\subset D_{M_{\wedge\wedge}}$). Without loss of generality we can assume that $\varkappa$ does not meet $\mathcal{S}_\circ$. Indeed, remove from $\varkappa$ a small (so small, in particular, that its closure is also contained in $M_{\wedge\wedge}$) neighbourhood of each intersection $\varkappa \cap \mathcal{S}_{\circ}$. This will transform $\varkappa$ into a finite (see Proposition 18) set of segments disjoint with $\mathcal{S}_{\circ}$. At least one of them being chosen to be the new $\varkappa$ will satisfy ($\star$).

According to Proposition 20, both ends of $\varkappa$ are in one of the three sets: M, $\tilde{G}_{\wedge}$, or $H_{\circ\!-\!\circ}$. But M and $\tilde{G}$ are convexly C-extendible and, therefore, the ends of $\varkappa$ that leaves $M_{\wedge\wedge}$ cannot lie in connected components of $D \cap M$, or $D \cap \tilde{G}_{\wedge}$. This excludes—on the strength of Corollary 22—their containment in $D_{M_{\wedge\wedge}} \cap M$, or $D_{M_{\wedge\wedge}} \cap \tilde{G}_{\wedge}$. Consequently,

$$a, b \in \left(D_{M_{\wedge\wedge}} \cap H_{\circ\!-\!\circ}\right).$$

This last possibility cannot be ruled out in the same way, because, first, $H_{\circ\!-\!\circ}$—in contrast to M and $\tilde{G}$—is *not* convexly C-extendible (this is easily verified by taking $\mathring{H}$ as the test C-extension and an arbitrary perfectly simple neighbourhood of a point of $\mathring{\Sigma}$—as the test perfectly simple set). And, second, for $H_{\circ\!-\!\circ}$ the connectedness of its intersections with $D_{M_{\wedge\wedge}}$ is not established. So, one can imagine that a and b lie in different components of that intersection, see the discussion above Corollary 22. Our method for solving these problems will depend on whether the component of $D \cap H_{\circ\!-\!\circ}$ containing a—we denote it $D_{H_{\circ\!-\!\circ}}$—intersects the surface $\mathcal{S}_{\circ}$, or, to be precise, whether there is a curve $\mu(s)\colon\ [-1, 0] \to D$ such that

$$\mu(-1) = p \in \mathcal{S}_{\frown}, \qquad \mu(s) \subset H_{\circ\!-\!\circ} \ \text{ at } -1 < s < 0, \qquad \mu(0) = a$$

(the 'dual' case, obtained by replacing $\mathcal{S}_{\frown} \leftrightarrow \mathcal{S}_{\smile}$, is perfectly analogous and we shall not consider it separately). If there is no such a curve, then, as we shall see, the missing $\mathring{\mathcal{S}}_{\frown}$ does not manifest itself. But if the curve do exist, then the position of D is severely restricted by the singularity formed by deleting $\mathring{\Sigma}$. The restriction is so strong that $\varkappa$—recall that it is future-directed and disjoint with $\mathcal{S}_{\frown}$—turns out to be confined to $H_{\circ\!-\!\circ}$.

(i) Thus, suppose that the above-defined μ does *not* exist. This would imply that $D_{M_{\wedge\wedge}}$ (as usual, this is the component of $D \cap M_{\wedge\wedge}$ containing a) lies wholly in $H_{\circ\!-\!\circ}$. Hence,

$$D_{M_{\wedge\wedge}} \cap H_{\circ\!-\!\circ} = D_{H_{\circ\!-\!\circ}} \tag{6}$$

(cf. Corollary 22). Thus, to obtain a contradiction with the assumption ($\star$) it would suffice to extend $W \leftrightharpoons D \cup H_{\circ\!-\!\circ}$ to the spacetime $\bar{\bar{W}} = D \cup \bar{\bar{H}}$, where $\bar{\bar{H}}$ is a convexly C-extendible extension of $H_{\circ\!-\!\circ}$ such that $\bar{\bar{H}} \cap D = H_{\circ\!-\!\circ} \cap D$.

The desired $\bar{\bar{W}}$ will be built by the 'cut-and-paste' method discussed in Sect. 6 in Chap. 1. For this purpose, first, cut the region W out of M^e, i.e. consider the former as a spacetime in its own right (rather than a region of a larger space). If $D \cap H_{\circ\!-\!\circ}$ is non-connected [which will be the case—in spite of (6)—if so is $D \cap M_{\wedge\wedge}$], change to W_{unst} by 'unsticking' all of its components but $D_{M_{\wedge\wedge}}$ (as before, we keep denoting the corresponding regions of W_{unst} by D and $H_{\circ\!-\!\circ}$). This will guarantee that

$$D \cap \mathcal{S}_{\frown} = \varnothing. \tag{7}$$

As a last step, we recall that $H_{\multimap}$ was created by deleting a surface in $\mathring{H}$, see Sect. 4.2,

$$H_{\multimap} = \pi_{\Diamond}(\mathring{H} - \mathrm{Cl}\mathring{\mathcal{S}}_{\frown}),$$

and close the slit, i.e. build a new space

$$\bar{\bar{W}} \leftrightharpoons W_{\mathrm{unst}} \cup_{\pi_{\Diamond}} \mathring{H}.$$

By Test 68 in Chap. 1, $\bar{\bar{W}}$ is a spacetime, since the boundary of W_{unst} in $\bar{\bar{W}}$ is $\mathring{\mathcal{S}}_{\frown}$, and thus, due to (7), is separated from the boundary of $\pi_{\Diamond}(\mathring{H})$ (this boundary belongs to D). As easy to verify the thus defined pair $\bar{\bar{W}}$, $\bar{\bar{H}} \leftrightharpoons \pi_{\Diamond}(\mathring{H})$ does satisfy all the requirements.

(ii) Assume now, that μ does exist. In considering this case it will be convenient to work with $\mathring{H}$, whose 'nice properties' are not spoiled by a slit. Keeping this in mind, we define ξ as a continuous map

$$\xi\colon \quad \left(H_{\multimap} \cup \mathcal{S}_{\frown}\right) \to \mathring{H},$$

coinciding with $\pi_{\Diamond}^{-1}$ in $H_{\multimap}$ (they both are isometries there) and recall that by convention, $\mathring{X}$ is an alias for $\xi(X)$, see Remark 11.

The contradiction to which, as we are going to show, the existence of μ leads, is that $\varkappa$, contrary to its definition ($\star$), cannot leave $M_{\curlywedge}$ or, to be specific, some $K \subset M_{\curlywedge}$. To define K we pick points $y \in D$ and $x \in H_{\multimap}$ such that

$$\varkappa, \mu \subset I^-_D(y), \qquad (\mu - p) \subset I^+_{H_{\multimap}}(x), \tag{$\star$}$$

see Fig. 7. The points y and $\mathring{x}$ (the latter is defined exactly as x, but with $^\circ$, placed over μ, p, and $H_{\multimap}$) can always be found, since D and $\mathring{H}$ are perfectly simple, while their subsets μ, $\varkappa$ and $\mathring{\mu}$ are compact. On the other hand, $I^-_{\mathring{H}}(\mathring{\mu}) \cap I^+_{\mathring{H}}(\mathring{x})$ is included in $\mathring{H}_{\curlyvee}$ (i.e. timelike curves from $\mathring{x}$ to $\mathring{\mu}$ do not meet $\mathring{\mathcal{S}}_{\frown}$) and hence also in the domain of $\pi_{\Diamond}$, so we can—in agreement with the notation—take x that enters equation ($\star$) to be $\pi_{\Diamond}(\mathring{x})$. Now we can at last define the afore mentioned set, which—as we shall prove—includes $\varkappa$:

$$K \leftrightharpoons I^-_{H_{\multimap}}(\mathcal{S}_{\frown}) \cap I^+_{H_{\multimap}}(x).$$

For future use, we need also a certain geodesic family. Specifically, consider the curve $\tilde{\mu}(s)$ obtained by translating every point $\mu(s)$, $s \neq 0$ along a past-directed path starting in this point and remaining in $I^-_D(y) \cup H_{\multimap}$ (the reason for doing so is that $\tilde{\mu}$, in contrast to μ, does not meet $\mathcal{S}_{\frown}$, so we can apply to it Corollary 19, see below). Let $\chi(s)$ be the curve resulting from joining $\tilde{\mu}$ to $\varkappa$:

$$\chi(s) \leftrightharpoons \begin{cases} \tilde{\mu}(s), & \text{at } s \in [-1, 0]; \\ \varkappa(s) & \text{at } s \in [0, 1]. \end{cases}$$

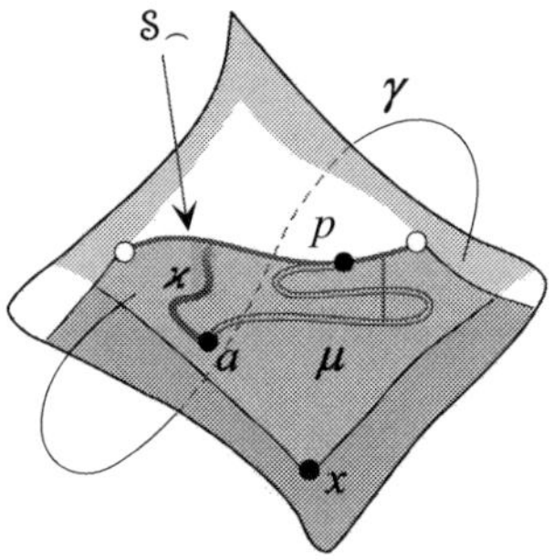

Fig. 7 The set $H_{\circ\!-\!\circ}$. The light grey region is $I^+_{H_{\circ\!-\!\circ}}(x)$. The dark grey—$I^-_{H_{\circ\!-\!\circ}}(\mathcal{S}_\frown)$. The intermediate is their intersection K. The thin vertical segment from $\mu(s)$ to $\mathcal{S}_\frown$ depicts $[\gamma_s)$, and γ is the geodesic that, as we must show, does not lie in D. Its past end point is in $I^-(\mathcal{S}_\frown) \cap H_{\circ\!-\!\circ}$, but not in $I^-_{H_{\circ\circ}}(\mathcal{S}_\frown)$

We focus on future-directed geodesics $\gamma_s \subset D$ connecting $\chi(s)$ with y. To begin with, note that γ_{-1} meets $\mathcal{S}_\frown$. This follows from Corollary 19, in which μ_0, μ_1, λ_{ξ_1}, and λ_{ξ_2} are taken to be, respectively, y, $\tilde{\mu}(-1)$, γ_{-1}, and the curve consisting of a timelike segment from y to p and the above- mentioned path from p to $\tilde{\mu}(-1)$.

Next, denote by $[\gamma_s)$ the segment of γ_s from $\chi(s)$ to the first meeting with $\mathcal{S}_\frown$. By convention, the intersection itself (that it exists can be verified by application of Corollary 19, this time with $\mu_1 \leftrightharpoons y$, $\mu_0 \leftrightharpoons \tilde{\mu}$, and $\lambda_\xi \leftrightharpoons \gamma_{\frac{\xi+1}{2}}$) is not contained in $[\gamma_s)$. Finally, define Σ (not to be confused with $\mathring{\Sigma}$) as the set of all s for which

$$[\gamma_s) \subset K. \qquad (*)$$

Then our assertion (to the effect that $\varkappa$ is imprisoned in K) is weaker than the assertion that $\Sigma = [-1, 1]$. We are going to prove this latter by showing that Σ is closed [it is open in $[-1, 1]$, as follows from the openness of K, and it is non-empty, because, by construction, $\chi(-1) = \tilde{\mu}(-1) \in K$].

Let $s_m \in \Sigma$, $m = 1, 2, \ldots$ be a sequence of points converging to some s_*. The geodesic γ_{s_*} is timelike (by the definition of y) and, as we just have discussed, meets $\mathcal{S}_\frown$. So, $\chi(s_*) \in \big(I^-(\mathcal{S}_\frown) \cap H_{\circ\!-\!\circ}\big)$. In combination with the—obvious—containment $\chi(s_*) \in I^+_{H_{\circ\circ}}(x)$, this *almost* proves our assertion [that $\chi(s_*) \in K$ and, hence, Σ is closed, whence $\varkappa$ is imprisoned in K]. The only problem is that we must prove the membership of $\chi(s_*)$ in $I^-_{H_{\circ\circ}}(\mathcal{S}_\frown)$, while the latter may differ from $I^-(\mathcal{S}_\frown) \cap H_{\circ\!-\!\circ}$. Thus, we are left with the task of excluding the possibility that γ_{s_*} leaves $H_{\circ\!-\!\circ}$ somewhere between $\mathcal{S}_\frown$ and $\chi(s_*)$, see Fig. 7.

On each γ_{s_m} pick a parameter τ so that

$$\gamma_{s_m}(0) = y, \qquad g(v_{s_m}(0), \boldsymbol{V}) = -1,$$

where $v_{s_m} \rightleftharpoons \partial_\tau(0)$ and $\boldsymbol{V}$ is an arbitrary future-directed timelike vector (the same for all m). The compactness of the ball

$$\{g(\boldsymbol{v}, \boldsymbol{v}) \leqslant 0,\ g(\boldsymbol{v}, \boldsymbol{V}) = -1\}$$

ensures the existence of a subsequence $\{s_j\} \subset \{s_m\}$ such that the velocities $\boldsymbol{v}_{s_j}$ converge to some $\boldsymbol{v}_*$. We denote by $\gamma_{\lim}$ the geodesic emanating from y with initial velocity $\boldsymbol{v}_*$. The solutions of the geodesic equation depend continuously on initial data, so for any τ

$$\gamma_{\lim}(\tau) = \lim_{j\to\infty} \gamma_{s_j}(\tau_j) \qquad \text{at} \quad \tau_j \to \tau,\ s_j \to s_*. \tag{$\circledast$}$$

Choosing τ_j so that $\gamma_{s_j}(\tau_j)$ lie between y and χ, we discover that the points in the right-hand side are in the compact set $\leqslant\chi, y\geqslant_D$ and, hence, $\gamma_{\lim}(\tau) \subset D$. But D is convex and y is connected to $\chi(s_*)$ by a *single* geodesic contained in D. This proves that $\gamma_{\lim} = \gamma_{s_*}$.

Likewise, $[\mathring{\gamma}_{\lim})$ is the limit—in the sense of ($\circledast$)—of the sequence of segments $[\mathring{\gamma}_{s_j})$. But each of them is in the compact set $\mathrm{Cl}\mathring{K}$, which, therefore, contains also $[\mathring{\gamma}_{\lim})$. Moreover, the latter segment lies in $\mathring{C} \rightleftharpoons \mathrm{Cl}\mathring{K} - \mathring{\mathbb{S}}$, since $\mathring{\mathbb{S}}$ is achronal in $\mathring{H}$, while $\mathring{\gamma}_{\lim}$ is timelike. $\mathring{C}$ is included in the domain of $\pi_\Diamond$ and, consequently, $[\gamma_{\lim}) = \pi_\Diamond([\mathring{\gamma}_{\lim})) \subset \pi_\Diamond(\mathring{C}) \subset H_{\circ\!-\!\circ}$, which proves our claim.

(2) Thus, it remains to prove that $M_{\wedge\!\wedge}$ satisfies also the second requirement of the proposition. Let $\ell \subset M_{\wedge\!\wedge}$ be a closed causal curve. It cannot be disjoint with M, because the projection $\pi_{\wedge\!\wedge}$ (see the very beginning of Sect. 5.1) would map it to a causal loop lying in H, which is impossible, since H is perfectly simple. But as ℓ is causal and M is a past set in $M_{\wedge\!\wedge}$, the existence of even a single common point implies the inclusion $\ell \subset M$, see Remark 34 in Chap. 1. □

6 Proof of the Theorem

Proposition 23 is quite close to what we wish to prove. The main differences are

1. The C-extension whose existence we just have proved may be convexly C-extendible, while the theorem asserts the existence a C-*maximal* one;
2. M is convexly C-extendible, while the theorem is formulated for an *arbitrary* U.

Correspondingly, we, first, duly strengthen Proposition 23 and then demonstrate that *any* U, being maximally extended to the past, becomes convexly C-extendible or C-maximal.

Proposition 24 *Every convexly* C*-extendible spacetime* M *has a* C*-maximal extension* $M^{\max}$ *with* $\overset{\looparrowright}{M}{}^{\max} = \overset{\looparrowright}{M}$.

The idea of the proof is close to that used by Choquet–Bruhat and Geroch [23]. We consider the set of *all* convexly C-extendible (or C-maximal) extensions of M in which the causality condition holds outside M and introduce a partial order on that set. A maximal, with respect to the corresponding relation, element turns out to be just the sought-for spacetime.

Proof For a given M, consider the set V of all pairs (V, ζ), where V is an extension of M that

(a) is C-maximal or convexly C-extendible,

(b) has $\overset{\looparrowright}{V} = \overset{\looparrowright}{M}$

and ζ is an isometric embedding of M into V. Each pair (V_1, ζ_1), (V_2, ζ_2) of sets contained in V defines the isometry $\zeta_2 \circ \zeta_1^{-1}$ mapping $\zeta_1(M)$ to $\zeta_2(M)$, cf. [76]. Let us introduce the following partial order in V: $(V_1, \zeta_1) \leqslant (V_2, \zeta_2)$, if $\zeta_2 \circ \zeta_1^{-1}$ can be extended to an isometry $\zeta_{1,2}$ napping V_1 to V_2.

Comment 25 The relation $(V_1, \zeta_1) \leqslant (V_2, \zeta_2)$ implies, of course, that V_2 is an extension of V_1. The converse, however, is not true. As discussed in Sect. 1 in Chap. 1, generally, M can be embedded into V in more than one way (some of embeddings may even send M to the same subset of V and be different, nevertheless). And it may happen that $(V_1, \zeta_1) \geqslant (V_2, \zeta_3)$. That is why the set of spacetimes cannot be ordered merely by inclusion.

Let $\{(V_\alpha, \zeta_\alpha)\}$, $\alpha \in A$ be an arbitrary chain in V. Our task is to demonstrate that it has an upper bound $(V^{\cup}, \pi\big|_M)$. By Zorn's lemma, see [87], this would imply the existence of a maximal element $V^* \in \mathsf{V}$. Thus, the existence of the mentioned upper bound would imply that M has a C-extension V^* which, first, being an element of V satisfies conditions (a), (b) and, second, (by the maximality in V) cannot be extended to a larger C-space satisfying them. And, hence, by Proposition 23, to any C-space *at all*. This means that V^* is C-maximal and can, therefore, be chosen as $M^{\max}$.

Warning The maximality of an element of V with respect to $\leqslant$ and the maximality (inextendibility) of a spacetime are different things.

Consider the set $V_A \leftrightharpoons \bigcup_A V_\alpha$ and introduce the following equivalence in it:

$$x \sim y \qquad \Leftrightarrow \qquad \exists \alpha_1, \alpha_2\colon \quad x = \zeta_{\alpha_1,\alpha_2}(y) \quad \text{or} \quad y = \zeta_{\alpha_1,\alpha_2}(x).$$

Define $V^{\cup}$ to be the quotient space $V^{\cup} \leftrightharpoons V_A \setminus \sim$ (i.e. to obtain $V^{\cup}$ we embed a spacetime into its extension, then this extension into *its* extension, etc.). By π, as usual, the projection is denoted that sends every point $p \in V_A$ to the corresponding point of $V^{\cup}$ (that is to the equivalence class containing p). Clearly, the restriction of π to any V_α is an isometry.

(1) That $V^{\cup}$ is a spacetime, is obvious (a formal proof can be inferred from Proposition 68 in Chap. 1). Moreover, C-spaces V_α form its open cover [from now on, we do not distinguish between V_α and their images $\pi(V_\alpha)$ in $V^{\cup}$] and, hence, $V^{\cup}$ is a C-space, see Test 3(c in Chap. 2).

(2) Next, let us prove that if $V^{\cup}$ is C-extendible, then it is also convexly C-extendible. To this end consider its arbitrary C-extension V^e and an arbitrary perfectly simple subset D thereof. Denote by φ a curve lying in $D \cap V^{\cup}$ (and, hence, in a connected component of this set), and by $\varkappa$—an arbitrary *timelike* curve with the same end points. Since φ and $\varkappa$ are compact, we can find a finite subchain $\{V_{\alpha_k}\}$

$$V_{\alpha_1} \dots \leqslant V_{\alpha_k} \leqslant \dots V_{\alpha_{k_0}}$$

such that both these curves lie in $\bigcup_{k=1}^{k_0} V_{\alpha_k}$ and, hence, in any V_β that is an upper bound of that subchain. But V_β, by hypothesis, is convexly C-extendible, and, consequently, $\varkappa \subset (D \cap V_\beta)$, i.e. $\varkappa$ lies wholly in a connected component D_{V_β} (which is the same component where φ lies) of the intersection $D \cap V_\beta$ and thereby in a connected component of the intersection $D \cap V^{\cup}$. By definition, this means that $V^{\cup}$ is convexly C-extendible. Thus, it satisfies condition 24(a).

(3) Let $\ell \subset V^{\cup}$ be a closed causal curve. Again, from its compactness we conclude that there is a subchain such that ℓ is contained in each V_β bounding that subchain from above. This means (since $\overset{\looparrowright}{V}_\beta = \overset{\looparrowright}{M}$) that there are no causal loops in $V^{\cup}$ besides those present in M. So, condition 24(b) holds too. □

Thus, we are left with the task of proving that any C-space U has a C-maximal or convexly C-extendible extension M with causality violations confined to U. As a first step, we consider the set W of all C-extensions of U that are of the form $V = I^-_V(U)$. Clearly, for any C-extension V^e of V

$$V \subset I^-_{V^e}(U).$$

Repeating the previous argument one checks that W contains a maximal, with respect to $\leqslant$, element M. Suppose, M^e is a C-extension of M. Then the inclusion above, combined with the maximality of M in W, implies the equality

$$M = I^-_{M^e}(U). \qquad (\circledast)$$

And this means, in particular, that it is impossible for a past-directed timelike curve in M^e to leave M. Consequently, M is causally convex *in each of its* C*-extensions*. Hence, M is a C-maximal or convexly C-extendible extension of U. And the inclusion $\overset{\looparrowright}{M} \subset I^-_M(U)$ follows immediately from ($\circledast$). In combination with Proposition 24 this proves Theorem 2.

Remark 26 We have imposed the requirement $\mathsf{W} \subset \mathsf{C}$ in order to ensure that c holds in M. But another consequence is that the maximality in W guarantees the equality (⊛) only for $M^e \in \mathsf{C}$. That is why in our proof convexly c-extendible spacetimes are used and not their (geometrically more natural) variety defined by dropping the condition $M^e \in \mathsf{C}$ in Definition 3.

Chapter 6
Time Travel Paradoxes

... Loads of them ended up killing
their past or future selves by mistake!
Hermiona in [158]

It seems appropriate now to turn attention to the most controversial issue related to the time machines—the time travel paradoxes. On the one hand, paradoxes seem to be something inherent to time machines (their main attribute, perhaps). And on the other hand, the (supposed) paradoxicalness of time travel is traditionally the main objection against it and a good pretext for dismissing causality violating spacetimes from consideration. Recall, however, that in studying physics one meets a lot of 'paradoxes' (Ehrenfest's, Gibbs', Olbers', etc.). Today they are just interesting and instructive toy problems. Our aim in this chapter is to examine the 'temporal paradoxes' and to reduce them to the same status. In particular, we are going to show that they do not increase the tension between the relativistic concept of spacetime and 'the simple notion of free will' [76]. As a by-product, we shall reveal, in the end of the chapter, a curious relation between the geometry of a spacetime and its matter content.

In our discussion, we follow [96]. A reader interested in alternative views on the paradoxes is referred to [38, 120, 138, 154, 172] and the literature therein.

1 The Essence of the Problem

1.1 The Two Kinds of Time Travel Paradoxes

The most known time travel paradox is undoubtedly the 'grandfather paradox' first proposed more than 70 years ago (perhaps, in *'Le Voyageur Imprudent'* by R. Barjavel), which may be formulated as follows.

S. Krasnikov, *Back-in-Time and Faster-than-Light Travel in General Relativity*, Fundamental Theories of Physics 193, https://doi.org/10.1007/978-3-319-72754-7_6

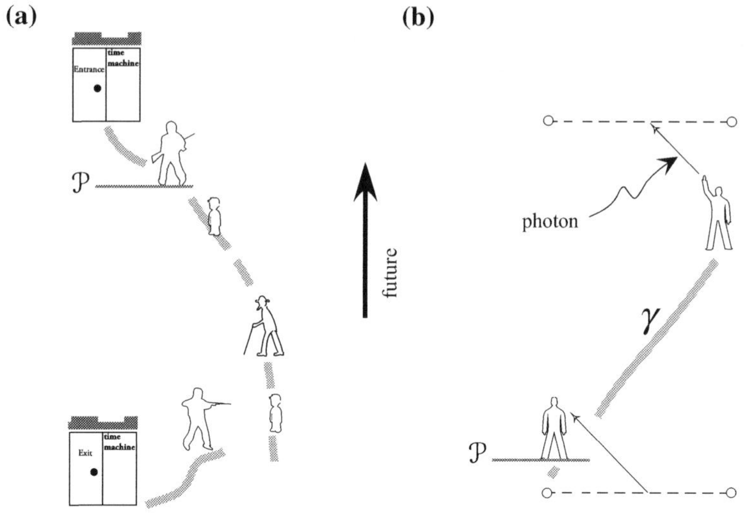

Fig. 1 **a** The classical grandfather paradox. The two rightmost silhouettes are the grandfather at two ages. **b** A free falling man in the DP space (the 'seams' depicted by the dashed lines play the same role as the entrance and exit doors of the time machine in the left picture). The man learns—merely sees—that in five minutes he will raise his right hand. After receiving this information, he tries to change the future by raising the 'wrong' hand, hence a paradox

Example 1 ('The Grandfather Paradox'). Suppose that at some moment (corresponding to the surface $\mathcal{P}$ in Fig. 1a) a man able to travel in time decides to kill his grandfather in infancy. He travels to the past, steals up to the baby and shoots. What will happen? We assume that the man under consideration is

1. intelligent enough not to confuse babies;
2. accurate enough not to miss the target;
3. motivated well enough not to suddenly change his mind;
4. ....

So, the baby *will* be killed. But on the other hand, it will *not* be killed, because otherwise the father of the would-be killer—and therefore also the killer himself—would never be conceived, so there would be nobody to pull the trigger.

The question that is usually asked in this relation is: Will or will not the baby be killed? Or, if the second alternative seems more convincing (after all, 'a person' and 'a person whose grandfather did not die in infancy' is *exactly the same*), one can ask: What saves the baby? or *Why didn't the time traveller shoot?*

In science fiction and popular literature, one can easily find hundreds of 'time travel paradoxes' (a voluminous bibliography on the subject can be found in [138]) including those that concern 'changing the past' (why cannot a time traveller save the dinosaurs or kill Hitler?). As far as I know *all* of them—except for trivial ones, see below—are just variants of the grandfather paradox.

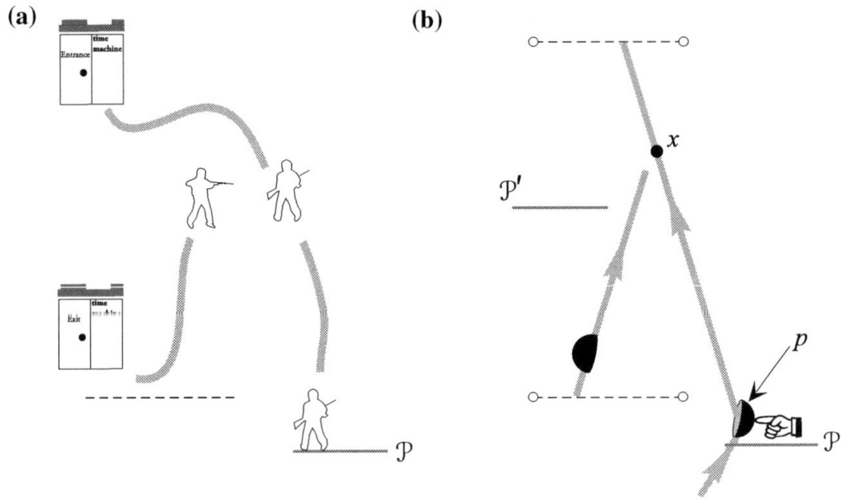

Fig. 2 The machine builder paradox. **a** The dashed line denotes the moment from which the time machine operates. Before crossing it the experimenter already ceases to make any decisions. **b** A simpler model of the same paradox

A paradox of another type was considered in [34]. In accordance with the homicidal tradition, it can be reformulated as a story of a self-sacrificing experimenter.

Example 2 ('The Machine Builder Paradox'). An inquisitive person plans to build a time machine and to use it in killing their younger self, see Fig. 2a. The plan is elaborate and the experimenter is resolved to stick to it no matter what happens. What will come out of it?

This looks very much like the grandfather paradox. The remarkable difference, however, is that now all preparations are done and the decisions are made *before* the time machine appears. In other words, even before crossing the Cauchy horizon the experimenter abandons the freedom of will.

As a simple model, consider an experiment performed in a point p, see Fig. 2b. The experiment consists in pushing a uranium lump with mass $m = \frac{2}{3}\times$(critical mass). The push is directed so that the lump moving on a geodesic must enter the time machine where that geodesic has a *self-intersection* in a point x. We know for sure that there must be an explosion in x, because the total mass of the two colliding lumps exceeds the critical. But at the same time, we can assert with equal confidence that there is *no* explosion there, because otherwise the 'younger' lump would collide not with an m-mass piece of uranium, but only with its remnants, so that the total mass of uranium in x would be less than critical. But the absence of explosion implies that the lumps in x are intact and *must* explode. Et cetera.

Thus, we have a typical time travel paradox with an important feature: the experimenter remains in the causal region. The time machine, in this case, may be occupied only by inanimate objects governed by a few simple laws, which relieves us of having to discuss freedom of will.

The difference between the two types of paradoxes sometimes may *seem* immaterial.

Example 3 ('Changing the Future'). A person sees (with the aid of a time machine, of course) that in five minutes they will raise the right hand, see Fig. 1b. What will happen if the person makes up their mind (out of contrariness) to raise the *other* hand? Obviously, this paradox is essentially the same as the previous ones. Which type of paradox it is depends on whether γ—the maximally extended world line of the experimenter—intersects the Cauchy horizon and when the decision to change the future is made.

1.2 Pseudoparadoxes

There are a lot of situations to which we shall not refer as time travel paradoxes, even though they are often called so. Let, for example, a certain Eckels travel to the past (or at least to what is *claimed* to be the past), step on a butterfly, and find on returning that the world differs from that he left [16]. Contrary to a widespread belief this story, however strange and exciting, has *nothing to do* with time machines as they are understood in this book and, in particular, with time travel paradoxes. No closed causal curves are involved in the story (precisely because the world to which Eckels returns is not the one from which he departed, his world line is not self-intersecting) and, correspondingly, no time machines. Eckels did not 'change the past' of the moment s in which he started the safari, he has never even been to that past, see Fig. 3.

Another kind of pseudoparadox is obtained by assuming that along with conventional macroscopic bodies, which we shall call *comotes*[1] in this context, there are objects, called *contramotes* [189]. Being perfectly usual in all other respects contramotes get younger with time. In other words, the entropy of each contramote decreases as time goes on.[2] An encounter of a comote with a contramote can lead to paradoxes similar to those considered above, because the latter *remembers the future* of the former, which is a variant of the grandfather paradox. It is also quite paradoxical that neither of the personae can kill the other. We shall not, however, consider paradoxes of this kind, because the spacetimes, in which they take place, do not have to be what we call time machines, see Definition 4.1 in Chap. 4 and comment 4.2.

---

[1]The term was coined by Tadasana.

[2]In fact, this assumption is not that extravagant. I am not aware of a single strong argument against it. Note, in particular, that the apparent lack of contramotes in the everyday life and in astronomical observations is not an argument: the contramotes must be practically invisible to us comotes. Indeed, they almost do not radiate light. Instead, a contramote star, say, absorbs a powerful flux of photons emitted (for some mysterious reason) towards the star by other bodies.

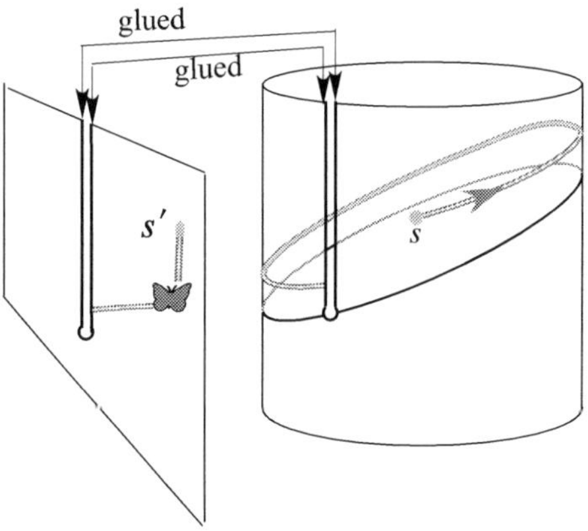

Fig. 3 A two-dimensional analogue of Eckels' universe (the thick curve from s to s' is his world line): a Minkowski plane (in which the vertical direction, against tradition, is null) and a Misner time machine (the ellipse is its horizon) are cut along a future-directed null ray each, and the edges of the cuts are glued crosswise. This universe contains *no* causal loops: the butterfly is in the past of s', but *not* of s

Yet another pseudoparadox[3] is a story of a man who receives a note (with a very helpful clue) from his older self, keeps this note in his wallet, and when (his) time comes hands it to the addressee—to his younger self [72]. Nothing *impossible* happened, but one might wonder who wrote the note and why it does not get dirtier with each cycle. The stories of this kind—'bootstrap paradoxes' in the terminology of [172]—can hardly be called paradoxes. In contrast to Examples 1 and 2, they do not contain plausible assumptions leading to unbelievable conclusions. If one finds the existence[4] of such a note incredible, all one needs to do (and, in fact, *must* do) is to deny the existence of the note (nothing prevents one from doing so—the said existence is just an assumption, indeed, not an implication from some other, very natural, say, premises, so the burden of answering the puzzled questions lies with Harrison).

A number of false paradoxes stem from a popular misunderstanding concerning energy conservation. In the initial globally hyperbolic region, i.e. before the time machine was put into operation, there had been exactly one m-mass lump at each moment of time (i.e. on each surface $\mathcal{P}$). But *after* the time machine started to work the total mass of uranium doubled, see Fig. 2b). Does not this violate energy conservation?

To answer this question note that there are two quite different laws which may equally well be called 'energy conservation':

[3]For a collection of such *pseudo*paradoxes see [138].

[4]We speak of the existence of the note and not of its *appearance*, because being a typical Cauchy demon, see Sect. 3 in Chap. 2, the note has always existed, without ever having come into being.

(1) One may call so the *local* property of matter fields to satisfy the equation $T^{ab}{}_{;b} = 0$. This equality must hold *universally* (in fact, by the definition of the stress–energy tensor T^{ab}). And it does hold in all time machines;
(2) Alternatively, one may require that the 'total energy' would be conserved, i.e. that the quantity $\int_{\mathcal{P}} T^{ab} n_a \mathrm{d}^3 V$ (here n is the unit normal to $\mathcal{P}$) would not depend on the choice of a spacelike edgeless surface $\mathcal{P}$. It is this requirement that fails in time machines. But there is nothing paradoxical or even unusual in this failure. We have no reason whatever to expect this law to hold anywhere except in the Minkowski space, where it follows from the local one.

2 Science Fiction

Science fiction offers a plenitude of ideas concerning time travel paradoxes [138] of which most interesting are the following two.

2.1 *The Banana Skin Principle*

According to a popular view, there are no time travel paradoxes at all: *any* reasonable initial state of a reasonable system gives rise to *some* evolution.

The situations traditionally considered in thought experiments with time machines (say, a human being in an unspecified causality violating spacetime) are, of course, too complex for a rigorous analysis.[5] So, one might conjecture that we just overlook something each time, while actually all contradictions are resolved by unforeseen contingencies. In Example 1 we compiled a list of precautions to be taken in sending a killer to the grandfather. The idea under discussion is that such a list can never be made exhaustive: no matter how many candidates we send to the past, each of them will get in an accident, or slip on a banana skin, etc. In general, each time the grandfather will be saved by natural causes. This approach especially suggests itself in the case of the machine builder paradox. Indeed, as soon as a system crosses the Cauchy horizon, it gets exposed to Cauchy demons (which are particles whose world lines are disjoint with M^{r}) with all their unpredictability, see Sect. 3.3 in Chap. 2.

Clearly, a wild demon behind the door of the time machine would easily reconcile the traveller's firm intention to shoot with the grandfather's invulnerability. Still:

1. Even the best plan can fail, and it is in the order of things that an experimenter suddenly fall sick or enter a wrong door. But if this happens a thousand times in a row, in spite of all precautions, the situation—paradoxical or not—would certainly need an explanation, however innocent each failure looks.
2. Similar paradoxes take place, as we shall see, even when *all* contingencies are provided for (which of course can be done only in toy models).

---

[5]In fact, they often are too complex even when consist of billiard balls, see [39, 127].

2.2 Restrictions on the Freedom of the Will

Sometimes the impossibility to shoot the would-be grandfather is explained on the hypothesis that the traveller will not even try to do so, which is interpreted as a restriction on their freedom of will. Here one is up against two problems:

1. The lack of the freedom of will is, by itself, quite an obscure notion. What exactly is it? How can it be registered? What mechanism enforces it?, etc. So, a puzzle of time travel paradoxes turns out to be 'explained' via an equally puzzling phenomenon. It is easy to imagine a person who—out of apathy—refuses to raise the required hand and to see what would happen. But if 100 of 100 test persons do the same, such a lack of curiosity might seem exceedingly strange (though not paradoxical, perhaps) and calling explanation, as we already discovered in the case of test persons in the causality violating region, see the previous subsection.
2. Whatever happens with the will of a traveller when they are in a time machine, this would not explain the machine builder paradox, because in this case all decisions (requiring presumably a free will) are taken in the region where the causality condition is satisfied and 'one is free to perform any experiment' [76].

3 Modelling and Demystification of the Paradoxes

Formally speaking, a time travel paradox is a proof (non-rigorous, of course, but in this chapter we omit such obvious provisos) of the fact that for a certain *feasible* system (e.g. for a maniacally curios grandson, in Example 1) there is a state that, on the one hand, is *inconsistent*[6] with the laws of motion governing that system—let us call such states *paradoxical*—and, on the other hand, is *feasible* (it seems easy to arm the grandson, instruct him, and bring face to face with his baby grandfather). That the feasibility is essential is clear from the following example, see also discussion on p. 65.

Example 4 Let a system $\mathscr{A}$ be a set of stable sterile massless pointlike particles, and let its state at the surface $\mathcal{P}$ be that shown in Fig. 4. Then we can ascertain that the system is in a paradoxical state. Indeed, the laws of motion require such a particle to pass through the points p' and p'' in some proper time after (respectively, before) it passed through p. Both requirements are violated by the initial conditions.

Thus, we have found a paradoxical state. This, however, does not yet constitute a 'brain-boggling logical paradox' [138], because (and only because) we have not proposed a plausible method of preparing the system in this state.

Based on the examples considered in Sect. 1.1 we separate the paradoxes into three types:

---

[6]As is known, '...either a tail *is* there or it isn't there. You can't make a mistake about it...' [128]. The same is true for evolutions. So, we shall not speak of 'self-inconsistent evolution' or 'trajectories with zero multiplicity'.

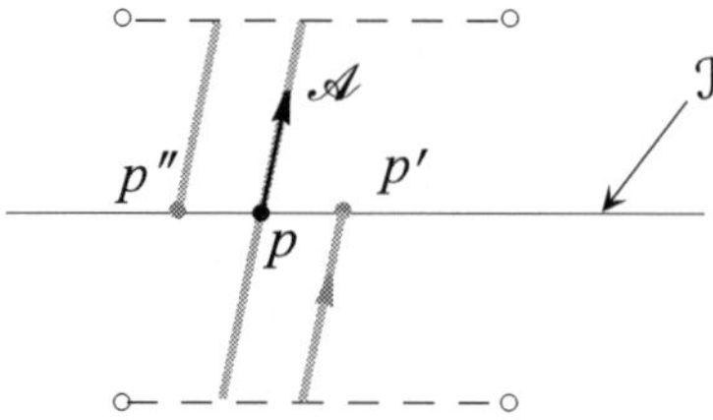

Fig. 4 The state of the system $\mathscr{A}$ at the surface $\mathcal{P}$ is a single particle in p moving slowly to the right (the black arrow from p is the particle's velocity and the grey line is a part of its world line determined by the laws of motion). When p is the origin in the DP space this state is paradoxical, since it does not contain particles in p' and p''

i. The Grandfather Paradox. The whole history of the relevant system lies in the non-causal region of the universe. We—inhabitants of M^{r}—play no role and may be considered missing as well;

ii. The Machine Builder Paradox. The state of the relevant system is fixed in the initial globally hyperbolic region. Cauchy demons play no role and may be considered missing as well;

iii. Mixed Paradoxes. Something intermediate. The paradoxes of this type do not exhibit any specific interesting features and we shall not consider them.

3.1 The Grandfather Paradox

With our definitions, the resolution of the grandfather paradox is trivially simple. It is convenient to demonstrate this with a concrete example based on the 'changing the future' paradox, see p. 150.

Assume that a time machine will be built one day and we shall be able to find 20 volunteers such that:

1. each of them will undertake to enter that time machine, to wait until their future self raises a hand—at some moment τ_0 (of proper time)—and
   (a) to raise the other hand in 5 minutes before τ_0;
   (b) to keep it (and only it) raised for 5 minutes.
2. they will be attacked by Cauchy demons and $\frac{9}{10}$ of them will fail to see their future;
3. half of those who succeeded in observing their future selves, will exercise their freedom of will and abandon the experiment.

To turn these assumptions into a paradox one only has to claim that they imply the existence of $\frac{1}{2} \cdot \frac{1}{10} \cdot 20 = 1$ feasible initial state (a volunteer seeing his future self) inconsistent with the laws of motion (which include the volunteer's determination to raise the wrong hand).

The assumptions listed above are disputable, but what makes the *fatal* flaw in the proposed paradox is the origin of the volunteers in question. Indeed, by construction they came from the initial causal region of the spacetime, while *by the definition* of the grandfather paradox, see p. 154, they must have been Cauchy demons. This observation resolves the paradox, since the existence of suitable volunteers, which seemed unquestionable, becomes highly doubtful, as soon as we require them to be Cauchy demons.

This resolution also clears up the logical status of the paradox. It turns out to be merely a proof—by contradiction—of an estimate on the number of demons of a certain type. In this capacity it goes roughly as follows: 'Let N be the number of two-handed demons in $\overset{\looparrowright}{M}$ that obey the laws 1–3. Then $N < 20$. For, suppose not. Then it follows from the said laws that at least one demon will have motive, means and opportunity to raise the wrong hand. Comparing the hands that are up at τ_0 and at $\tau_0 - 5'$ we see that by 1(a) they are different, but by 1(b) they are the same, a contradiction. Q.E.D.'

3.2 *The Machine Builder Paradox [96]*

Before resolving the paradox let us verify that—in contrast to the previous one—it does exist. To this end, we must produce an example of a system such that it can be prepared at $\mathcal{P} \subset M^{\mathrm{r}}$ in a state inconsistent with its laws of motion. These laws must be local geometrical (see the beginning of Sect. 1 in Chap. 2) so that we could speak of the same laws in different spacetimes. They also must be consistent with any feasible initial state in causality respecting spacetimes, for otherwise it is the laws, not causality violations, that should be blamed for the paradox.

The paradox is built in the twisted DP space. This space can be visualized[7] as a space obtained by deleting the horizontal segments $x_1 \in [-1, 1]$, $x_0 = \pm 1$ in the Minkowski plane and gluing the lower side of either cut to the upper side of the other. The sides are glued so as to identify diametrically opposite points

$$(1 \mp 0, c) \mapsto (-1 \pm 0, -c), \qquad \forall c \in (-1, 1),$$

see Fig. 5, not the points with the same abscissas x_1 (which would result in the regular, non-twisted, DP space). Except in two 'seams' (the two segments that appeared when we glued the sides of the cuts together), we can use in M the coordinates $x_{0,1}$ inherited from the Minkowski plane, cf. Example 73 in Chap. 1. This is supplemented with the following rules:

- any smooth causal curve approaching from below (from above) a point of the upper (lower) seam continues from the symmetric—with respect to the origin of

[7] For a technical description see Example 74 in Chap. 1.

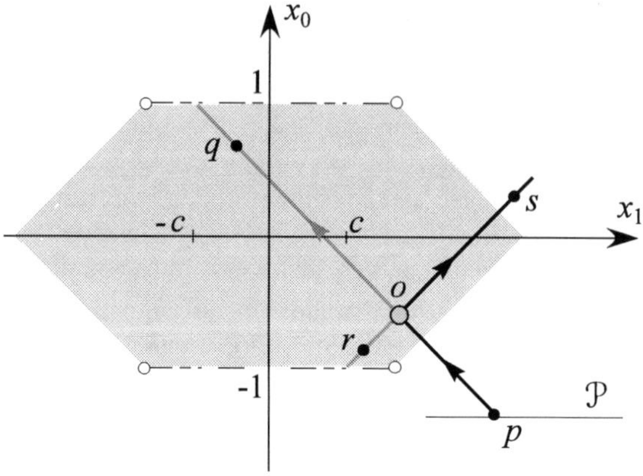

Fig. 5 The two-dimensional twisted Deutsch–Politzer space M. The shadowed region is $\overset{\looparrowright}{M}$. The self-intersecting line *poqros*, which actually is straight, is a null geodesic. Its dark gray (closed) part may be the world line of a Cauchy demon. The seams are shown by the dashed segments

the coordinates—point of the lower (respectively, upper) seam. The *velocity* $\boldsymbol{v}$ of the curve jumps at this moment (v_{x_1} changes the sign), see Fig. 5;

- the curves that, loosely speaking, terminate at a 'corner point' $x_0, c = \pm 1$ are inextendible.

In this spacetime consider a set of particles that obey the following laws:

(*a*) the world line of each particle is an inextendible broken line in M whose edges are null geodesic segments. The vertices are the points where two particles collide;

(*b*) after a collision, either of the particles changes the direction of its motion.

Physically speaking, we consider pointlike perfectly elastic massless particles which interact with, and only with, their like.

Every null geodesic that enters $\overset{\looparrowright}{M}$ has a self-intersection. Thus, at first glance, it might seem that already the initial data shown in Fig. 5—a single particle ready to fly into the time machine—constitute the desired paradox. Indeed, if the particle enters the time machine and freely flies through o, then later it will have to hit its younger self in that same point. But in such a case the right ('younger') particle will have to change the direction of motion and fly away on the ray os. Then it will never pass through q and get in o. Correspondingly, there will be no collision and the younger particle will arrive at o to meet its older self, cf. [154]. Et cetera.

In fact, however, this situation is a *pseudo*paradox like those mentioned in Sect. 1. Indeed, the entire reasoning is based on the implicit assumption that the number of particles is a conserved quantity: if there was a single particle on the surface $\mathcal{S}$, then there must be only one particle on the surface $x_0 = 0$ too. But in the case under consideration, there is no such a law, cf. the discussion of the energy conservation in the end of Sect. 1. It does not follow from our laws (*a,b*), nor any reasons are seen to impose it as a separate additional law (especially, as we require all laws to be local).

Another drawback of this candidate paradox is that it does not take into account the possible interaction between the particle and Cauchy demons. Demons never meet

$\mathcal{P}$, so, their evolution is not determined by the data at that surface: they are guided only by dynamical laws [(a,b) in this case] and the urge to provide the existence—in the entire M—of a solution to the Cauchy problem.

Taking both of these issues into consideration, we immediately find a solution to the paradox in question, i.e. a solution of laws (a,b) that is consistent with the given initial state. This solution contains *two* different particles. One of them is a demon whose world line is the closed curve $oqro$ in Fig. 5. The other—its world line is pos—collides with the demon in o and bounces away from the time machine.

To exclude such solutions let us supplement the physics of the particles with the following law:

(c) Each particle is characterized at every instant by a 'charge' with the possible values ± 1. The charge of each particle change after every collision, see Fig. 6a, and in no other case.

Remark 5 The introduction of 'charge' is inspired by the uranium lump story, see Fig. 2b. The two values of charge correspond to the two states of the uranium—exploded and intact. The reaction *exploded* + *exploded* → *intact* + *intact* may seem strange, but we *must* introduce it insofar as we require the dynamics to be geometric and local. This is the price for modelling such a complex non-equilibrium object as a bomb by a mere pointlike particle.

Now the evolution depicted in Fig. 5 is impossible: on the one hand, the charge of the particle must have the same sign on the segments (o, q) and (r, o) (because on the segment (q, r) the particle does not collide with anything). But on the other hand, the signs must differ because of the collision in o, cf. Figure 6a. And still, self-consistent evolutions, satisfying all requirements, exist. An example is shown in Fig. 6b. It consists of two particles again: one of them starts from $\mathcal{P}$, collides with a demon in o and flies away to the right. The other is a demon appearing 'from the singularity'. It collides consecutively with its older self at o', then with the former particle at o, and finally with its (demon's) younger self at o', after which it vanishes in the singularity.

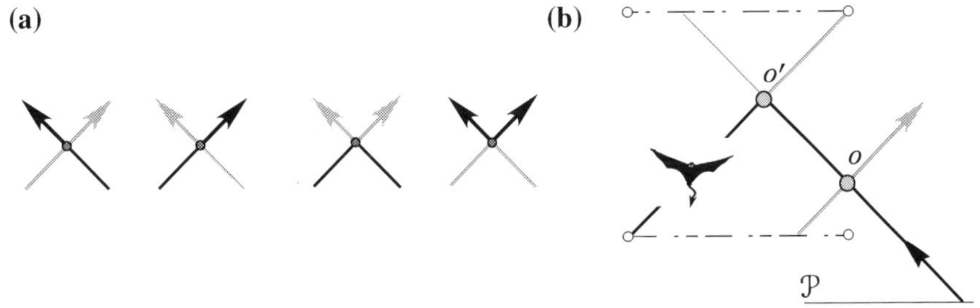

Fig. 6 **a** All possible collisions satisfying (a–c). The world lines of 'positive' and 'negative' particles are drawn in black and in light grey, respectively. **b** An evolution consistent with both the laws (a–c) and the initial data at $\mathcal{P}$

To exclude this last type of permitted solution, we assign, in addition to charge, one more characteristic (constant, this time) to every particle. This characteristic, called *colour*, is subject to the following law.

(d) There are three colours and the particles of different colours do not interact.

To summarize, interaction in the world under consideration reduces to the following rule. The world lines of all particles taken together form a set of inextendible null geodesics (though each geodesic may include segments of world lines of different particles). If γ is such a geodesic then its colour is constant and the value of charge in $a \in \gamma$ differs from that in $b \in \gamma$ if and only if between the points a and b, γ has an odd number of intersections with geodesics of the same colour.

Now we are in a position to produce a paradox. Suppose, in the region $x_0 < -1$ of the twisted DP space there are exactly three particles of different colours moving in the same direction towards $\overset{\looparrowright}{M}$, see Fig. 7. This initial state is consistent with *no* solution of dynamical laws $(a - c)$.

Proof Indeed, any such solution must contain all three self-intersecting geodesics depicted by the solid—thin, thick, and double—lines. Besides, it *may* contain a few Cauchy demons. That is in addition to the solid lines the set of world lines may contain a few *dashed* lines. Each of the solid lines has a self-intersection and hence includes a loop [cf, the loop $oqro$ in Fig. 5]. But there are at most *two* demons, $\eth_2$ and $\eth_3$, intersecting the *three* loops. So, there is at least one particle that experiences no collisions between the two moments of its history in which it passes through the same point. And as we have already discussed, see just below Remark 5, it is impossible to assign consistently any charge to such a particle. □

Thus, we have demonstrated that, at least in a simple model, feasible paradoxical states—contrary to the banana skin principle—do exist. Now the question of what *actually* will happen if the system under discussion will be prepared in such a state becomes meaningful.

The answer proposed here is based on the observation that the geometry of a spacetime is related to its matter content not only via the Einstein equations. For

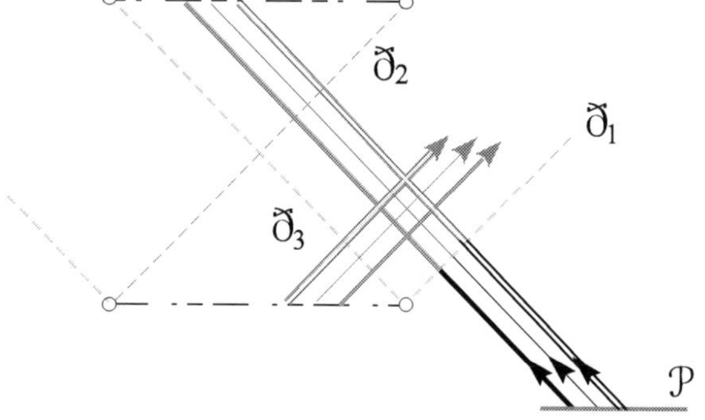

Fig. 7 Any solution consistent with the conditions at $\mathcal{P}$ must contain all three self-intersecting geodesics. It also may contain some of the geodesics depicted by the thin dashed lines

example,[8] in a spacetime of the form $\mathbb{L}^1 \times \mathcal{K}$, where $\mathcal{K}$ is a compact two-dimensional surface, a field of non-zero null vectors *by its very existence* would imply that $\mathcal{K}$ is a torus. But such a relation makes our approach—where backreaction is neglected and the geometry of the spacetime is taken the same for all candidate solutions—unjustified. And as soon as we admit the dependence of the geometry to the future of $\mathcal{P}$ on the state of the particles at $\mathcal{P}$, the paradox vanishes. Now the Minkowski plane with three parallel null lines makes a legitimate solution of the evolution laws consistent with the—former paradoxical—initial data depicted in Fig. 7.

The same reasoning applies to *any* machine builder paradox. Indeed, in all of them, by definition, the initial state is fixed at an acausal subset $\mathcal{P}$ of the initial causal region M^{r}. But according to Theorem 2 in Chap. 5, M^{r} can always be extended to a causality respecting maximal spacetime $M^{\max}$, where *all* initial data—as long as they are fixed at an acausal surface—are allowed.

Remark 6 How general paradoxes are is still a question. An interesting possibility is that any 'realistic' state is paradoxical in any 'realistic' causality violating extension of M^{r}. This would mean that time machines are impossible after all and it is the paradoxes that protect causality.

[8] For a less trivial one see [65].

Part II
Semiclassical Effects

Chapter 7
Quantum Corrections

So far our consideration was purely classical. There are no reasons, however, to expect such consideration to be exhaustive even in the initial globally hyperbolic region M^{r}. Indeed, every causal curve leaving a sufficiently small neighbourhood U of a point in M^{r} will never return back to U. In this sense the physics in U is the same as in the Minkowski space. However, the closer is U to the Cauchy horizon, the smaller[1] must it be to avoid causal loops, cf. Sect. 4 in Chap. 4. And when the relevant distances approach the Planck scale, one expects quantum effects to become noticeable (cf. [173]). Thus, quantum gravity may be pertinent to our subject. Unfortunately, it is not clear, how exactly it should be taken into account. The situation is better with 'semi-classical gravity'—the approximation (presumably justified when U is large in Planck units) in which matter fields are quantized and gravitation is not. In semi-classical gravity, see [14, 70, 178], it is assumed that the metric keeps adhering the Einstein equations, while the contribution to their right-hand side of a field in a pure quantum state $|Q\rangle$ is the (renormalized) expectation value of its stress–energy tensor operator $\langle Q|T_{ab}|Q\rangle$ (to simplify notation I shall often write $\langle\, T_{ab}\rangle_Q$ for $\langle Q|T_{ab}|Q\rangle$). Correspondingly, the semi-classical Einstein equations have the form

$$G_{ab} = 8\pi T^{\mathrm{C}}_{ab} + 8\pi \langle\, T_{ab}\rangle_Q, \qquad (1)$$

where T^{C}_{ab} is the contribution of the classical matter, that is the matter for which quantum effects can be neglected. $\langle Q|T_{ab}|Q\rangle^{\mathrm{ren}}$ depends on the type of the field and its state, but also on the geometry of the spacetime, which makes Eq. (1) by far more complex than Eq. (1) in Chap. 2, only a few solutions are known. Much can be said pro and contra semi-classical gravity, see e.g. [14]. In what follows, we take the validity of the semi-classical Einstein equations for granted and confine discussion to

[1] Of course, in the *pseudo*-Riemannian case one should exercise certain caution in speaking of 'sizes', see, in particular item 2 below.

S. Krasnikov, *Back-in-Time and Faster-than-Light Travel in General Relativity*, Fundamental Theories of Physics 193, https://doi.org/10.1007/978-3-319-72754-7_7

its consequences. It should be emphasized that even in macroscopic problems (which are our concern), the transition from the classical to semi-classical approximation may not reduce to small corrections:

1. In some important cases, the quantum effects may give rise to *qualitatively* new phenomena. For example, it is believed, see Hypothesis 11 in Chap. 3 and the 'chronology protection theorem' discussed just below Corollary 20 in Chap. 4, that the WEC must break down on compactly generated Cauchy horizons and in the traversable wormholes.

Remark 1 As before, we understand the WEC as the condition

$$G_{ab}v^a v^b \geqslant 0 \qquad \forall\, \boldsymbol{v}\colon\ v^a v_a \leqslant 0.$$

For solutions of the classical Einstein equations (1) in Chap. 2, it is equivalent to (2) in Chap. 2, and in the semi-classical case, it takes the form

$$T^{\mathrm{C}}_{ab}v^a v^b + \langle\, T_{ab}\rangle_Q v^a v^b \geqslant 0 \qquad \forall\, \boldsymbol{v}\colon\ v^a v_a \leqslant 0. \tag{2}$$

It is matter violating this condition that will be called exotic. Alternatively, one could reserve the term WEC for the inequality (2) in Chap. 2 and interpret the appearance of a new term in the right-hand side of the Einstein equations an evidence that in the semi-classical theory exotic matter is not necessary for creating time machines, traversable wormholes, etc., rather than that usual matter can become exotic under the action of quantum effects.

The need for purely classical matter to violate the WEC would mean, in fact, that the corresponding spaces are *forbidden*. However, for *quantum* fields, the violations of all energy conditions are commonplace, for explicit examples see Sect. 1.3 in Chap. 8 and the beginning of Chap. 10. Irrespective of their strength (as we shall see below there is no natural way, as of now, to characterize the 'strength' of the WEC violations) they disarm the mentioned forbidenness of time machines and shortcuts. Indeed, in Chap. 9, we show that quantum effects do make traversable a wormhole that classically is known to be a model of non-traversability.

2. There are no reasons to expect the quantum effects to be weak at, say, Planck scale. And due to the Lorentzian signature, such scales sometimes turn out to be involved in apparently macroscopic situations. For example, what can be called the 'distance between the mouths' of a wormhole, tends to zero even for a macroscopic wormhole in the course of its transformation into a time machine, see Sect. 1.3 in Chap. 8.

Remark 2 In semi-classical gravity, a vacuum is *not* necessarily the lowest energy state or the most symmetric one. Roughly speaking, a field is represented as a sum of excitations and a vacuum is a state with as few excitations as possible (note that speaking of vacuum solutions in this context one does not necessarily mean that $T^{\mathrm{C}}_{ab} = 0$—the word 'vacuum' may refer only to the *quantized* fields). So, it is not surprising that often $\langle\, T_{ab}\rangle_Q$ is non-zero even when $|Q\rangle$ is a vacuum. This phenomenon is known as *vacuum polarization*. The vacuum stress–energy tensor $\langle T_{ab} \rangle_Q$ affects—via the Einstein equations—the spacetime geometry. That is why the vacuum energy density is so important in semi-classical gravity, even though in the rest of physics, it can safely be set to zero. This density, as a rule, is so much greater than the contribution to the full $\langle T_{ab} \rangle_Q$ of separate particles, that by $|Q\rangle$ in (1) a vacuum is tacitly understood.

Quantum fields are a subject of a mature and sophisticated science. Of all its results and concepts, we shall only need the estimates for $\langle T_{ab} \rangle_Q$ in a few simplest situations. So, in this chapter instead of presenting (the basics of) quantum field theory, we restrict ourselves to the discussion—as brief as possible, for details and substantiation we refer the reader to textbooks such as [14, 70, 178]—of relevant methods for calculating $\langle T_{ab} \rangle_Q$. Also some notations and formulas are established for future use.

1 Direct Calculation

In this section, we recapitulate the basic procedure prescribed by quantum field theory for finding the vacuum expectation of the stress–energy tensor of the scalar field in a globally hyperbolic spacetime.

We confine ourselves to the free scalar field. In the classical theory, it would be described by a smooth function ϕ obeying a certain equation of motion. Its stress–energy tensor would be a quadratic combination of ϕ and its derivatives. For example, in the case of the 'minimally coupled' field the equation of motion is

$$(\Box - m^2)\,\phi = 0, \tag{3}$$

and the stress–energy tensor is given by the expression

$$T_{ab} = \phi_{,a}\,\phi_{,b} - \tfrac{1}{2} g_{ab}(g^{cd}\phi_{,c}\,\phi_{,d} + m^2\phi^2). \tag{4}$$

In *quantum* case, the field is described by an 'operator-valued distribution' $\boldsymbol{\phi}$, which is an analogue of ϕ and solves the Klein–Gordon equation (3) too. The operators satisfy certain commutation relations, see below, and to specify the state of the field one must build a representation for these relations. The space of this representation (the Fock space) can be built, for example, as follows. Find the set $\mathcal{U} = \{u_k\}$ of functions u_k (they are called *modes*) which adhere to the conditions of 'orthonormality'

$$y(u_l, u_m) = \delta_{lm}, \quad y(u_l, u_m^*) = 0, \qquad \text{where} \quad y(f, g) \leftrightharpoons i \int_{\mathcal{S}} (g^* \nabla^a f - f \nabla^a g^*)\, \mathrm{d}\mathcal{S}_a$$

($\mathcal{S}$ is a Cauchy surface) and 'completeness'. The latter means that every smooth complex-valued solution of the equation of motion is a sum of two functions, of which one is in $H_{\mathcal{U}}$ and the other—in $H^*_{\mathcal{U}}$. Here, $H_{\mathcal{U}}$ denotes the Hilbert space with basis $\mathcal{U} \leftrightharpoons \{u_n\}$ and internal product y (the choice of y depends on the dynamics, the expression given above is specific to the Klein–Gordon case). The basis $\mathcal{U}$ is chosen arbitrary and every choice defines the expansion of $\boldsymbol{\phi}$ in terms of annihilation and creation operators:

$$\boldsymbol{\phi} = \sum_k [\boldsymbol{a}_k u_k + \boldsymbol{a}^\dagger_k u^*_k]$$

(recall that $\boldsymbol{\phi}$ is a distribution, so convergence is understood in the distributional sense). The operators obey the commutation relations

$$[\boldsymbol{a}_k, \boldsymbol{a}_{k'}] = [\boldsymbol{a}^\dagger_k, \boldsymbol{a}^\dagger_{k'}] = 0, \quad [\boldsymbol{a}_k, \boldsymbol{a}^\dagger_{k'}] = \delta_{kk'}, \tag{5}$$

which are either postulated, or are obtained from relations imposed on $\boldsymbol{\phi}$ (stemming, as a rule, from the correspondence 'Poisson brackets $\mapsto$ commutators'). The vacuum $|\mathcal{U}\rangle$ is defined now by the equality

$$\boldsymbol{a}_k|\mathcal{U}\rangle = 0, \qquad \forall k,$$

and the Fock space $\mathbb{F}_{\mathcal{U}}$ is the result of acting on $|\mathcal{U}\rangle$ by all possible combinations of creation operators $\boldsymbol{a}^\dagger_k$.

Remark 3 We consider vacuums corresponding to different choices of the basis as different (that is why vacuums are labelled by letters referring to the relevant bases: a vacuum $|\mathcal{U}\rangle$ corresponds to the basis $\mathcal{U}$). In fact, however, bases generating the same Hilbert spaces give rise to equivalent theories.

It might seem natural to define the stress–energy tensor as the result of substituting $\phi \to \boldsymbol{\phi}$ into the classical expression. However, $\boldsymbol{\phi}$ in contrast to ϕ are distributions, so the quadratic combinations like $\boldsymbol{\phi}^2$ are not defined. A possible way out is defining the stress–energy tensor as a *limit* of some[2] well-defined quantity. For example, in the classical case

$$T_{ab}(p) = \lim_{p' \to p} \mathfrak{D}_{ab}\big(\phi(p)\phi(p')\big), \tag{6}$$

where $\mathfrak{D}_{ab}$ is a certain differential operator. It is easy to check, in particular, that for the field (4)

$$\mathfrak{D}_{ab} \leftrightharpoons \nabla_{x^a}\nabla_{x'^b} - \tfrac{1}{2} g_{ab}(\nabla_{x^c}\nabla^{x'^c} + m^2) \tag{7}$$

(x^a and x'^a are understood to be the coordinates of points p and p', respectively). So, one could expect that $\langle T_{ab}\rangle_{\mathcal{U}}$, where $|\mathcal{U}\rangle$ is a vacuum, is obtained by replacing $\phi(p)\phi(p')$ in (6) by $\langle \boldsymbol{\phi}(p)\boldsymbol{\phi}(p')\rangle_{\mathcal{U}}$ or, equivalently, by $\frac{1}{2}G^{(1)}(\mathcal{U}; p, p')$. Here $G^{(1)}$ is the *Hadamard function*

[2]There are different regularizations.

$$G^{(1)}(\mathcal{U};\, p,\, p') \leftrightharpoons \big\langle |\phi(p)\phi(p') + \phi(p')\phi(p)| \big\rangle_{\mathcal{U}},$$

which is easily found[3] from the commutation relations (5),

$$G^{(1)}(\mathcal{U};\, p,\, p') = \sum_n u_n(p)u_n^*(p') + \text{complex conjugate}, \tag{8}$$

However, $\mathfrak{D}_{ab}G^{(1)}$ diverges at $p' \to p$ and in reality $\langle T_{ab} \rangle_{\mathcal{U}}$ is defined as the result of subtracting from $\frac{1}{2}\mathfrak{D}_{ab}G^{(1)}$ a certain T_{ab}^{div}, which also diverges in this limit:

$$\langle \mathcal{U}|T_{ab}|\mathcal{U}\rangle^{\text{ren}}(p) \leftrightharpoons \lim_{p' \to p} [\tfrac{1}{2}\mathfrak{D}_{ab}G^{(1)}(\mathcal{U};\, p,\, p') - T_{ab}^{\text{div}}(p,\, p')]. \tag{9}$$

The art of finding and justifying a suitable expression for T_{ab}^{div} is one of the central issues of semi-classical gravity. We shall not go into it, but two remarks are to be made:

1. T_{ab}^{div} does not depend on the state of the field, so there is no need to know it in calculating differences like $\langle T_{ab} \rangle_{\mathcal{U}} - \langle T_{ab} \rangle_{\mathcal{V}}$.

Remark 4 The just mentioned independence combined with Eq. (9) implies that $\langle T_{ab} \rangle_{\mathcal{U}}$ depends on the choice of $\mathcal{U}$ (or, to be more accurate, of the Hilbert space $H_{\mathcal{U}}$, cf. Remark 3), because so does the Hadamard function. So, it should be emphasized that generally there are infinitely many different bases and, correspondingly, different vacuums. *None of them is preferable or 'natural'*. Thus, the vacuum expectation of the stress–energy tensor is not defined until it is specified exactly which vacuum is meant. In the Minkowski case, it is customary to speak about *the* vacuum, or the 'standard' vacuum referring to the vacuum $|0\rangle$ associated with the modes $\sim e^{ik^a x_a - i\omega t}$, where $\omega = \sqrt{k^a k_a + m^2}$. By obvious reasons, it is agreed that $\langle T_{ab} \rangle_0 = 0$ (though actually it is a requirement on T_{ab}^{div}).

2. T_{ab}^{div} does depend on the geometry of spacetime and the dependence is not quite known. Fortunately, there are a few cases in which the ambiguity under discussion is overcome owing to the high symmetry of the relevant spacetimes or to 'accidental' cancellations (cf. Remark 6).

2 Auxiliary Spacetimes

Direct calculations of $\langle \mathcal{U}|T_{ab}|\mathcal{U}\rangle^{\text{ren}}$ are, as a rule, very difficult. A trick to sometimes facilitate the problem is the use of auxiliary spacetimes. One, first, finds relevant quantities in spacetimes where the calculations are easy and then relates the result to that in the 'physical' spacetime. In this section, two such examples are discussed.

[3]Note that $G^{(1)}(\mathcal{U})$ may not be a regular function. It often suffices to define it as a 'bi-distribution', that is a functional that acts on $\mathcal{D}(M) \times \mathcal{D}(M)$ [where $\mathcal{D}(M)$ is the space of test functions on M] and is linear and continuous separately in each of the two arguments.

2.1 Conformal Coupling

Consider a pair of spacetimes, $(M, \mathring{g})$ and (M, g), where $g = \Omega^2\mathring{g}$ and Ω is a smooth positive function on M. Generally, no simple relation exists between matter fields in these two universes. There is an important exception, however. This is the case of the *conformally invariant* fields and, in particular,[4] of the scalar field governed by the equation

$$\left(\Box - \tfrac{n-2}{4(n-1)}R\right)\phi = 0, \tag{10}$$

where R is the scalar curvature and n is the dimension of spacetime. It turns out (see Sect. 3.1 in [14]) that $\phi(x)$ satisfies Eq. (10) in (M, g) if and only if $\mathring{\phi}(x) \rightleftharpoons \Omega^{n/2-1}\phi(x)$ satisfies it in $(M, \mathring{g})$.

Definition 5 For a given functional $S[\phi, g]$, denote by S_Ω the functional defined by the relation

$$S_\Omega\big[\mathring{\phi}, \mathring{g}, \Omega\big] = S\big[\phi = \Omega^{1-n/2}\mathring{\phi}, g = \Omega^2\mathring{g}\big].$$

The field $\phi(x)$ with action $S[\phi, g]$ is called *conformally invariant*, if $\delta S_\Omega/\delta\Omega = 0$.

The fact that S_Ω does not depend on Ω implies, among other things, that the stress–energy tensor is traceless:

$$0 = \frac{\delta S_\Omega}{\delta(\ln\Omega)} = 2g^{ac}\frac{\delta S}{\delta g^{ac}} + (1 - n/2)\phi\frac{\delta S}{\delta\phi} = 2g^{ac}\frac{\delta S}{\delta g^{ac}} = \sqrt{-g}\,T^a{}_a$$

(the last but one equality in this chain is the equation of motion of the field ϕ and the last one is the definition of the metric stress–energy tensor [111, Sect. 94]). At the same time, direct calculations show that, generally, $\langle T^a{}_a\rangle_{\mathcal{U}} \neq 0$. This mysterious at first glance phenomenon is called *conformal anomaly*. The anomalous trace $\langle T^a{}_a\rangle_{\mathcal{U}}$ depends on transformation properties of the field and on the geometry of the spacetime, but not on the state $|\,\mathcal{U}\rangle$.

For a vacuum $|\,\mathring{\mathcal{U}}\rangle$, defined in $(M, \mathring{g})$ by the modes $\{\mathring{u}_k\}$, there is a corresponding quantum state $|\,\mathcal{U}\rangle$ in the universe (M, g), defined by the set $\{u_k \rightleftharpoons \Omega^{1-n/2}\mathring{u}_k\}$. The states $|\,\mathring{\mathcal{U}}\rangle$ and $|\,\mathcal{U}\rangle$ are said to be *conformally related*. Since the quantities $\langle T_{ab}\rangle_{\mathcal{U}}$ and $\langle T_{ab}\rangle_{\mathring{\mathcal{U}}}$ are defined on the same manifold, they can be pointwise compared. It turns out [144] that in the *four-dimensional* case

$$\begin{aligned}\langle T_{\hat{a}\hat{c}}\rangle_{\mathcal{U}} = \Omega^{-4}\langle T_{\hat{a}\hat{c}}\rangle_{\mathring{\mathcal{U}}} - c_1\Omega^{-4}\big[(\mathring{C}^j{}_{\hat{a}i\hat{c}}\ln\Omega)_{;j}{}^{i} + \tfrac{1}{2}\mathring{R}^{ji}\mathring{C}_{j\hat{a}i\hat{c}}\ln\Omega\big]\\ + c_2\big[(2R^{ji}C_{j\hat{a}i\hat{c}} - H_{\hat{a}\hat{c}}) - \Omega^{-4}(2\mathring{R}^{ji}\mathring{C}_{j\hat{a}i\hat{c}} - \mathring{H}_{\hat{a}\hat{c}})\big] + c_3\big[I_{\hat{a}\hat{c}} - \Omega^{-4}\mathring{I}_{\hat{a}\hat{c}}\big],\end{aligned} \tag{11}$$

where

$$H_{\hat{a}\hat{c}} \rightleftharpoons -R_{\hat{a}}{}^{i}R_{i\hat{c}} + \tfrac{2}{3}RR_{\hat{a}\hat{c}} + \left(\tfrac{1}{2}R_i{}^{d}R_d{}^{i} - \tfrac{1}{4}R^2\right)g_{\hat{a}\hat{c}},$$

[4] Another example is the electromagnetic field (though, only at $n = 4$).

$$I_{\hat{a}\hat{c}} \leftrightharpoons 2R_{;\hat{a}\hat{c}} - 2RR_{\hat{a}\hat{c}} + (\tfrac{1}{2}R^2 - 2\Box R)g_{\hat{a}\hat{c}},$$

the covariant derivatives are taken with respect to the metric $\mathring{g}$, and c_k are constants, characterizing the field. The accent $^\circ$ over a tensor shows that it refers to $\mathring{g}$, and hats over a few indices mean that the relevant components are found in a (fixed once and forever) orthonormal basis [the elements of the bases, in $(M, \mathring{g})$ and in (M, g), are connected: the former are obtained from the latter by multiplying them by Ω]. In the special case, when

$$G_{ab} = 0 \quad \text{and} \quad \Omega = const,$$

we have

$$R_{;ac} = \mathring{R}_{;ac} = 0, \qquad R = \mathring{R} = 0, \qquad \mathring{C}^j{}_{aic\,;j} = \mathring{R}_{a[c\,;i]} - \tfrac{1}{6}\mathring{g}_{a[c}\mathring{R}_{;i]} = 0,$$

and relation (11) takes the remarkably simple form

$$\langle T_{ac}\rangle_{\mathcal{U}} = \Omega^{-2}\langle T_{ac}\rangle_{\mathring{\mathcal{U}}} \tag{12}$$

(the power is 2 and not 4 because in this expression we use the components without hats).

Remark 6 Equation (12) may seem trivial. Indeed, a conformal transformation with constant Ω is equivalent to transition to a new unit of length (and hence of energy), and one might think that (12) can be obtained merely for dimensional reasons. This, however, is not the case, as seen from (11). Generally, and, in particular, when $G_{ab} \neq 0$, the equality (12) does not take place. The origin of such a counter-intuitive property of $\langle T_{ac}\rangle_{\mathcal{U}}$ is the same as that of the conformal anomaly: fixing the ambiguity in $\langle T_{ac}\rangle_{\mathcal{U}}$ one has to introduce a constant of dimension of mass, which makes the resulting (quantum) theory non-scale invariant. Fortunately, that constant is multiplied by a factor that vanishes in the Einstein spaces, cf. the end of Sect. 1.

The connection between $\langle T_{\hat{a}\hat{c}}\rangle_{\mathcal{U}}$ and $\langle T_{\hat{a}\hat{c}}\rangle_{\mathring{\mathcal{U}}}$ has an important consequence in the two-dimensional case too. The point is that in this case, all spacetimes are conformally flat. So, we always can choose $(M, \mathring{g})$ to be flat. Correspondingly, in suitable coordinates, its metric will have the form (locally, at least)

$$\mathring{g}: \qquad ds^2 = -d\tau^2 + d\psi^2. \tag{13}$$

It follows that for the conformally invariant field, see Eq. (6.136) in [14],[5]

$$\langle T_{ab}\rangle_{\mathcal{U}} = \langle T_{ab}\rangle_{\mathring{\mathcal{U}}} - \vartheta_{ab} + \tfrac{1}{48\pi}Rg_{ab}, \tag{14}$$

[5]The last term there has a different sign from ours because so does the metric signature.

where ϑ_{ab} is a tensor with the following components in the coordinate basis

$$\vartheta_{\tau\psi} = \vartheta_{\psi\tau} = \tfrac{1}{12\pi}\Omega\,\partial_\psi\partial_\tau\Omega^{-1}, \qquad \vartheta_{\tau\tau} = \vartheta_{\psi\psi} = \tfrac{1}{24\pi}\Omega\,(\partial^2_\psi + \partial^2_\tau)\Omega^{-1}.$$

In the early 90s, the idea was popular that to prove the impossibility of a particular time machine, one has only to prove that the energy density is unbounded in the frame of a fiducial observer located at the Cauchy horizon. The (apparently, fatal) drawback of this idea is that the quantity $\langle T_{ab}\rangle_{\mathcal{U}}$ characterizes the *pair*: the background spacetime *and* the state $|\mathcal{U}\rangle$. In spacetimes with time machines, there are states with *both* bounded and unbounded energy densities (examples will be presented in Chap. 10). But exactly the same is true even for Minkowski space, and thus cannot serve as evidence that the corresponding spacetime is 'flawed'. The relation (14) enables us to illustrate this assertion with a very simple example.

Example 7 Pick a smooth function f and on the plane with coordinates ψ, τ define new coordinates $\alpha \leftrightharpoons \psi + \tau$, $\beta \leftrightharpoons f(\psi - \tau)$. Then, the flat metric $\eta_{(\psi,\tau)}$ on this plane will take the form

$$\eta_{(\psi,\tau)} = \Omega^2(\beta)\eta_{(\alpha,\beta)}, \quad \text{where} \quad \Omega \leftrightharpoons (f')^{-1/2} \text{ and } \quad \eta_{(\alpha,\beta)}\colon \quad \mathrm{d}s^2 = \mathrm{d}\alpha\mathrm{d}\beta.$$

For the state $\mathcal{U}$, conformally related to the standard vacuum in the plane (α, β), the first and third terms in (14) vanish, while the second does not. We see that even in the Minkowski space the vacuum expectation of the stress–energy tensor may be non-zero (in a suitable vacuum state). Moreover, f can be chosen so that Ω *diverges* at $\psi - \tau = c < \infty$. So, obviously, does $\langle T_{ab}\rangle_{\mathcal{U}}$. This, apparently, points to the fact that for such f, $|\mathcal{U}\rangle$ as a quantum state in Minkowski space[6] is pathological, but by no means this compromises the space itself.

2.2 The Method of Images

Let M be a non-simply connected spacetime and represent it as the quotient space $M = \widetilde{M}/\overset{\Gamma}{\sim}$, where $\widetilde{M}$ is the universal covering of M, and Γ is the relevant group of isometries $\widetilde{M} \to \widetilde{M}$ (cf. Sect. 2 in Chap. 4). As $\widetilde{M}$ is simply connected, it may happen that it is easier to establish a property of a quantum field $\tilde{\phi}$ in $\widetilde{M}$ than of the same (i.e. obeying the same equation of motion) field ϕ in M. Our task in this section is to discuss what relation (if any) exists between the two fields.

Let $|\widetilde{\mathcal{U}}\rangle$ be a vacuum state of the field $\tilde{\phi}$. Suppose, that the Hadamard function $G^{(1)}(\widetilde{\mathcal{U}}; q, q')$ corresponding to this state respects the symmetries of $\widetilde{M}$, i.e. that

$$G^{(1)}(\gamma q, q') = G^{(1)}(q, \gamma^{-1}q'), \qquad \forall\gamma \in \Gamma, \quad q, q' \in \widetilde{M}. \tag{15}$$

[6]Another—better known, but also more complex—example is the Boulware vacuum in Schwarzschild space.

Then—assuming that the relevant series converges—the natural projection π mapping $\widetilde{M}$ to M generates the function $G^{\Sigma}(p, p')$:

$$G^{\Sigma}(p, p') \leftrightharpoons \sum_{\gamma\in\Gamma} G^{(1)}(\widetilde{\mathcal{U}}; q, \gamma q'), \qquad p \leftrightharpoons \pi(q), \quad p' \leftrightharpoons \pi(q'). \tag{16}$$

Like a Hadamard function, G^{Σ} is symmetric and in both its variables solves the—linear, since we discuss only free fields—equation of motion of $\boldsymbol{\phi}$. Denote by $T^{?}_{ab}$ the result of replacing $G^{(1)}(\mathcal{U}; p, p')$ in (9) by G^{Σ}. The analogy with electrostatics (which is responsible for the name of the method) suggests that $T^{?}_{ab}$ is just the vacuum expectation of the stress–energy tensor in M. The 'method of images' is the approach to the calculation of quantum effects based on the belief that the mentioned analogy can be converted into specific rigorous results and that, in particular, G^{Σ} and $T^{?}_{ab}$ are, respectively, the Hadamard function and the stress–energy tensor of a certain vacuum. This belief goes back to [35]. To time machines, the method was applied for the first time in [78], where the spacetimes $\widetilde{M}$ and M were taken to be, respectively, Minkowski and Misner spaces. The (scalar) field $\tilde{\boldsymbol{\phi}}$ was massless and its state $|\widetilde{\mathcal{U}}\rangle$ was the standard Minkowski vacuum $|0\rangle$ (see Remark 4). It turns out that in this case $T^{?}_{ab}$ diverges at the Cauchy horizon, from which the conclusion was drawn [78] that in the full-fledged quantum gravity the horizon will prove to be a singularity (and thus will protect causality). Later, approximately the same results were obtained for many more spacetimes. There are a few questions, however, putting these results in doubt.

Questions 8 (1) Does $|\mathcal{U}\rangle$, such that $T^{?}_{ab} = \langle T_{ab}\rangle_{\mathcal{U}}$, exist for *all* $|\widetilde{\mathcal{U}}\rangle$?
(2) If the answer is no, then what is $T^{?}_{ab}$ and what does its unboundedness prove in the cases when the relevant $|\mathcal{U}\rangle$ does *not* exist?

It seems that the answer to the first question is no, indeed. In defining G^{Σ} it is essential that (15) be valid. The full inverse image $\pi^{-1}(p)$ consists of the infinite number of points and it is (15) that guarantees the independence of the sum of the series in (16) on which of them is taken as q. (To avoid this problem one could have considered, instead of G^{Σ}, a more symmetric object $\sum_{\gamma,\gamma'\in\Gamma} G^{(1)}(\widetilde{\mathcal{U}}; \gamma q, \gamma' q')$, but, when $|\Gamma| = \infty$, such a series will hardly converge.) However, (15) need not always hold even in the simplest case of a flat two-dimensional space with $|\Gamma| < \infty$.

Example 9 Let $\widetilde{M}$ be a cylinder (we drop the requirement of simple connectedness of $\widetilde{M}$, in order to make $|\Gamma|$ finite) with coordinates τ, ψ, metric (13), and identification rule $(\tau, \psi) = (\tau, \psi + \psi_0)$. Let, further, $|\widetilde{\mathcal{F}}\rangle$ be a vacuum (cf. the state $|\mathring{\mathcal{U}}\rangle$ built on p. 204), defined by a basis $\{u_k\}$ with

$$u_2 = \tfrac{1}{\sqrt{8\pi}} e^{-2i\beta}, \qquad \text{where } \beta \leftrightharpoons \tfrac{2\pi}{\psi_0}(\tau - \psi).$$

Another vacuum, $|\widetilde{\mathcal{V}}\rangle$, is associated with a different set of modes $\{v_k\}$. Namely, these are the modes defined by the following relations:

$$v_k \leftrightharpoons u_k, \quad k \neq 2, \qquad v_2 \leftrightharpoons \sqrt{2}u_2 + u_2^*.$$

Then, by Eq. (8),

$$\Delta G(q, q') \leftrightharpoons G^{(1)}(\widetilde{\mathcal{V}}; q, q') - G^{(1)}(\widetilde{\mathcal{F}}; q, q') =$$
$$= \frac{1}{4\pi}\Big(\cos 2(\beta - \beta') + \sqrt{2}\cos 2(\beta + \beta')\Big).$$

Now, choosing Γ to be the group generated by the translation

$$\gamma\colon \quad \psi \mapsto \psi + \psi_0/8$$

(so that M is merely a cylinder eight times thinner than $\widetilde{M}$), we immediately see that either $G^{(1)}(\widetilde{\mathcal{V}})$, or $G^{(1)}(\widetilde{\mathcal{F}})$ violates (15), since

$$\Delta G(\gamma q, q') - \Delta G(q, \gamma^{-1} q') = \frac{\sqrt{2}}{2\pi}\cos 2\Big(\beta + \beta' + \tfrac{1}{4}\pi\Big) \neq 0.$$

Questions 10 Suppose the answer to question 8(1) is positive. Then, which *exactly* vacuum $|\widetilde{\mathcal{U}}\rangle$ must be used in calculating $T^{?}_{ab}$ in order to judge from its boundedness the regularity of the Cauchy horizon in the hypothetical quantum gravity.

Indeed, the physical system under study is the field $\boldsymbol{\phi}$ in the spacetime M, while $\widetilde{M}$, $|\widetilde{\mathcal{U}}\rangle$, and $\tilde{\boldsymbol{\phi}}$ play the role of fictitious auxiliary entities. They have no specific physical meaning, so it is totally unclear how to choose them. For example, the vacuum $|0\rangle$ used in [78] stands out, because it possesses all the symmetries of Minkowski space, but the 'physical' spacetime M (which is the Misner space) does *not* have some of them. In particular, $|0\rangle$ has some properties that may be desirable for a vacuum in a static spacetime, but the Misner space is *not* static.

The answer to question 10 is of crucial importance: as we have seen, the divergence of the energy density in *some* vacuums says *nothing* of whether the spacetime under consideration is feasible: there are states with such a divergence even in the Minkowski space, see Example 7. Of course, the problem would be solved, were it shown that in a given spacetime the divergence is present *irrespective* of the choice of the vacuum. Such a strategy was adopted in [55]. Assuming (implicitly), that for every vacuum $|\mathcal{U}\rangle$ in the physical space M, there is a vacuum $|\widetilde{\mathcal{U}}\rangle$ in the universal covering $\widetilde{M}$, such that the stress–energy tensor for $|\mathcal{U}\rangle$ is just $T^{?}_{ab}(\widetilde{\mathcal{U}})$

$$\forall \mathcal{U}\, \exists \widetilde{\mathcal{U}}\colon \quad \langle T_{ab}\rangle_{\mathcal{U}} = T^{?}_{ab}(\widetilde{\mathcal{U}})$$

(this is *not* the positive answer to question 8(1): the quantors here stand in a 'wrong' order), Frolov made an attempt to show that $T^{?}_{ab}$ diverges at the Cauchy horizon of any time machine for any choice of $\widetilde{\mathcal{U}}$ and, hence, of $\mathcal{U}$. The fallacy of the relevant derivation will be demonstrated in Chap. 10 with an explicit example.[7]

[7]For a detailed critical analysis of the approach developed in [55], see [90].

A lot of results are obtained by now employing the method of images. However, to my knowledge, none of the problems pointed out above have been solved (though some consideration was given to the case of the *finite* $|\Gamma|$ [7]). So, we shall neither discuss, nor review those results.

Remark 11 Two theories have been considered above: one of them describes the field ϕ (or its quantum analogue $\boldsymbol{\phi}$) in the spacetime M, the other deals with the field $\tilde{\phi}$ in the universal covering $\widetilde{M}$. The former theory is, obviously, obtained from the latter, by defining ϕ via the natural projection:

$$\phi(p) = \tilde{\phi}[\pi^{-1}(p)],$$

and imposing an additional condition on $\tilde{\phi}$

$$\pi(q) = \pi(q') \quad \Rightarrow \quad \tilde{\phi}(q) = \tilde{\phi}(q') \qquad \forall q, q' \in \widetilde{M}. \tag{$*$}$$

This condition is necessary for the self-consistency of the theory: if ϕ is required to be single-valued, it must not change after travelling around a closed contour. One could adopt, however, a different point of view and allow ϕ to be only defined *locally*. Then the former theory has a curious generalization. It is obtained by replacing the condition ($*$) with

$$\pi(q) = \pi(q') \quad \Rightarrow \quad |\tilde{\phi}(q)| = |\tilde{\phi}(q')| \qquad \forall q, q' \in \widetilde{M}.$$

Now the phase of ϕ defined by ($*$) may change after travelling around a closed curve. The fields described by such functions are called *automorphic*. Locally, they resemble the 'usual' fields, but the theory contains some non-trivial *global* effects, see [14]. These effects contribute to the vacuum polarization in the vicinity of the Cauchy horizon in non-simply connected time machines [164].

Chapter 8
WEC-Related Quantum Restrictions

So far, we have been focused on kinematical possibility of time machines or shortcuts, and the Einstein equations played no role in our consideration. As a next step, it is natural to find out whether among such spaces there are solutions of the Einstein equations with realistic matter sources. In general, any *necessary* property of matter filling these spacetimes would be of interest. The only such property found so far is that the geometries under consideration often[1] violate the weak energy condition.

In the literature, different candidates were proposed for exotic matter, but in this book I stand on the conservative position that they all are *too* exotic and the necessary WEC violations are provided (if at all) only by the term $\langle T_{ab}\rangle_Q$ in (2) in Chap. 7. Our task in this chapter is to explore the ensuing consequences and to check whether they prohibit exotic spaces as the lack of exotic matter prohibits such spaces the classical case.

There are strong grounds for expecting that they do. It was shown in some assumptions (below we analyse them thoroughly) that to support an Alcubierre bubble 100 m in diameter, one needs -10^{67} g $\approx -\frac{1}{2}10^{34}M_\odot$ of exotic matter [148]. A similar result was obtained [47] for the Krasnikov tube and, as we argue below, can be similarly obtained for the traversable wormhole as well [98]. Such a figure looks absolutely discouraging and for all practical purposes can be viewed as a *prohibition* of shortcuts.

Our point, however, is that these estimates must not be taken too seriously, because:

(*a*) the assumptions mentioned above are *quite* disputable, see below;
(*b*) the spacetimes under discussion were brought forward as *illustrations* to the concept of superluminal travel. Accordingly, they were intended to be as simple as possible. So, it does not seem impossible that it is their simplicity that is responsible for the undesirable properties of the matter sources.

Nevertheless, the arguments leading to such impressive estimates merit detailed consideration and demolition.

[1] What 'often' means is a separate question; as is stated above, the relevant rigorous assertions are yet to be formulated.

S. Krasnikov, *Back-in-Time and Faster-than-Light Travel in General Relativity*, Fundamental Theories of Physics 193, https://doi.org/10.1007/978-3-319-72754-7_8

1 Quantum Restrictions on Shortcuts

1.1 The Quantum Inequality

Suppose, a freely falling observer in a spacetime M measures the renormalized vacuum expectation value of the energy density $\varrho \leftrightharpoons \langle T_{\hat{0}\hat{0}}\rangle_{\mathcal{V}}$ of a quantum field. Pick a non-negative function f normalized by the condition

$$\int_{-\infty}^{\infty} f(\tau)\,\mathrm{d}\tau = 1$$

and define the 'weighted average' of the energy density as

$$\varrho_f(\mathcal{V};\tau_0) \leftrightharpoons \int_{-\infty}^{\infty} \varrho(\mathcal{V};\tau) f(\tau-\tau_0)\,\mathrm{d}\tau. \tag{1}$$

Here, τ is the proper time of the observer and the integral is taken along their world line. To formulate the 'quantum inequality', consider a timelike geodesic segment $\gamma\colon\ (\tau_1,\tau_2) \to M$ and denote

$$\mathcal{T} \leftrightharpoons \left(\max |R_{\hat{a}\hat{b}\hat{c}\hat{d}}|\right)^{-1/2},$$

where the maximum is taken over all $\tau \in (\tau_1,\tau_2)$ and all sets of indices (the hats over the indices mean that the components are found in the proper reference system of the observer γ, i.e. in an orthonormal basis with one of the basis vectors tangent to γ). Set γ to be so short that

$$|\tau_2-\tau_1| \lesssim \mathcal{T}, \tag{2a}$$

$$J^+(\gamma)\cap J^-(\gamma) \text{ is a ball (topologically)}. \tag{2b}$$

A central role in this chapter is played by the following assertion.

Quantum inequality 1 There exists $f \in \mathscr{D}([\tau_1,\tau_2])$ such that in any M for any γ satisfying (2) and any reasonable $\mathcal{V}$

$$-\varrho_f(\mathcal{V}) \lesssim c|\tau_2-\tau_1|^{-4}, \tag{3}$$

where c is a constant of the order of unity.

A clear distinction should be made between this quantum inequality and any of the other *similar* statements, which can often be met in the literature under—sometimes—the same name (see [157] or [149] for a review). Some of these namesakes are correct and proven. Still, we shall not consider them, the only exception being a brief discussion in Remark 2. The reason is that even when the difference in formulation is minor, it makes the corresponding statement either useless in deriving restrictions on the shortcuts, or even wrong, see below. This is partly true even for the statement 1 itself: as of today it is proven only when M is the Minkowski space,[2] but in application to the shortcuts this case is obviously unusable. At the same time, no easy way is seen to strengthen this result. It is known, for example to be false [50], if either of the conditions (2) is dropped, or (in the two-dimensional case [99], at least) if $f\big|_{(\tau_1,\tau_2)} = const.$

1.2 Connection with Shortcuts

Once the validity of the quantum inequality is assumed it is used for estimating the value of the WEC violations. The reasoning goes, roughly, as follows. Consider a point p through which there is a timelike geodesic segment $\gamma(\tau)$ satisfying (2). Suppose, that

$$\max|R_{\hat{a}\hat{b}\hat{c}\hat{d}}(p)| \approx \max|T_{\hat{e}\hat{f}}(p)| \approx -\varrho(p) \tag{4}$$

(p satisfying all these requirements happens to exist in all three above-mentioned shortcuts). Then, it follows from (3), when the possibility of $|\tau_2 - \tau_1| \ll \mathcal{T}$ is neglected, that

$$|\varrho_f| \lesssim c\mathcal{T}^{-4} = c\big(\max|R_{\hat{a}\hat{b}\hat{c}\hat{d}}|\big)^2 \approx c\varrho^2(p), \tag{5}$$

or

$$|\varrho(p)| \gtrsim c^{-1}\varrho_f(p)/\varrho(p) \approx 1. \tag{6}$$

Thus, the energy density in p must be Planckian!

All the prohibitive estimates mentioned above stem merely from (6) (recall that $1 \approx 5 \times 10^{93}$ g/cm^3). Indeed, pick a spacelike surface $\mathcal{N}$ in the region where the weak energy condition fails and suppose that all points of $\mathcal{N}$ satisfy all the requirements imposed on p above. Then one can define the 'total amount of negative energy'

$$E^-_{\text{tot}} \leftrightharpoons \int_{\mathcal{N}} |\varrho|\,\mathrm{d}^3x \gtrsim V^3(\mathcal{N}), \tag{7}$$

where $V^3(\mathcal{N})$ is the volume of $\mathcal{N}$. In both Alcubierre and Krasnikov spaces, $\mathcal{N}$ is chosen to be a spherical layer of diameter $\rlap{-}p$ and thickness δ surrounding the domain $\mathcal{V}$ of the 'false' flat metric. For a spherically symmetric wormhole, $\mathcal{N}$ is essentially the throat of the wormhole, that is also a spherical layer of diameter $\rlap{-}p$ and thickness δ. The volume of $\mathcal{N}$ can be estimated as $V^3(\mathcal{N}) \gtrsim A_{\mathrm{i}}\delta$, where A_{i} is the area of its inner surface [this is a quite rough bound from below: the area of the outer surface may be much greater than A_{i}, even though δ is small]. To enable a human to travel by

---

[2] If $n = 4$. In the conformally trivial [14] *two-dimensional* case, it is proven for curved spaces too [48].

the shortcut, $\not{b}$, apparently, must be at least $\sim$ 1 m, which means that $A_{\text{i}} \approx \not{b}^2 \gtrsim 10^{70}$. So, even if the thickness of the layer is $\delta \sim 1$ (recall that in our units $1 = l_{\text{Pl}}$), one concludes that it would take at least

$$E^-_{\text{tot}} \sim |\varrho \not{b}^2 \delta| \approx 10^{32} M_\odot \tag{8}$$

of exotic matter to support a practical shortcut.

Remark 2 As mentioned above, there are a number of assertions resembling the quantum inequality that—in contrast to the latter are *proven*, but that do not imply estimates like (8). Replace, for example, $\varrho(\mathcal{V}; \tau)$ in the definition (1) by $\varrho(\mathcal{V}; \tau) - \varrho(\mathcal{O}; \tau)$, where $|\mathcal{O}\rangle$ is some 'reference' state. The inequalities (3) with the thus obtained $\varrho_f(\mathcal{V})$—they are proven in [51]—do not yield the estimate in question, because there is no reason now to believe that $\varrho_f(p) \approx \varrho(p)$, so, the right-hand side of (6) may well be small. Likewise, any inequalities with an unknown (and thus potentially big) c, or those valid in the *limit* $\tau_2 \to \tau_1$ are useless for us here, because the sign $\leqslant$ in (6) has no justification in this limit.

1.3 The Meaning of the Restrictions

...physicists are comfortable with little huge numbers, but not with big ones.

B. S. DeWitt [36]

The quantity E^-_{tot} is usually understood to be a quantifier of feasibility: spacetimes with such astronomical E^-_{tot} as in (8) are considered obviously impossible. It is not improbable, however, that this interpretation is only due to the hypnotic effect of large numbers. Indeed, the physical meaning of E^-_{tot} is rather obscure. The surface $\mathcal{N}$ and the congruence of geodesics γ can be chosen in infinitely many ways, and no choice is preferred. At the same time, the value of E^-_{tot} clearly depends on that choice. Moreover, as the following example shows this dependence is so strong that even with a fixed $\mathcal{N}$ simply by choosing a suitable γ one can make E^-_{tot} *arbitrary*.

Let U be the region $0 < x_a < c_i$, $a = 1, 2, 3$ of the Minkowski space. Here, x_a are the Cartesian coordinates with x_0 being the time, and c_i are some constants (thus U is the world tube of a freely falling parallelepiped, it is such a parallelepiped that we are going to choose as $\mathcal{N}$, but it takes some preparation, because the foliation of U by parallelepipeds may be done differently). Consider two families of geodesic observers: some moving in the x_1- and the others—in x_2-directions. The speeds of all observers are the same (in absolute value) and, correspondingly, their four velocities are

$$\boldsymbol{v}_1 = \tfrac{1}{\sqrt{1-v^2}}(1, v, 0, 0) \quad \text{and} \quad \boldsymbol{v}_2 = \tfrac{1}{\sqrt{1-v^2}}(1, 0, v, 0).$$

We define $\mathcal{N}_i$, $i = 1, 2$ to be spacelike sections of U normal to $\boldsymbol{v}_i$.

Now, suppose that the stress–energy tensor in U has the form

$$\langle T_{ab}\rangle = \mathrm{diag}(-1, -3, 1, 1) \tag{9}$$

and, thus, violates the WEC. It is easy to check that E^{-}_{tot} are *different* for $i = 1$ and $i = 2$. Indeed, though the volumes are the same $V^3(\mathcal{N}_i) = c_1c_2c_3\sqrt{1-v^2}$ in these two cases, the energy densities ϱ_i are not:

$$\varrho_1 = \langle T_{ab}v_1^a v_1^b\rangle = -\frac{1+3v^2}{1-v^2}, \qquad \varrho_2 = \langle T_{ab}v_2^a v_2^b\rangle = -1.$$

As a result,

$$E^{-}_{\mathrm{tot}\,1} = c_\Pi \frac{1+3v^2}{\sqrt{1-v^2}}, \qquad E^{-}_{\mathrm{tot}\,2} = c_\Pi\sqrt{1-v^2}, \qquad \text{where } c_\Pi \leftrightharpoons c_1c_2c_3,$$

and we see that merely by the choice of a suitable observer, E^{-}_{tot} can be made arbitrarily large/small. Does this make the matter distribution under consideration unphysical?

Remark 3 The stress–energy tensor (9) coincides, up to a factor, with that generated by the Casimir effect (see [14, Eq. (4.39)]), i.e. with that which would be observed (up to edge effects), if the parallelepiped $\mathcal{N}$ were bounded in the x_1-direction by superconducting plates. Also, (9) is the vacuum stress–energy tensor of the scalar field in the spacetime obtained from $\mathbb{L}^4$ by identifying $x_1 = x_1 + c_1$.

2 Counterexamples

Whatever is the meaning of E^{-}_{tot}, it seems likely that its *moderate* value for some 'naturally chosen' observers is a merit of a shortcut. Let us check, therefore, that the (presumed) validity of the quantum inequality does not generally imply the estimate (8).

2.1 The Weyl Tensor

In deriving (8), we assumed that in the relevant region the components of the Riemann and the Einstein tensors are roughly of the same order, see (4). So, one does not expect (8) to hold, even approximately, unless

$$\max|C_{\hat{a}\hat{b}\hat{c}\hat{d}}|/\max|R_{\hat{a}\hat{b}}| \lesssim 1$$

for the relevant observers, where

$$C_{abcd} \leftrightharpoons R_{abcd} + g_{a[d}R_{c]b} + g_{b[c}R_{d]a}$$

is the so-called Weyl tensor. But this condition fails more often than not. For example, in *any* curved (and, hence, having a non-zero Riemann tensor), but empty (and, hence, having a zero Ricci tensor) region—in particular, in the vicinity of each star—for *any* observer

$$\max |C_{\hat{a}\hat{b}\hat{c}\hat{d}}| / \max |R_{\hat{a}\hat{b}}| = \infty.$$

For this reason alone, the whole argumentation resulting in the estimate (8) is wrong in the general case. And, indeed, in Chap. 9, we produce an example of a traversable wormhole which is prevented from collapse by the vacuum polarization just in the Schwarzschild space.

2.2 The Non-trivial Topology

The inequality (3) becomes increasingly more restrictive as $|\tau_2 - \tau_1|$ grows, so it is important that the length of γ is bounded by the conditions (2). The main reason for imposing (2)—even at the cost of the power of the quantum inequality—is that in the Minkowski space the inequality (3) does hold, while any region is believed to be 'almost a portion of the Minkowski space', if it is small enough. Of course, this idea may be employed in different ways and the condition (2b) may be reformulated one day, but it cannot be dropped completely: there are static spacetimes in which $\varrho = const < 0$, see Remarks 3 and 1 in Chap. 10 for example. In such spacetimes, there are γ-s that are sufficiently long to violate (3). In the absence of restrictions like (2b), these γ-s would violate the quantum inequality too.

At the same time, (2b) prevents the quantum inequality from yielding a restriction like (6) and thus (8) for the most promising (as shortcuts) wormholes. Indeed, consider a timelike geodesic in the region between the mouths of a static wormhole: such a geodesic is depicted by a vertical segment γ in Fig. 3b in Chap. 3. The segment may be arbitrarily short, but if c_τ is sufficiently close to d—and as discussed in Remark 12 in Chap. 3, it is the wormholes with $c_\tau \approx d$ that are best for interstellar travel—the end points of γ are connected by a causal curve λ. The latter passes through the throat of the wormhole and is, therefore, non-homotopic to γ, see Fig. 3b in Chap. 3. The existence of such a λ guarantees that $J^+(\gamma) \cap J^-(\gamma)$ is *not* a ball and the condition (2b) does not hold. Thus, choosing a suitable c_τ one can make the maximal length allowed by condition (2b) arbitrarily small. This implies a loophole in the derivation of (6), which leans heavily on the assumption that $|\tau_2 - \tau_1| \approx \mathcal{T}$, see the provision between (4) and (5).

An additional merit of wormholes with $c_\tau \sim d$ is that—as is reasonable to assume by analogy with the case $\mathbb{L}^3 \times \mathbb{S}^1$, see Eq. (9) and Remark 3—the vacuum polarization generates large [of the order of $(c_\tau^2 - d^2)^{-2}$] negative energy density, which relieves one of having to seek additional sources of exotic matter. So, the better is the wormhole as shortcut, the less severe is the Quantum inequality.

2.3 'Economical' Shortcuts

Now let us check that Planck-scale densities by themselves do not imply estimates like (8). Even if all the assumptions of Sect. 1 are valid, a macroscopic wormhole can nevertheless be maintained by just $E^{-}_{\text{tot}} \approx 10^{-2}M_{\odot}$. Moreover, contrary to naive expectations [cf. the reasoning above formula (8)] a macroscopic body may, in principle, be transported through a *micro*scopic wormhole (or another shortcut), which makes it possible to further reduce the required E^{-}_{tot} to a modest value of $\approx 10^{-4}$ g.

'Portal'

In any wormhole, there is a region where a converging congruence of null geodesics becomes diverging. Correspondingly, the violation of the weak energy condition is inevitable there, see the end of Sect. 2.1 in Chap. 3 for discussion. However, the *magnitude* of this violation depends dramatically on the geometry of the wormhole. Thus, for example, in the wormhole W_{8}, the violation takes place in the throat. Here, the sphere Σ_t is identified with $\Sigma'_{t'}$ 'turned inside out', cf. Fig. 4 in Chap. 3. It seems reasonable to assume that the least exotic wormholes must be those where Σ in the process of turning inside out changes its shape as little as possible.

With this guess in mind, we now turn to a special type of wormhole called *portals*. These are static spacetimes with the Cauchy surfaces like those depicted in Figs. 1 and 2. In the (2+1)-dimensional case, it is a surface obtained as the space W in Example 3 in Chap. 3 with the only difference that now $\mathcal{B}$ and $\mathcal{B}'$ are ovals rather than circles, that is the long sides of the holes depicted in Fig. 1 are straight. Cauchy

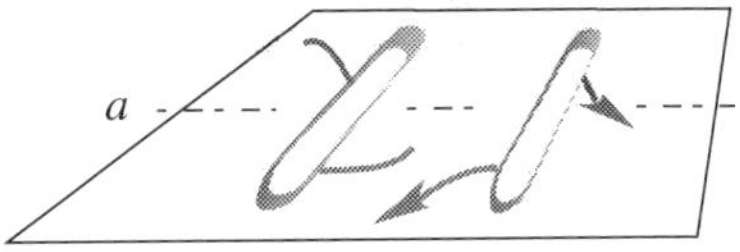

Fig. 1 A surface of simultaneity of a (2+1)-dimensional portal

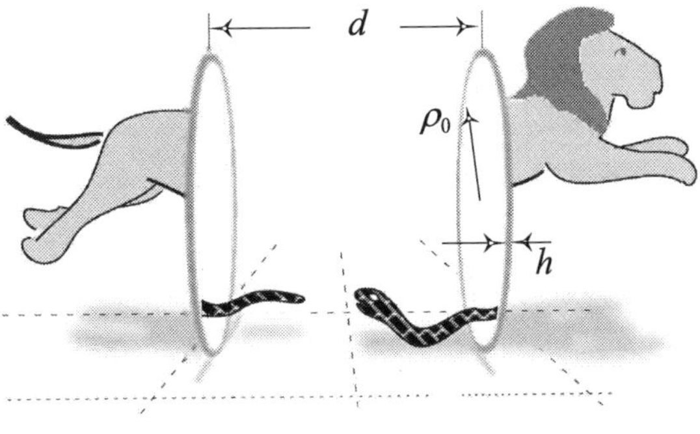

Fig. 2 The 'distance' between the hoops (defined, for example, as the length of the snake) is d. The radius of each of them is ρ_0 and the thickness is h

surfaces in the (3+1)-dimensional case are built from the two-dimensional ones by rotation with respect to the a-axis or, equivalently, by replacing the *balls* $\mathcal{B}$ and $\mathcal{B}'$ from the description of $W_{\aleph}$, see Sect. 2.1 and Fig. 4 in Chap. 3, by *cylinders* (the bases of the cylinders are flat, but their side walls are not, this is where the spacetime is curved).

The portal may be considered as an approximation to the limit case, which is the spacetime H—a *dihedral wormhole* in the terminology of [172]—built as follows. From the Euclidean space $\mathbb{E}^{(n-1)}$ remove two equal $(n-2)$-dimensional disks perpendicular to the line connecting their centres. Then glue the left bank of either slit to the right bank of the other one (in perfect analogy to how the Deutsch–Politzer space was built). The resulting space $\mathcal{A}$ is just a spacelike section of $H = \mathbb{L}^1 \times \mathcal{A}$. Though H resembles an ordinary wormhole, it differs from the latter in two respects:

a. H is everywhere flat. Thus, it does not need exotic matter *at all*: $E^-_{\text{tot}} = 0$;
b. H is non-globally hyperbolic: it has a closed stringlike singularity.[3] So, by definition, H is not a shortcut.

Conversely, H may be modified so as to obtain from it a proper shortcut (the explicit expression for the relevant metric is presented in the Appendix). To this end cut out some narrow neighbourhood of the singularity in $\mathcal{A}$ and replace it by a solid torus (similarly to how a cone singularity is regularized sometimes by cutting out the sharp tip and replacing it with a smooth blunted ‘cap’). The torus of course is not flat, but its thickness h is small and E^-_{tot}, though non-zero, turns out to be moderate. It can be roughly estimated from the fact that when h tends to zero, $\varrho = \frac{1}{8\pi} G_{00}$ grows as h^{-2} (since the Einstein tensor includes the second derivatives of metric). At the same time, the volume of a solid torus with fixed length falls as h^2. Hence, the total amount of energy concentrated in the hoop remains approximately constant in this limit and, consequently, can be found by setting $h \approx \varrho \approx 1$. Thus, to support a human-sized wormhole of this type it would suffice $E^-_{\text{tot}} \approx 1\,\text{m} \approx 10^{-3} \text{M}_\odot$ of exotic matter. This tiny [in relation to (8)] quantity is comparable with the energy of a supernova.

Van Den Broeck’s Trick

In fact, E^-_{tot} can be reduced further yet by tens of orders of magnitude. The idea [170] is to use a capsule which would be, on the one hand, large enough to accommodate a human and, on the other hand, small enough to be transported through a microscopic [i.e. with $\not{p} \approx 1$, see (8)] shortcut. Strange it might seem, both requirements can be satisfied simultaneously and this would not require large amounts of exotic matter.

Consider the spacetime

$$\mathrm{d}s^2 = -\mathrm{d}t^2 + \mathrm{d}l^2 + r(l)^2(\mathrm{d}\vartheta^2 + \sin^2\vartheta\,\mathrm{d}\varphi^2), \qquad l \geqslant 0. \tag{10}$$

[3] Due to the exceptional simplicity of this case one can assign a particular shape to the singularity [101].

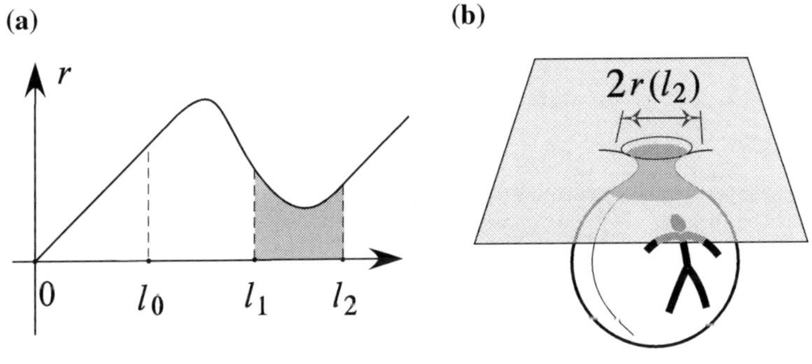

Fig. 3 **a** Variation of the capsule's radius with the distance from the centre. **b** By using just $\approx M_{\rm Pl}$ of exotic matter, a Planck-size (to an external observer) capsule can be made arbitrarily roomy

(It is meant that the underlying manifold is merely $\mathbb{R}^4$, rather than a wormhole: at $l = 0$ points that have the same t are identified even when their ϑ and φ differ.) The Einstein tensor for the metric (10) can be easily found, see e. g. [135, Eq. (14.52)]:

$$G_{\hat{t}\hat{t}} = \frac{1 - r'^2 - 2rr''}{r^2}, \quad G_{\hat{t}\hat{t}} + G_{\hat{r}\hat{r}} = -\frac{2r''}{r}, \quad G_{\hat{t}\hat{t}} + G_{\hat{\nu}\hat{\nu}} = \frac{1 - r'^2 - rr''}{r^2},$$

where $\nu \leftrightharpoons \varphi, \vartheta$.

Pick three positive numbers $l_0 < l_1 < l_2$ and let $r(l)$ be a smooth non-negative function which is concave at and only at $l \in (l_1, l_2)$ and which has a—unique—root at $l = 0$. Impose the additional condition

$$|r'| \leqslant 1 \quad \text{at } l \in (l_0, l_1), \qquad r' = 1 \quad \text{at } l \notin (l_0, l_2),$$

see Fig. 3b. Then, it is easy to check that the WEC is violated in and only in the spherical layer $\mathcal{N}$: $l \in (l_1, l_2)$. Obviously, the 'external' radius $r(l_2)$ of that layer and its volume $V(\mathcal{N})$ may be of Planck size, even when the 'internal' radius l_0 is macroscopic. In the simplest case, it takes only $E^-_{\rm tot} \approx M_{\rm Pl} \approx 10^{-7}$ kg of exotic matter to support the spacetime in discussion [98] (though, of course, the practicability of such a spacetime is out of question). Moreover, if we relax for a moment the requirement that the spacetime would be flat at large l, then a similar 'pocket' can be built without exotic matter *at all* [82].

Remark We have used the Van Den Broeck pocket not as a shortcut, but only as a tool for reducing the amount of negative energy required for a regular shortcut. However, a similar trick was applied directly to the Krasnikov tube and enabled the relevant $E^-_{\rm tot}$ to be reduced to $\approx 10^{30}$ g [69].

Chapter 9
Primordial Wormhole

1 Introduction

In discussing the subject matter of this book, one of central points is the existence of traversable wormholes. It turns out that approaches to this problem differ depending on whether we consider the creation of a wormhole by a hypothetical advanced civilization as a possibility. If we do, then it only remains to find a method for stabilizing such a wormhole, i.e. for making it traversable. There are a number of—speculative—proposals of how this can be done. Thus, for example, one could use well-aimed 'streams of pure phantom radiation' [77] to this end.

We can, however, leave advanced civilizations aside[1] and specify the question as follows: consider a 'primordial' (i.e. that which appeared at the same time with the rest of the universe and for the same reason) wormhole. Would it be traversable for some time, at least? But even this question, as it turns out, is too general. There is a plethora of more or less innovative theories in which the WEC does not hold. If any of them is valid, the wormhole will have a chance to exist for sufficiently long time to be traversed. For example, static wormholes are, apparently, possible in theories with classical 'ghosts' [44] or scalar fields [8], in a 'brane world' [19], etc. In my view, however, this fact does not make traversable wormholes any more plausible.

In this chapter, we adopt the approach of 'maximal banality' (cf. 'the boring physics conjecture' [172]) and, in the spirit of [161], formulate our question as follows: will a primordial wormhole be traversable within ordinary semi-classical gravity involving nothing exotic in the least? Clearly, the answer must depend on the initial shape of the wormhole, the state of the matter filling it, etc. Today it is hard to tell what initial state is more and what is less realistic. So we choose it to be as simple as possible in the hope that the simpler a wormhole is the more probably it

[1] In a globally hyperbolic spacetime no new wormhole can appear, see Proposition 51 in Chap. 1, while the evolution of *non*-globally hyperbolic spacetimes is a rather obscure matter, see Sect. 3 in Chap. 2 for discussion.

S. Krasnikov, *Back-in-Time and Faster-than-Light Travel in General Relativity*, Fundamental Theories of Physics 193, https://doi.org/10.1007/978-3-319-72754-7_9

resembles a real one. Specifically, we assume that the spacetime $M_{\rm wh}$ containing the wormhole in question is:

(1) spherically symmetric,[2]
(2) empty (in the classical sense, i.e. $T^{\rm C}_{ac} = 0$), and
(3) similar to that which would evolve from the same initial state in the purely classical case (i.e. to the Schwarzschild space in this case).
Classically, such a wormhole would be non-traversable, see Sect. 2.1, and our task is only to find out how this property is affected by quantum effects.

Remark 1 Our choice of simplifying assumptions is, of course, disputable. Indeed, there are at least two works in which the solution was required to be *static*, while conditions (3) or (2, 3) were, on the contrary, weakened. Neither of these attempts, however, was successful. The length of the wormhole built in [79] is $\approx l_{\rm Pl}/2$, the radius of its throat is $\approx 67 l_{\rm Pl}$. And the second work [95] contains a serious mathematical error.

To describe the *appearance* of the wormhole, we introduce—as initial condition—a surface $\mathcal{E}$ dividing $M_{\rm wh}$ and assume that the part of $M_{\rm wh}$ to the future of $\mathcal{E}$ solves the semi-classical Einstein equations $G_{ab} = 8\pi \langle T_{ab} \rangle_{\rm Q}$. The remaining part of $M_{\rm wh}$ is *terra incognita* and we do not consider it. At $\mathcal{E}$ proper the geometry is taken to be *exactly* the same as at the corresponding surface in the Schwarzschild space. By requiring $\mathcal{E}$ to obey a few seemingly natural conditions, see page 187, we reduce the multitude of wormholes to a family parameterized by the initial mass m_0, the times (measured in a special coordinate system) at which the mouths appeared, and, finally, some quantity $\mathfrak{h}$. The last-named parameter shows how close to collapse the new-born wormhole would be if it were classical. As we shall see, for our model to be self-consistent a wormhole must appear already *almost* torn apart by the singularity. Thus, it is clear in advance that the 'traversability time' $\mathscr{T}^{\rm trav}$ of the wormholes under consideration is very small (if non-zero). It should be stressed that this is not a non-trivial property of primordial wormholes, but rather a criterion of eligibility for being considered within our model. It turns out, however,—and this is one of the main results of this book—that there are wormholes whose $\mathscr{T}^{\rm trav}$, though small, are *macroscopic*. Thus, the intra-universe version of $M_{\rm wh}$, we discuss it in Sect. 4, may well endanger causality.

2 The Model and Assumptions

2.1 *Schwarzschild Spacetime*

The question of traversability of a spherically symmetric empty wormhole reduces in the classical case to studying the Schwarzschild space, merely because by the

[2] As a next step it would be natural to consider a rotating wormhole, see [85].

Birkhoff theorem it is the only (maximal, globally hyperbolic) spherically symmetric solution of the Einstein equations $G_{ab} = 0$. Since our approach is based on the idea that $M_{\rm wh}$ is a perturbation of the classical wormhole, we begin with recapitulating some basic facts concerning the geometry of the latter (for a detailed geometrical consideration see [103, 135], for a discussion of quantum fields in Schwarzschild space see [14, 20, 56]). These facts are then used in formulating the assumptions concerning the geometry of $M_{\rm wh}$.

The Geometry

The Kruskal or the (maximally extended) Schwarzschild space $M_{\rm Sc}$ is the spacetime with topology $\mathbb{R}^2 \times \mathbb{S}^2$ and metric

$$\begin{gathered} \mathrm{d}s^2 = -\mathring{F}^2 \mathrm{d}u \mathrm{d}v + \mathring{r}^2(\mathrm{d}\vartheta^2 + \cos^2\vartheta\, \mathrm{d}\varphi) \\ u, v \in \mathbb{R}, \quad \mathring{r} > 0, \end{gathered} \tag{1}$$

where φ and ϑ coordinatize the sphere and

$$\mathring{F}^2 \leftrightharpoons 16 m_0^2 x^{-1} e^{-x}, \qquad \mathring{r} \leftrightharpoons 2m_0 x. \tag{2}$$

m_0 is a positive[3] parameter called the *mass* and $x(u, v)$ is (implicitly) defined by the equation

$$uv = (1 - x)e^x. \tag{3}$$

$M_{\rm Sc}$ includes two globally hyperbolic regions with radius $\mathring{r}$ (see Remark 8 in Chap. 3) varying from $2m_0$ to infinity in either. These are II $\leftrightharpoons \{u > 0, v < 0\}$ and IV $\leftrightharpoons \{u < 0, v > 0\}$, see Fig. 1a.

In region IV, consider the coordinate transformation from u, v to $\mathring{r}$ and $t_S \leftrightharpoons 2m_0 \ln(-v/u)$. In the new coordinates, the metric takes the customary form[4]

$$\begin{gathered} \mathrm{d}s^2 = -(1 - \tfrac{2m_0}{\mathring{r}})\mathrm{d}t_S^2 + (1 - \tfrac{2m_0}{\mathring{r}})^{-1}\mathrm{d}\mathring{r}^2 + \mathring{r}^2(\mathrm{d}\vartheta^2 + \cos^2\vartheta\, \mathrm{d}\varphi) \\ t_S \in \mathbb{R}, \quad \mathring{r} > 2m_0, \end{gathered}$$

and we see, in particular, that region IV is asymptotically flat. It is static and each of its spacelike sections $t_S = const$ is a (slightly deformed) Euclidean space $\mathbb{E}^3$, from which a ball of radius $2m_0$ is removed. A maximal extension of the region under discussion is obtained by replacing this ball with a singular spacetime or with a collapsar, or, as in the Schwarzschild case, with a throat connecting IV to the region II. So, Schwarzschild space is a wormhole, indeed—two static asymptotically flat universes are connected by an evolving throat (which forms the union of I $\leftrightharpoons \{u > 0, v > 0\}$

[3] Solutions corresponding to $m_0 \leqslant 0$ exist, but they have a completely different structure and will not be considered here.

[4] A detailed discussion of how Schwarzschild space looks in different coordinates can be found, for example, in [135].

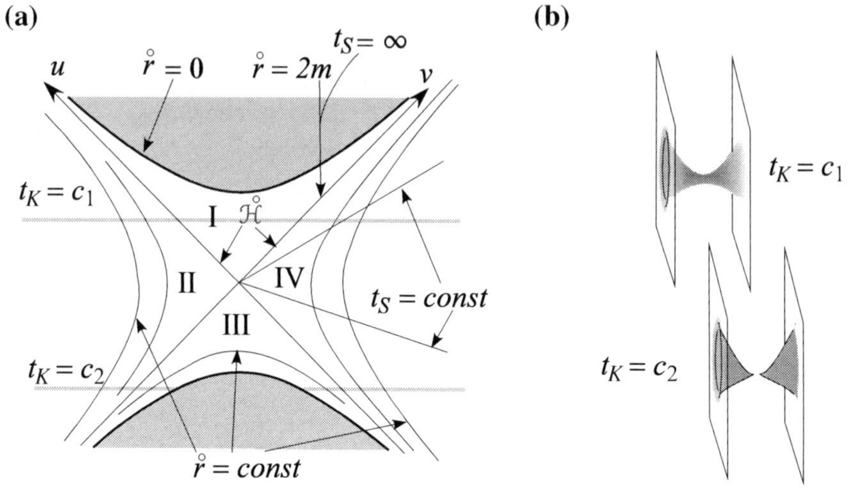

Fig. 1 **a** The section $\varphi = const$, $\vartheta = const$ of Kruscal spacetime. A causal curve in each of its points lies inside the angle with sides parallel to the u- and v-axis. So, the quadrants IV and II are causally non-connected. **b** The sections $t_K = const$, $\varphi = const$ of the same space. c_1 is in $(-1, 1)$. As is easily seen, the depicted spacetime is actually a (non-static) wormhole

and III $\leftrightharpoons \{u < 0, v < 0\}$). In this capacity, Schwarzschild space is often called the *Einstein–Rosen bridge*.

Remark 2 M_{Sc} is non-static. Though there *is* a group of isometries acting on it

$$p \mapsto \breve{p}, \qquad \text{where } u(\breve{p}) = Cu(p),\ v(\breve{p}) = C^{-1}v(p) \qquad \forall C > 0, \tag{4}$$

in regions I and III the corresponding Killing vectors are spacelike.

To visualize the geometry of M_{Sc} consider its sections $t_K \leftrightharpoons \frac{1}{2}(u + v) = const$. At first, when t_K is small, each of them is a pair of cylinders $\mathbb{R}^1 \times \mathbb{S}^2$ with radiuses of the 2D spheres taking all positive values, see the section $t_K = c_2$ in Fig. 1b. Then, at $t_K = -1$ [this value is easily found from (3)] the cylinders merge into a wormhole, see section $t_K = c_1$. The radius of its throat grows from 0 to $2m_0$ (the latter value is taken at $t_K = 0$) and then begins to decrease. The throat becomes more and more narrow, its radius tending to zero as $t_K \to 1$. At $t_K = 1$ it collapses and we again have a pair of disjoint singular surfaces. Thus, the Einstein–Rosen bridge is a 'transient' wormhole. Moreover, it is non-traversable, because it is destroyed *before* any signal has time to traverse it [58, 136]. This is easy to check by inspection of Fig. 1a: $\mathring{r}$ in region I decreases along any future-directed causal curve. So, a signal which escapes to spatial infinity must become superluminal somewhere. We denote the boundary of region I by $\mathring{\mathcal{H}}$ and call it the *horizon* (implying the event or apparent, but not, of course, Cauchy horizon).

As for specific relations, later we shall need a few equalities following immediately from definitions (2) and (3):

$$\mathring{r}_{,v} = -\frac{2m_0 u}{x e^x}\,, \tag{5a}$$

$$\mathring{r}_{,u} = -\frac{2m_0 v}{x e^x} = 2m_0\frac{x-1}{ux}\,, \tag{5b}$$

$$\mathring{r}_{,uv} = -\frac{2m_0}{x^3 e^x}\,, \tag{5c}$$

$$\mathring{f}_{,u} = -\tfrac{1}{2}(\ln x + x)_{,u} = -\frac{1+x}{2x}x_{,u}\,, \qquad \text{where} \quad \mathring{f} \leftrightharpoons \ln \mathring{F}. \tag{5d}$$

Vacuum Polarization

The matter in our simple model will be represented by the conformal scalar field, see Sect. 2.1 in Chap. 7, in a vacuum state. For the maximal Schwarzschild space, there are three different vacuums that are regarded 'natural'—Boulware's, Hartle-Hawking's, and Unruh's (see [14] and the literature cited there), but in the first state the expected stress–energy tensor diverges at the future horizon $u = 0,\ v > 0$ and the second contains radiation incoming from infinity. Either feature is undesirable in modelling a regular evaporating wormhole, so we are left with the Unruh vacuum. We shall denote it $|\mathring{\mathscr{Q}}\rangle$, but sometimes drop the brackets, for simplicity, and write $\mathring{T}_{ab}$ for $\langle\, T_{ab}\rangle_{\mathring{\mathscr{Q}}}$. The Unruh vacuum respects all the symmetries of the Kruskal space, which by itself puts serious restrictions on the structure of $\mathring{T}_{ab}$:

(1) Consider the two-dimensional space S_p tangent[5] in a point $p \in M_{\rm Sc}$ to the sphere $u = v = const.$ S_p is a subspace of T_p and this induces a Euclidean metric g^R in the former. The linear operator

$$\boldsymbol{A}\colon \qquad x^c \mapsto \mathring{T}^b_a x^a, \qquad \forall \boldsymbol{x} \in S_p.$$

is self-adjoint with respect to this metric:

$$g^R(\boldsymbol{y}, \boldsymbol{A}\boldsymbol{x}) = y_b \mathring{T}^b_a x^a = y^b \mathring{T}_{ba} x^a = g^R(\boldsymbol{A}\boldsymbol{y}, \boldsymbol{x})$$

and hence its eigenvectors make up an orthogonal basis, i.e. there is a pair of orthogonal unit vectors $\boldsymbol{s}_1, \boldsymbol{s}_2 \in S_p$ such that

$$\mathring{T}^b_a x^a y_b = \lambda_1 s_{1a} s_1{}^b x^a y_b + \lambda_2 s_{2a} s_2{}^b x^a y_b, \qquad \forall \boldsymbol{x}, \boldsymbol{y} \in S_p.$$

[5] Since the metric is given, we shall not pedantically distinguish co- and contravariant vectors.

The spherical symmetry forces $\lambda_1 = \lambda_2 = \lambda$ and therefore

$$\mathring{T}_a^b x^a y_b = \lambda \delta^b{}_a x^a y_b\,, \quad \text{whence} \quad \mathring{T}_{ab} x^a y^b = \lambda g_{ab} x^a y^b \qquad \forall \boldsymbol{x}, \boldsymbol{y} \in S_p.$$

Thus,

$$\mathring{T}_{\vartheta\varphi} = 0 \quad \mathring{T}_{\varphi\varphi} = \cos^2\vartheta\, \mathring{T}_{\vartheta\vartheta}.$$

(2) Assume, that in a point p

$$0 \neq \mathring{T}_{v\varphi} = (\partial_v)^a \mathring{T}_{ab} (\partial_\varphi)^b,$$

and, correspondingly, the projection of the vector $\mathring{T}_{ab}(p)(\partial_v)^b$ on S_p is non-zero. This defines a preferential direction in S_p, which contradicts the spherical symmetry of the problem. Applying the same reasoning to the cases $u \to v$ and/or $\varphi \to \vartheta$, we conclude that

$$\mathring{T}_{w\vartheta} = \mathring{T}_{w\varphi} = 0, \qquad w = u, v.$$

(3) Any two points $p, \breve{p}$ with the same $x \neq 1$ and the same sign of v are related by a combination of a rotation and an isometry (4). Hence,

$$\mathring{T}_{uu}(\breve{p}) = C^{-2}\mathring{T}_{uu}(p), \quad \mathring{T}_{uv}(\breve{p}) = \mathring{T}_{uv}(p), \quad \mathring{T}_{vv}(\breve{p}) = C^2\mathring{T}_{vv}(p),$$

where $C = x_{,u}(p)/x_{,u}(\breve{p})$ and, consequently,

$$\mathring{T}_{uu}(p) = x^2{}_{,u}(p) c_{uu}, \quad \mathring{T}_{vv}(p) = x^{-2}{}_{,u}(p) c_{vv},$$
$$\text{and} \quad \mathring{T}_{ab}(p) = c_{ab} \quad \text{for the rest pairs } ab.$$

c_{ab} in the last expression stand for some quantities that do not vary with coordinates as long as $\mathring{r}$ is fixed.

(4) Finally, note that the metrics (1) with different masses are conformally related and, hence, by (7.12)

$$\mathring{T}_{w_1 w_2}(m_0, p) = (\breve{m}/m_0)^2 \mathring{T}_{w_1 w_2}(\breve{m}, \breve{p}).$$

The right and left-hand sides of this equality refer to points with equal coordinates[6] (the coordinate systems are understood to be those in which one metric is proportional to the other). Thus, $x(\breve{p}) = x(p)$, but, for example, $\mathring{r}(\breve{p}) = x(\breve{p})/(2\breve{m}) \neq x(p)/(2m_0) = \mathring{r}(p)$.

Thus, summarizing the results of items (1)–(4), we can state that in the coordinates v, u, ϑ, φ (exactly in this order)

[6]They are *different* points, nevertheless, because they lie in different spacetimes.

$$\mathring{T}_{ab} = \frac{m_0^{-2}}{4\pi}\begin{pmatrix} (\mathring{r},_u/m_0)^{-2}\tau_1(x) & \tau_3(x) & & \\ \tau_3(x) & (\mathring{r},_u/m_0)^2\tau_2(x) & & \\ & & \tau_4(x) & \\ & & & \tau_4(x)\cos^2\vartheta \end{pmatrix},$$

where $\tau_i, i = 1, \dots 4$ are some functions of x (but not of u, v, or m_0 separately). These functions are connected via energy conservation and the value of the anomalous trace [24]. This, however, is not sufficient for finding all four of them and some calculations have to be done numerically, which, indeed, was done in [45]. The results that we shall use in this chapter are (see Sect. A.5 in Appendix) as follows:

a. The vacuum polarization is weak (now, that it is described by *scalar* functions τ_i this is a meaningful statement). In particular, it follows from (A.57), (A.52), (A.54), (A.58) that $|\tau_i(1)| \lesssim 10^{-3}$;
b. At the horizon, the component $\mathring{T}_{vv}$ is negative.

More specifically, we shall need the following estimates for the components $\mathring{T}_{ab}$ at the horizon:

$$\mathring{r},_u^2 \mathring{T}_{vv}\big|_{\mathcal{H}} = -\frac{\mathring{F}^4(1)K}{16m_0^4}, \qquad \text{where} \quad K \leftrightharpoons \frac{9}{40\cdot 8^4\pi^2} \approx 6\times 10^{-6}. \tag{6a}$$

$$\text{Accordingly,} \quad \tau_1(1) = -64\pi K e^{-2} \approx -10^{-3},$$

$$-\mathring{r},_u^{-2}\mathring{T}_{uu}\big|_{\mathcal{H}} = \frac{\tau_2(1)}{4\pi m_0^4} \approx 2\cdot 10^{-5} m_0^{-4} \ll m_0^{-4} \tag{6b}$$

and, as follows from the comparison with (2) and (A.54), at any macroscopic m_0

$$|\mathring{T}_{uv}|\big|_{\mathcal{H}} = \frac{\tau_3(1)}{4\pi m_0^2} \ll \frac{\mathring{F}^2(1)}{64\pi m_0^2} \tag{6c}$$

(we do not substitute here the numerical value of $\mathring{F}$ in order to simplify the transition from $\mathring{F}$ to F).

2.2 *The Geometry of the Evaporating Wormhole*

The wormhole M_{wh} that we are studying is spherically symmetric and, correspondingly, has the metric

$$ds^2 = -F^2(u,v)\,dudv + r^2(u,v)(d\vartheta^2 + \cos^2\vartheta\, d\varphi), \tag{7}$$

where F and r (note the absence of $^\circ$ over them) are the functions to be found. Our next task is to express mathematically the idea that the wormhole is 'initially

Schwarzschild'. To this end, we pick a hypersurface $\mathring{\mathcal{E}}$ dividing the Schwarzschild space (later we shall restrict the possible choice) and require that there exist a surface $\mathcal{E} \subset M_{\text{wh}}$ such that F, r and their first derivatives coincide on $\mathcal{E}$ with, respectively, $\mathring{F}$, $\mathring{r}$ and their derivatives (the values are compared in points of $\mathcal{E}$ and $\mathring{\mathcal{E}}$ with the same coordinates; we shall not repeat this trivial stipulation any more). In particular, relations (5) with accents $^\circ$dropped must hold on $\mathcal{E}$. We require $\mathcal{E}$ to be spherically symmetric—in the sense that with every point p it also contains all points q such that $u(q) = u(p)$, $v(q) = v(p)$—and to satisfy the following conditions:

(i) It is spacelike at $r < 2m_0$;
(ii) In quadrant IV, its section $\varphi = \vartheta = 0$ is a graph of a smooth positive function $v = V(u)$ which has no maximum. The same must also hold with the substitution IV $\to$ II, $V \to U$, $v \leftrightarrow u$;
(iii) Far from the wormhole (i.e. at $r \gg m_0$), $\mathcal{E}$ is required to be merely a surface of constant Schwarzschild time, that is to satisfy the equation $u/v = const$. Thus for some positive constants κ_L, κ_R and any point $p \in \mathcal{E}$ with $r(p) \gg m_0$

$$\begin{aligned} v(p) < u(p) \quad &\Rightarrow \quad v(p) = -\kappa_L u(p), \\ v(p) > u(p) \quad &\Rightarrow \quad u(p) = -\kappa_R v(p). \end{aligned}$$

Condition (i) restricts substantially the class of wormholes under examination, in contrast to (ii), which is of minor importance and can be easily weakened, if desired. The idea behind (iii) is that far from the wormhole mouths the Schwarzschild time becomes the 'usual' time and that the Planck era ended—by that usual time—simultaneously in different regions of the universe. Though, remarkably, (iii) does not affect the relevant *geometrical* properties of M_{wh}, it proves to be very useful in their interpretation. In particular, it enables us to assign in an intuitive manner the 'time' $\mathscr{T}$ to any event p' near the throat of the wormhole. Namely, p' happens at the moment when it is reached by the photon emitted in the end of Planck era from a point p (or p'') located on the left (respectively, right) asymptotically flat region, see Fig. 2. The distance—when it is large enough—from this point to the wormhole is approximately

$$r = 2m_0 x \approx 2m_0 \ln[-u(p')v(p')] = 2m_0 \ln[u^2(p')\kappa_L]$$

[x here is expressed in terms of uv via (3)], or $\approx 2m_0 \ln[v^2(p')\kappa_R]$. Taking this distance to be the measure of the time elapsed from the end of the Planck era, we define

$$\mathscr{T}_{\text{L}}(p') \leftrightharpoons 2m_0 \ln[u^2(p')\kappa_L], \qquad \mathscr{T}_{\text{R}}(p') \leftrightharpoons 2m_0 \ln[v^2(p')\kappa_R] \tag{8}$$

and *interpret* $\mathscr{T}_{\text{L(R)}}$ as time, even though $\nabla \mathscr{T}_{\text{L(R)}}$ is null (as is the case with the 'advanced' and 'retarded' times in M_{Sc}). Note that as long as we consider the two asymptotically flat regions of M_{wh} as totally independent (i.e. up to Sect. 4), there is no relation between κ_R and κ_L, nor there is a preferred value for either of them.

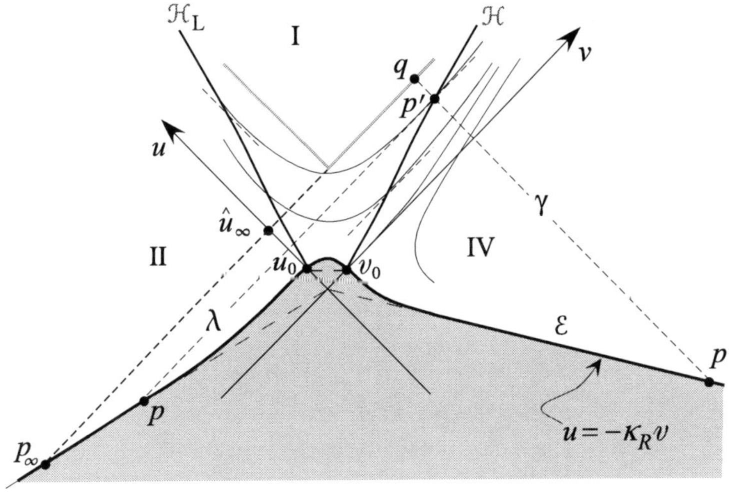

Fig. 2 The section $\varphi = \vartheta = 0$ of an evaporating wormhole. The thinnest solid curves are the surfaces $r = const$. The grey angle containing q is the horizon

Among other things, the choice of $\mathcal{E}$ [combined with the requirement that (1) holds on it] fixes the coordinates u and v up to a transformation

$$u \mapsto u' = Cu, \qquad v \mapsto v' = C^{-1}v.$$

To fix this remaining arbitrariness and thus to make formulas more compact, we require the u- and v- intercepts of $\mathcal{E}$—let us denote them u_0 and v_0—be positive and equal, see Fig. 2. The remaining quantity, v_0, is a free parameter of the model. And though no reasons are seen to think that wormholes with some particular values of v_0 are more common than with any other, we restrict our consideration to those with

$$1 + \sqrt{\mathfrak{c}} < \mathfrak{h} < \tfrac{\sqrt{5}+1}{2}, \qquad \text{where} \quad \mathfrak{h} \leftrightharpoons e\mathfrak{c}/v_0^2, \qquad \mathfrak{c} \leftrightharpoons 16\pi K m_0^{-2} \tag{9}$$

[K is defined by (6a)]. As we shall see, wormholes with smaller $\mathfrak{h}$ may be non-traversable, while those with larger $\mathfrak{h}$ evaporate too intensely and cannot be studied within our simple model. The lower bound of $\mathfrak{h}$ differs from 1 by an extremely small quantity

$$\mathfrak{c} \approx 3 \times 10^{-73} \left(\frac{2m_0}{1\,\mathrm{m}}\right)^{-2},$$

which is chosen non-zero for a purely technical reason, see the derivation of (45). To summarize, we have four independent parameters m_0, $\mathfrak{h}$, and $\kappa_{R(L)}$, all values of which are considered equally possible as long as $m_0 \gg 1$, $\mathfrak{h} \in (1 + \sqrt{\mathfrak{c}}, \frac{\sqrt{5}+1}{2})$ and $\kappa_{R(L)} > 0$.

The class of wormholes under consideration is restricted further by requiring that in the semi-classical region of the universe (i.e. above $\mathcal{E}$ in Fig. 2), the following inequalities would hold

$$r,_{uv} < 0, \tag{10a}$$

$$\nabla r \neq 0. \tag{10b}$$

This requirement means that the difference between $M_{\rm wh}$ and Schwarzschild space must not be so strong as to change the sign in (5c), or to make the throat transit from contraction to expansion.

Our subject will be the (right, for definiteness) horizon, by which I understand the curve $\mathcal{H}$ lying in the (u, v)-plane and defined by the condition

$$r,_v (u, v)\big|_{\mathcal{H}} = 0. \tag{11}$$

By (5a), $r,_v$ is negative in all points of $\mathcal{E}$ with positive u-coordinates and vanishes in the point $(0, v_0)$. It is from this point that the horizon starts. $\mathcal{H}$ cannot have an end point, being a level line of the function $r,_v$, which has a non-zero [by condition (10a)] gradient. Neither can it return to $\mathcal{E}$, because v can only *increase* along $\mathcal{H}$ [again by condition (10a), which would fail in a point of maximal v]. Thus, $\mathcal{H}$ goes from $\mathcal{E}$ to infinity dividing the region of the (u, v)-plane lying above $\mathcal{E}$ into two parts: $r,_v$ is strictly negative to the left of $\mathcal{H}$ and strictly positive to the right. So the horizon exists in $M_{\rm wh}$ and is unique. Loosely speaking, any small segment of $\mathcal{H}$ shows where the event horizon would pass if the evolution stopped at this moment and the metric remained (approximately, at least) Schwarzschild with mass

$$m(v) \leftrightharpoons \tfrac{1}{2} r(\mathcal{H}(v)) \tag{12}$$

(that $\mathcal{H}$ can be parameterized by v, as implied in this expression, follows from the—already mentioned—fact that v grows monotonically on the horizon. Alternatively, it can be parameterized by m). It is the behaviour of the horizon that defines whether the wormhole is traversable. Obviously, a photon can leave the region bounded by the horizon only if the horizon becomes timelike somewhere, see Sect. 2.4.

Notation. Given a function $y(u, v)$, we shall write $\hat{y}$ for its restriction to $\mathcal{H}$. In doing so, we view $\hat{y}$ as a function of v or m depending on which parameterization is chosen for $\mathcal{H}$ (this is a—slight—abuse of notation, because strictly speaking $\hat{y}(v)$ and $\hat{y}(m)$ are different functions, but it must not cause confusion). Partial derivatives are, of course, understood to act on y, not on $\hat{y}$. Thus, for example,

$$m \leftrightharpoons \tfrac{1}{2}\hat{r}, \qquad \frac{\partial}{\partial v}\hat{r},_u \leftrightharpoons r,_{uv} (\mathcal{H}(v)), \qquad \hat{f},_{uv} (m) \leftrightharpoons f,_{uv} (\mathcal{H}(m)) \qquad \text{etc.}$$

In conformity with this notation, the function $v \to u$ whose graph is $\mathcal{H}$ will be denoted by $\hat{u}(v)$, while $\hat{u}(m)$ is a shorthand notation for $\hat{u}(v(m))$. On the intervals where $\mathcal{H}$ is not parallel to the v-axis (i.e. on the entire $\mathcal{H}$ in our case, as we shall see), the function $\hat{v}(u)$ is defined likewise.

2.3 Weak Evaporation Assumption

The physical assumption lying in the heart of the whole analysis is the 'evaporation stability' of the wormhole under consideration: we assume that in the region above $\mathcal{E}$ there is a solution of the system (Einstein equations + field equations) having the following property: the geometry in a small neighbourhood of each point p is similar to that in a point $\mathring{p}$ of Schwarzschild space with some mass $\mathring{m}$ (of course $\mathring{p}$ and $\mathring{m}$ depend on p), while the stress–energy tensor in p is small (in the sense that will be specified in a moment) and close to $\mathring{T}_{ab}(\mathring{m}, x(\mathring{p}))$. This main assumption is realized in the form of a set of (in)equalities. Let us list them.

The requirement that T_{ab} on the horizon be close to $\mathring{T}_{ab}(m, 1)$ is embodied in the assumption that the relations (6) remain valid when the sign $^\circ$ is removed in the left-hand side, while on the right-hand side $\mathring{F}(1)$ and m_0 are replaced with, respectively, $\hat{F}$ and m:

$$\hat{r},_u^2\, \hat{T}_{vv} = -\tfrac{K}{16}\hat{F}^4 m^{-4}, \qquad K \approx 6\times 10^{-6}; \tag{13a}$$

$$\hat{r},_u^{-2}\, \hat{T}_{uu} = c m^{-4}, \qquad 0 > c \gg -1; \tag{13b}$$

$$|\hat{T}_{vu}| \ll \tfrac{1}{64\pi}\hat{F}^2 m^{-2}. \tag{13c}$$

Two more assumptions concern the 'long-distance' behaviour of the stress–energy tensor. Let γ be a segment of a null geodesic $v = const$ from some $p'' \in \mathcal{E}$ to $p' \in \mathcal{H}$, see Fig. 2. We shall assume that for all p'

$$\left|\int_\gamma 4\pi r,_u^{-2}\, T_{uu} r\, \mathrm{d}r\right| \ll 1. \tag{13d}$$

This assumption seems fairly reasonable. Indeed, as follows from (6b), in the Schwarzschild case the combination $4\pi\mathring{r},_u^{-2}\,\mathring{T}_{uu}$ is approximately $10^{-4}m_0^{-4}$ on the horizon and falls with x [at least, for $x > 2$, see [45], where that combination corresponds to the expression $\frac{4\pi x}{x-1}(\mu + p_r + 2s^{\mathrm{U}})$]. At large r it falls as $1/r^2$ [24], so the left-hand side is presumably of the order of $10^{-4}m_0^{-4}\ln(r(p'')/m_0)$. Since $r(p'')$ is less than the age of the universe, that quantity is small for any macroscopic m_0 (that we shall extend the assumption to small masses is simply a matter of convenience: instead of the Planck mass, we could introduce some macroscopic minimal mass—$m_{\min} = 10^3$, say—which would only result in exponentially small corrections).

Finally, we assume that

$$|T_{\vartheta\vartheta}| \ll \tfrac{1}{2\pi} r|r,_{vu}|F^{-2}. \tag{13e}$$

Again, in the Schwarzschild case the corresponding inequality—which is $\tau_4 \ll 2m_0^2/x$, see (2) and (5c)—holds both on the horizon and at large x, see (A.52) and (A.50). And, again, we actually do not need (13e) to be true *pointwise*. It would suffice that the relevant integral be small, see (30).

2.4 Preliminary Discussion

The traversability of the wormhole is determined by the fact that in the course of evaporation $\hat{u}(m)$ does not remain constant (as in the Schwarzschild case), but tends to $\hat{u}_\infty > \hat{u}_0$ as $m \to 1$ (what happens at smaller m is, of course, anybody's guess). Indeed, draw through $p' \in \mathcal{H}$ a future-directed null geodesic λ parallel to the v-axis. In our model $\hat{r}_{,vv}$ is strictly positive, see (20) below, and hence in the point where λ meets the horizon, $r\big|_\lambda(v)$ reaches its global minimum. So, λ intersects $\mathcal{H}$ once only. Summarizing, λ is the world line of the photon emanating from $p = \lambda \cap \mathcal{E}$, traversing, in $p' \in \mathcal{H}$, the throat of the wormhole and escaping to infinity.

As we move from p to the left, see Fig. 2, the same reasoning applies to all photons as long as their u-coordinates are small enough to ensure the intersection of $\mathcal{H}$ and λ. The boundary of the part of $\mathcal{E}$ from which the right infinity is reachable is generated by the points p_∞ with

$$u(p_\infty) = \hat{u}_\infty \leftrightharpoons \sup_{m\in[1,m_0]} \hat{u}(m)$$

(as we shall see the supremum is provided, in fact, by $m = 1$). Correspondingly, we define the *closure time* as the moment when the wormhole ceases to be traversable for a traveller wishing to trip from the left asymptotically flat region to the other one:

$$\mathcal{T}_\mathrm{L}^\mathrm{cl} \leftrightharpoons 2m_0 \ln \hat{u}_\infty^2 \kappa_L.$$

Similarly, the *opening time* is defined as $\mathcal{T}_\mathrm{L}^\mathrm{op} \leftrightharpoons 2m_0 \ln v_0^2 \kappa_L$. Photons with $u < v_0$, that is with $\mathcal{T}_\mathrm{L} < \mathcal{T}_\mathrm{L}^\mathrm{op}$ (such photons exist unless $\mathcal{E}$ is everywhere spacelike, which is uninteresting) also cannot traverse the wormhole: on their way out they get into the Planck region. For all practical purposes, this means that they 'vanish into the singularity'. Finally, we define the *traversability time* as

$$\mathcal{T}_\mathrm{L}^\mathrm{trav} \leftrightharpoons \mathcal{T}_\mathrm{L}^\mathrm{cl} - \mathcal{T}_\mathrm{L}^\mathrm{op} = 4m_0 \ln \frac{\hat{u}_\infty}{v_0}. \tag{14}$$

Obviously,

$$\text{A wormhole is traversable} \quad \Leftrightarrow \quad \mathcal{T}_\mathrm{L}^\mathrm{trav} > 0 \quad \Leftrightarrow \quad \hat{u}_\infty > v_0 \tag{$\star$}$$

and our goal in this chapter is to estimate $\hat{u}_\infty/v_0$. This must be done accurately enough, because there is a drastic difference between really traversable wormholes and those that are traversable only nominally and have $T^{\text{trav}} \approx 1$, say.

Remark 3 The behaviour of the apparent horizon under assumptions very similar to ours was studied back in the 1980s (see e.g. [20, 56] for reviews). The spacetime being considered, was not a wormhole, though, but a black hole originating from gravitational collapse (which means, in particular, that it is not empty even classically). Still, the problems are closely related. It should be noted therefore that according to a widespread opinion (see [73], though), the backreaction results only in the shift of the event horizon to a radius that is smaller than $2m$ by $\delta \sim m^{-2}$, which is physically negligible [9]. This inference, however, is logically flawed: the smallness of δ does not at all imply that the effect is weak. The situation here is similar to that with the 'total amount of negative energy', see Sect. 1 in Chap. 8—the huge value of a quantity (for instance, $M_\odot^2 \approx 10^{76}$) obscures the disputableness of its interpretation. Indeed, the event horizon is a *null* surface, but there is simply no such thing as the distance between a point and a null surface.[7] And even if it were properly defined, the distance between spheres with radiuses r_1 and r_2 would hardly be uniquely determined by the difference $\delta \leftrightharpoons r_1 - r_2$, which depends only on the *areas* of the spheres, see Remark 8 in Chap. 3.

3 The Evolution of the Horizon

The Einstein equations for metric (7) read

$$4\pi T_{vu} = (\tfrac{1}{4}F^2 + rr,_{vu} + r,_v r,_u)r^{-2}, \tag{15}$$

$$4\pi T_{vv} = (2r,_v f,_v - r,_{vv})r^{-1}, \qquad \text{where} \quad f \leftrightharpoons \ln F, \tag{16}$$

$$4\pi T_{uu} = (2r,_u f,_u - r,_{uu})r^{-1} = \tag{17}$$

$$= -\frac{F^2}{r}\left(\frac{r,_u}{F^2}\right),_u, \tag{17$'$}$$

$$4\pi T_{\vartheta\vartheta} = -\frac{2r^2}{F^2}(r,_{vu}/r + f,_{vu}). \tag{18}$$

Under the weak evaporation assumption, they can be simplified. In particular, on the horizon, the left-hand side of (15) can be neglected by (13c), while $r,_v$ vanishes there by definition. So, we have

---

[7] Consider, for example, the surface $t = x + \Delta$ in Minkowski plane. Is the distance from the origin of the coordinates to that surface large or small? Apparently, neither: Δ can be made arbitrary merely by a coordinate transformation $t' = t\operatorname{ch}\gamma + x\operatorname{sh}\gamma$, $x' = t\operatorname{sh}\gamma + x\operatorname{ch}\gamma$ with a suitable γ.

$$\hat{r},_{vu} = -\frac{\hat{F}^2}{8m}. \tag{19}$$

Equations (16) and (13a) yield

$$\hat{r},_{vv} = \frac{\pi K \hat{F}^4}{2m^3 \hat{r},_u^2}. \tag{20}$$

Likewise, (17) in the approximation (13b) gives

$$\hat{r},_{uu} = 2\hat{r},_u \hat{f},_u - 4\pi c\hat{r},_u^2 m^{-3}. \tag{21}$$

Finally, Eq. (18) with (13e) taken into consideration reduces to

$$f,_{vu} = -r,_{vu} / r. \tag{22}$$

3.1 Evaporation

In this subsection, we establish the relation between the 'mass' m (or the *normalized* mass $\mu = m/m_0$) and the 'time' v (this relation is used not to establish the traversability of M_{wh}, but to check whether the model is self-consistent). As we shall see m falls with v and it is this process that is referred to as *evaporation*.

As already mentioned, the horizon can be parameterized both by the mass m and the coordinates u or v. The three parameterizations are related by the obvious formulas:

$$2\frac{\mathrm{d}m}{\mathrm{d}u} \equiv \frac{\mathrm{d}\hat{r}}{\mathrm{d}u} = \hat{r},_u + \hat{r},_v \frac{\mathrm{d}\hat{v}}{\mathrm{d}u} = \hat{r},_u \tag{23}$$

and

$$\frac{\mathrm{d}\hat{v}}{\mathrm{d}u} = -\frac{\hat{r},_{vu}}{\hat{r},_{vv}}, \tag{24}$$

of which the former follows strictly from the definitions (11), (12) and the latter from the fact that $0 = \mathrm{d}\hat{r},_v = \hat{r},_{vu}\,\mathrm{d}u + \hat{r},_{vv}\,\mathrm{d}\hat{v}$.

Corollary 4 *By substituting* (19) *and* (20) *in* (24), *we deduce that* $\mathrm{d}\hat{v}/\mathrm{d}u$ *is positive. This means that the horizon* $\mathcal{H}$ *is* timelike *in our model.*

Now substitute the same formulas in the expression for $\mathrm{d}\hat{v}/\mathrm{d}m$ obtained by combining (23) and (24):

$$\frac{\mathrm{d}\hat{v}}{\mathrm{d}m} = -2\hat{r},_u^{-1} \frac{\hat{r},_{vu}}{\hat{r},_{vv}} = \frac{\hat{r},_u m^2}{2\pi K} \hat{F}^{-2},$$

rewrite the resulting equation as

$$\frac{\mathrm{d}\hat{v}}{\mathrm{d}\mu^3} = \frac{8m_0}{3\mathfrak{c}}\frac{\hat{r},_u}{\hat{F}^2}, \qquad \text{where } \mu \leftrightharpoons m/m_0, \tag{25}$$

and estimate its right-hand side. To this end consider the geodesic segment γ between the points p' and p'', see Fig. 2 and the passage above condition (13d). On this segment

$$\left(\frac{r,_u}{F^2}\right),_u \mathrm{d}u = -\frac{4\pi r}{F^2} T_{uu}\mathrm{d}u = -\frac{4\pi r}{r,_u^2} T_{uu}\left(\frac{r,_u}{F^2}\right)\mathrm{d}r$$

(the former equality follows from (17′) and the latter—from the fact that $\mathrm{d}r = r,_u\,\mathrm{d}u$ on γ) and, hence,

$$\frac{\hat{r},_u}{\hat{F}^2}(p') \equiv \frac{r,_u}{F^2}(p'') \cdot \exp\left\{\int_\gamma \left(\ln\frac{r,_u}{F^2}\right),_u \mathrm{d}u\right\} = \\ = -\frac{v}{8m_0}\exp\left\{-\int_\gamma \frac{4\pi r T_{uu}\,\mathrm{d}r}{r,_u^2}\right\} \approx -\frac{v}{8m_0}. \tag{26}$$

The approximate equality here is due to (13d), and the factor in front of the exponent is simplified with the use of the first equalities in formulas (2) and (5b). Thus, we see that the expression (5b), which in the Schwarzschild case is valid in the entire spacetime, remains valid—at least, on the horizon—in M_{wh} too.

Substituting (26) in (25) and integrating the resulting equation (recall that v, and $\hat{v}$, as functions of p', are the same) gives the sought-for relation

$$\hat{v}(\mu) = v_0 \exp\left\{\frac{1}{3\mathfrak{c}}(1-\mu^3)\right\}. \tag{27}$$

As an immediate application, we find the limit

$$v_\infty \leftrightharpoons \hat{v}(m=0) = v_0 e^{\frac{1}{3\mathfrak{c}}},$$

which enables us to estimate the time [in the sense of (14)] that it takes for a wormhole to evaporate. Namely, take the beginning of the evaporation to be the very moment $\mathcal{T}_{\mathrm{R}}^{\mathrm{st}}$ when the wormhole appeared, i.e. when $\hat{v} = v_0$. The evaporation (and the existence of the wormhole, in general) ceases at the moment $\mathcal{T}_{\mathrm{R}}^{\mathrm{fi}}$ when $\hat{v} = v_\infty$. Thus, the evaporation time is

$$\mathcal{T}_{\mathrm{R}}^{\mathrm{ev}} \leftrightharpoons \mathcal{T}_{\mathrm{R}}^{\mathrm{fi}} - \mathcal{T}_{\mathrm{R}}^{\mathrm{st}} = 2m_0\ln(v_\infty^2\kappa_R) - 2m_0\ln(v_0^2\kappa_R) = \frac{4m_0}{3\mathfrak{c}} \approx 6\cdot 10^{66}\left(\frac{m_0}{M_\odot}\right)^3 \text{yr}.$$

It is instructive to compare this result with that in [143, (26)].

Remark 5 So far, nothing has suggested that a wormhole has time to evaporate completely, i.e. that μ and v have time to reach the zero and v_∞, respectively. One could imagine that the evolving horizon tends to some asymptote $v = v_{\mathcal{H}}$, where $v_{\mathcal{H}}$ is a constant smaller than v_∞. In fact, however, such a behaviour is excluded in our model. Indeed, by symmetry the left horizon would have to have an asymptote $u = u_{\mathcal{H}}$. This would imply that the horizons intersect (recall that they are timelike, by Corollary 4), which contradicts (10b).

Remark 6 The coordinate transformation $(u, v) \to (r, \tilde{v})$, where $\tilde{v} \leftrightharpoons 4m_0 \ln v$, casts the metric in the following form:

$$\begin{aligned} ds^2 &= -F^2 r,_u^{-1} dv(-r,_v dv + dr) + r^2(d\vartheta^2 + \cos^2\vartheta\, d\varphi) = \\ &= \frac{F^2 v}{8r,_u m_0}\Big[\tfrac{1}{2m_0} v r,_v d\tilde{v}^2 - 2dr d\tilde{v}\Big] + r^2(d\vartheta^2 + \cos^2\vartheta\, d\varphi) = \\ &= (2m_V/r - 1)d\tilde{v}^2 + 2dr d\tilde{v} + r^2(d\vartheta^2 + \cos^2\vartheta\, d\varphi), \quad m_V \leftrightharpoons r\frac{2m_0 - vr,_v}{4m_0}, \end{aligned}$$

the last equality being obtained by the use of (26). In the vicinity of the horizon this, in fact, is the Vaidya metric [71, (9.32)], because the chain of equalities

$$4m_0 m_V,_u = 2m_0 r,_u - v(r,_u r,_v + r r,_{uv})\big|_{\mathcal{H}} = 2m_0 \hat{r},_u + \frac{1}{4} v\hat{F}^2 = 0$$

[the second one follows from (11), (19), and the last one—from (26)] proves that m_V depends only on $\tilde{v}$.

3.2 *The Shift of the Horizon*

In this section we solve a, quite cumbersome, technical problem—we find the function $\hat{u}(m)$. To this end we, first, combine the relations (19)–(22) into an ordinary differential equation defining $\hat{r},_u$ [this is Eq. (32)], and then solve the Eq. (23) with the thus found right-hand side.

We begin with writing down the following consequences of relations (23) and (24):

$$\frac{d}{dm}\hat{r},_u = \frac{du}{dm}\left(\frac{\partial}{\partial u} + \frac{d\hat{v}}{du}\frac{\partial}{\partial v}\right)\hat{r},_u = 2\hat{r},_u^{-1}\left(\hat{r},_{uu} - \frac{\hat{r},_{vu}^2}{\hat{r},_{vv}}\right). \tag{28}$$

Then, using (21) and the equation

$$\frac{\hat{r},_{vu}^2}{\hat{r},_u^2 \hat{r},_{vv}} = \frac{m}{32\pi K}$$

[it follows from (19) and (20)], we rewrite (28) as

$$\hat{r},_u^{-1}\frac{\mathrm{d}\hat{r},_u}{\mathrm{d}m} = 4\frac{\hat{f},_u}{\hat{r},_u} - 4\pi c m^{-3} - \frac{m}{16\pi K}. \tag{29}$$

The next step is to estimate the first term on the right-hand side. Consider the segment λ of the null geodesic $u = const$ between the points $p \in \mathcal{E}$ and $p' \in \mathcal{H}$. On this segment, (5d) and (22) force the equality

$$\hat{f},_u \equiv f,_u(p') \equiv f,_u(p) + \int_\lambda f,_{uv}\,\mathrm{d}v = -\frac{1+\bar{x}}{2\bar{x}}\bar{x},_u - \int_\lambda \frac{r,_{uv}}{r}\,\mathrm{d}v \tag{30}$$

[from now on for the sake of brevity I write $\bar{r}$, $\bar{x},_u$, etc. instead of $r(p)$, $x,_u(p)$, etc.; note that in this notation $\bar{u}(v) = \hat{u}(v)$]. The sign of $r,_{uv}$ is constant by (10a), while r (which, as shown in Sect. 2.2, monotonically falls on λ) changes from $\bar{r}$ to $2m$. Hence, on the strength of the corresponding theorem [193, n° 304], one gets

$$\hat{f},_u = \Big(\frac{1}{2m_\blacktriangledown} - \frac{1+1/\bar{x}}{4m_0}\Big)\bar{r},_u - \frac{1}{2m_\blacktriangledown}\Big(\bar{r},_u + \int_\lambda r,_{uv}\,\mathrm{d}v\Big) =$$
$$= \Big(\frac{1}{2m_\blacktriangledown} - \frac{1+1/\bar{x}}{4m_0}\Big)\bar{r},_u - \frac{1}{2m_\blacktriangledown}\hat{r},_u\,,$$

where $m_\blacktriangledown$ is a constant (for a given λ) lying between m and $\bar{r}/2$ (put differently

$$\frac{1}{\bar{x}} \leqslant \frac{m_0}{m_\blacktriangledown} \leqslant \frac{1}{\mu}, \tag{31}$$

which will be used in a moment). Substituting the just derived expression for $\hat{f},_u$ in (29) and neglecting the terms $\sim m_\blacktriangledown^{-1}$, m^{-3} in comparison with the last one we, finally, obtain the differential equation mentioned above:

$$\hat{r},_u^{-1}\frac{\mathrm{d}\hat{r},_u}{\mathrm{d}m} = \frac{2\xi\bar{x},_u}{\hat{r},_u} - \frac{m}{16\pi K}, \qquad \xi \leftrightharpoons \Big(\frac{2m_0}{m_\blacktriangledown} - \frac{1}{\bar{x}} - 1\Big). \tag{32a}$$

Its initial condition is the equality

$$\hat{r},_u(m_0) = -2m_0 v_0/e \tag{32b}$$

following from (5b). And the solution is

$$\hat{r},_u(\mu) = -2\frac{m_0 v_0}{e}[1 + \Xi(\mu)]\,y(\mu), \qquad \text{where}$$
$$y(\mu) \leftrightharpoons e^{\frac{1-\mu^2}{2c}}, \qquad \Xi(\mu) \leftrightharpoons \frac{e}{v_0}\int_\mu^1 \frac{\xi}{y}\Big(\frac{\bar{x}-1}{\bar{u}\bar{x}}\Big)\,\mathrm{d}\mu'. \tag{33}$$

Proof Rewrite the equation in question as

$$\frac{\mathrm{d}\hat{r}_{,u}}{\mathrm{d}\mu} + \frac{\mu}{\mathrm{c}}\hat{r}_{,u} = 2\xi\bar{x}_{,u}\, m_0,$$

or, equivalently, as

$$e^{-\frac{\mu^2}{2\mathrm{c}}}\frac{\mathrm{d}}{\mathrm{d}\mu}\big(e^{\frac{\mu^2}{2\mathrm{c}}}\hat{r}_{,u}\big) = 2\xi\bar{x}_{,u}\, m_0.$$

Now it is evident that

$$\hat{r}_{,u}(\mu) = \left(\hat{r}_{,u}(1) + \int_1^{\mu}\frac{2\xi\bar{x}_{,u}\, m_0}{y}\,\mathrm{d}\mu'\right)y.$$

(33) is obtained by applying (5b) to both terms in the parentheses. □

In the remainder of this subsection, we turn to our main task, which is to explore $\hat{u}(m)$. For this purpose, using, first, (23), then the fact that $\hat{u}(1) = 0$ (which is obvious from Fig. 2) and, finally, the expression (33), we represent $\hat{u}$ as

$$\hat{u}(\mu) = 2m_0\int_1^{\mu}\frac{\mathrm{d}\mu'}{\hat{r}_{,u}(\mu')} = \frac{e}{v_0}\int_{\mu}^{1}\frac{\mathrm{d}\mu'}{y(\mu')[1+\Xi(\mu')]}. \tag{34}$$

The plan is to find a constant Ξ^{m} (its value will depend on which stage is considered) restricting $|\Xi|$

$$|\Xi| \leqslant \Xi^{\mathrm{m}} < 1$$

and thus to obtain an estimate

$$\frac{e}{v_0(1+\Xi^{\mathrm{m}})}\int_{\mu}^{1}\frac{\mathrm{d}\mu'}{y(\mu')} \leqslant \hat{u}(\mu) \leqslant \frac{e}{v_0(1-\Xi^{\mathrm{m}})}\int_{\mu}^{1}\frac{\mathrm{d}\mu'}{y(\mu')}.$$

To simplify the task, we introduce an auxiliary quantity

$$\mu_\star\colon\quad \hat{u}(\mu_\star) = v_0. \tag{35}$$

and consider separately the cases $m > \mu_\star m_0$ and $m < \mu_\star m_0$. Physically, this threshold mass $\mu_\star m_0$ is distinguished by the fact that, as seen on Fig. 2, it is the mass of the wormhole in the moment when an observer in the region IV for the first time can see the light of the 'other universe' at the end of the throat.

The Early Stage

On the segment $\hat{u} < v_0$, the ray λ can meet $\mathcal{E}$ in one, two or three points, see properties (i) and (ii) on page 187. But (exactly) one of them always lies between the horizons. It is this point that we take to be the point p appearing in the definition of $m_\blacktriangledown$ and thereby of ξ, see Eq. (32a). This choice ensures, in particular, that $\bar{x} < 1$

and $\bar{v} < v_0$. Then [the first equality follows from (3)]

$$(1-\bar{x})/\bar{u} = \bar{v}e^{-\bar{x}} < v_0,$$
$$\bar{x} = 1-(1-\bar{x}) > 1-(1-\bar{x})e^{\bar{x}} = 1-\bar{u}\bar{v} > 1-v_0^2$$

and, hence [recall that according to (9), $\mathfrak{c}$ and, thereby, v_0 are much less than 1]

$$\frac{|\bar{x}-1|}{\bar{u}\bar{x}} < \frac{v_0}{1-v_0^2} < 2v_0. \tag{36}$$

Now note that in our case, i.e. at $\bar{x} < 1$, (31) implies

$$1 \leqslant \frac{1}{\bar{x}} \leqslant \frac{m_0}{m_\blacktriangledown} \leqslant \frac{1}{\mu},$$

which gives the estimate

$$0 < \xi \equiv \left(\tfrac{m_0}{m_\blacktriangledown} - 1\right) + \left(\tfrac{m_0}{m_\blacktriangledown} - \tfrac{1}{\bar{x}}\right) \leqslant 2\frac{1-\mu}{\mu}. \tag{37}$$

Substituting it in the definition of Ξ, see (33), and using then the inequality (36), we get

$$|\Xi| \leqslant \frac{2e}{v_0}\int_\mu^1 \frac{1-\acute{\mu}}{\acute{\mu}y(\acute{\mu})}\frac{|\bar{x}-1|}{\bar{u}\bar{x}}\,\mathrm{d}\acute{\mu} < 4eZ, \qquad \text{where } Z \leftrightharpoons \int_\mu^1 \frac{1-\acute{\mu}}{\acute{\mu}y(\acute{\mu})}\,\mathrm{d}\acute{\mu}. \tag{38}$$

To assess Z, take the relevant integral by parts

$$Z \equiv \mathfrak{c}\int_\mu^1 \frac{\acute{\mu}}{\mathfrak{c}}\frac{1-\acute{\mu}}{\acute{\mu}^2}e^{\frac{\acute{\mu}^2-1}{2\mathfrak{c}}}\,\mathrm{d}\acute{\mu} = \mathfrak{c}\left[\frac{\mu-1}{\mu^2}e^{\frac{\mu^2-1}{2\mathfrak{c}}} + e^{-\frac{1}{2\mathfrak{c}}}\int_\mu^1\left(\frac{2}{\acute{\mu}^3}-\frac{1}{\acute{\mu}^2}\right)e^{\frac{\acute{\mu}^2}{2\mathfrak{c}}}\,\mathrm{d}\acute{\mu}\right]$$

and note that the integrand in the right-hand side is positive (this is obvious) and grows monotonically at $1 > \acute{\mu} \geqslant 1/m_0$ (i.e. as long as the wormhole remains macroscopic).

Proof Indeed, the extrema of the integrand are the roots of the equation

$$\acute{\mu}^3 - 2\acute{\mu}^2 - 2\mathfrak{c}\acute{\mu} + 6\mathfrak{c} = 0. \tag{$*$}$$

The function in the left-hand side is obtained from the function $\acute{\mu}^3 - 2\acute{\mu}^2$, whose roots are 0 and 2, by adding a small ($\sim\mathfrak{c}$) function positive on the whole interval [0, 1]. So, Eq. ($*$) has a single (non-degenerate) root on that interval. It is $\approx\sqrt{3\mathfrak{c}}$, which can be easily established by solving the equation $\acute{\mu}^2 + \mathfrak{c}\acute{\mu} - 3\mathfrak{c} = 0$ [obtained by neglecting the term $\acute{\mu}^3$ in ($*$)]. But $\sqrt{3\mathfrak{c}} < 1/m_0$, see (9), so, there is no roots in the interval of interest. □

Expanding the integration range to $(\sqrt{3\mathfrak{c}}, 1)$ (what happens at smaller μ is irrelevant), dividing it by the point $\acute{\mu} = 1 - 100\mathfrak{c}$, and, finally, replacing the integrand by its maxima on either side of $\acute{\mu}$, we have

$$e^{-\frac{1}{2\mathfrak{c}}} \int_{\mu}^{1} \Big(\frac{2}{\acute{\mu}^3} - \frac{1}{\acute{\mu}^2}\Big) e^{\frac{\acute{\mu}^2}{2\mathfrak{c}}}\, \mathrm{d}\acute{\mu} \leqslant \Big(1 - 100\mathfrak{c}\Big)\Big(\frac{2}{\acute{\mu}^3} - \frac{1}{\acute{\mu}^2}\Big) e^{\frac{-1+\acute{\mu}^2}{2\mathfrak{c}}}\Big|_{\acute{\mu}=1-100\mathfrak{c}} + \\ + 100\mathfrak{c} \cdot 1 \approx e^{-100} + 100\mathfrak{c}.$$

Hence (taking into consideration that in the definition of Z the first term in the square brackets is negative)

$$Z \leqslant e^{-100}\mathfrak{c} + 100\mathfrak{c}^2, \qquad \forall \mu \geqslant 1/m_0. \tag{39}$$

Being substituted into (38) this gives

$$|\Xi(\mu)| \ll 1, \qquad \text{at } \mu \gtrsim \mu_\star \tag{40}$$

and, in consequence,

$$\hat{u}(\mu) \approx \frac{e}{v_0} \int_{\mu}^{1} \frac{\mathrm{d}\mu'}{y(\mu')}, \qquad \text{at } \mu \gtrsim \mu_\star. \tag{41}$$

It is clear, see Remark 5, that the (normalized) mass of the wormhole does reach $\mu_\star$ and keeps decreasing (that stage is considered below). However, we also need to know the specific value of $\mu_\star$. To find it, substitute in (41) the formula

$$\int_0^{\mu} \frac{\mathrm{d}\acute{\mu}}{y(\acute{\mu})} = \sqrt{2\mathfrak{c}}\, e^{-\frac{1}{2\mathfrak{c}}} \int_0^{\frac{\mu}{\sqrt{2\mathfrak{c}}}} e^{\grave{\mu}^2} \mathrm{d}\grave{\mu} \approx \frac{\mathfrak{c}}{\mu}\, e^{\frac{\mu^2-1}{2\mathfrak{c}}}, \tag{42}$$

which is (asymptotically) correct at large $\frac{\mu}{\sqrt{2\mathfrak{c}}}$ (see, for example, [191, Theorem 2.6]). The result is

$$v_0 = \hat{u}(\mu_\star) = \frac{e}{v_0} \int_{\mu_\star}^{1} \frac{\mathrm{d}\acute{\mu}}{y(\acute{\mu})} = \frac{e}{v_0} \Big(\int_0^{1} \frac{\mathrm{d}\acute{\mu}}{y(\acute{\mu})} - \int_0^{\mu_\star} \frac{\mathrm{d}\acute{\mu}}{y(\acute{\mu})} \Big) \approx \frac{e\mathfrak{c}}{v_0} (1 - \tfrac{1}{\mu_\star} e^{\frac{\mu_\star^2-1}{2\mathfrak{c}}}), \tag{43}$$

whence

$$\tfrac{1}{\mu_\star} e^{\frac{\mu_\star^2-1}{2\mathfrak{c}}} \approx 1 - \frac{v_0^2}{e\mathfrak{c}} = 1 - \mathfrak{h}^{-1} \gtrsim \sqrt{\mathfrak{c}}. \tag{44}$$

This means, in particular, that $\mu_\star$ is *very* close to 1:

$$1 - \mu_\star < \tfrac{1}{2}\mathfrak{c}\, |\ln \mathfrak{c}| \tag{45}$$

[it is to make this, quite useful, estimate valid even at the minimal relevant $\mathfrak{h}$, that we added to it an—extremely small—quantity $\sqrt{\mathfrak{c}}$, see (9)] and the approximation (42) was legitimate.

The Late Stage

Consider the evaporation of the horizon at $\hat{u} > v_0$. Here, in contrast to the previous case, $\bar{x} > 1$. So, instead of (36) we have

$$\left|\frac{\bar{x}-1}{\bar{u}\bar{x}}\right| - \frac{1}{\bar{u}}(1 - 1/\bar{x}) < \frac{1}{v_0} = \frac{\mathfrak{h}}{e\mathfrak{c}} v_0. \tag{46}$$

On the other hand, combining (31) with the definition of ξ, see (32a), one gets

$$-1 \leqslant \left(\frac{2}{\bar{x}} - \frac{1}{\bar{x}} - 1\right) \leqslant \xi \leqslant \left(\frac{2}{\mu} - \frac{1}{\bar{x}} - 1\right) \leqslant \frac{2-\mu}{\mu},$$

whence

$$|\xi| \leqslant \max\{2/\mu - 1, 1\} = 2/\mu - 1. \tag{47}$$

In the definition (33), neglect the contribution of the segment $(\mu_\star, 1)$ to Ξ [we may do so owing to (40)] and substitute the inequalities (46), (47) in the remaining integral to obtain

$$|\Xi| \leqslant \frac{e}{v_0}\int_{\mu}^{\mu_\star} \frac{|\xi|}{y}\left|\frac{\bar{x}-1}{\bar{u}\bar{x}}\right| \mathrm{d}\acute{\mu} \leqslant \frac{\mathfrak{h}}{\mathfrak{c}}\int_{\mu}^{\mu_\star} y^{-1}\left(\frac{1-\acute{\mu}}{\acute{\mu}} + \frac{1}{\acute{\mu}}\right)\mathrm{d}\acute{\mu} \leqslant \frac{\mathfrak{h}}{\mathfrak{c}}Z + \frac{\mathfrak{h}}{\mathfrak{c}}\int_{\mu}^{\mu_\star}\frac{\mathrm{d}\acute{\mu}}{\acute{\mu}y}.$$

The estimate (39) enables one to neglect also the first term in the right-hand side

$$|\Xi| \leqslant \frac{\mathfrak{h}}{\mathfrak{c}}\int_{1/m_0}^{\mu_\star} e^{\frac{\acute{\mu}^2-1}{2\mathfrak{c}}}\acute{\mu}^{-1}\mathrm{d}\acute{\mu} = \frac{\mathfrak{h}}{2\mathfrak{c}}e^{-\frac{1}{2\mathfrak{c}}}\int_{\frac{1}{2\mathfrak{c}m_0^2}}^{\frac{\mu_\star^2}{2\mathfrak{c}}} e^{\zeta}\zeta^{-1}\mathrm{d}\zeta \approx \frac{\mathfrak{h}}{\mu_\star^2}e^{\frac{\mu_\star^2-1}{2\mathfrak{c}}} \approx \mathfrak{h} - 1. \tag{48}$$

The last equality here follows from (45) and (44), while the penultimate one—from the fact that asymptotically

$$\int_a^b \frac{e^{\acute{\mu}}\mathrm{d}\acute{\mu}}{\acute{\mu}} \approx \frac{e^b}{b}, \qquad \text{at } a \approx 1 \text{ and large } b.$$

4 The Traversability of the Wormhole

The farther a point of the horizon is from $\mathcal{E}$, the greater its u-coordinate is, see Corollary 4. Therefore, the quantity $\hat{u}_\infty$ mentioned in Sect. 2.4 is merely $\hat{u}(\mu = 0)$. The latter can be easily assessed by (34), (48):

$$v_0 + \frac{e}{\mathfrak{h} v_0} \int_0^{\mu_\star} \frac{\mathrm{d}\mu'}{y(\mu')} \leqslant \hat{u}_\infty = \hat{u}(\mu = 0) \leqslant v_0 + \frac{e}{(2 - \mathfrak{h}) v_0} \int_0^{\mu_\star} \frac{\mathrm{d}\mu'}{y(\mu')}$$

[v_0 is the contribution of $(\mu_\star, 1)$], or, by (42) and (44),

$$v_0(2 - \mathfrak{h}^{-1}) \leqslant \hat{u}_\infty \leqslant v_0(2 - \mathfrak{h})^{-1}. \tag{49}$$

As discussed above, our subject is only those wormholes that satisfy (9). Consequently, in our case $\mathfrak{h} > 1$ and, hence, (49) implies $\hat{u}_\infty > v_0$. Thus, according to the criterion ($\star$), see page 192, *the wormhole under discussion is traversable.*

Depending on the value of $\mathfrak{h}$, its traversability time varies [see (14) and (49)] from

$$\mathscr{T}_{\mathrm{L}}^{\mathrm{trav}} = 0 \qquad \text{at} \quad \mathfrak{h} = 1$$

(we have neglected $\sqrt{\mathfrak{c}}$) to

$$\mathscr{T}_{\mathrm{L}}^{\mathrm{trav}} = \alpha m_0, \quad 1.3 \leqslant \alpha \leqslant 3.8 \qquad \text{at} \quad \mathfrak{h} = \tfrac{\sqrt{5}+1}{2}.$$

The question of whether the obtained result is reliable, i.e. whether the model is self-consistent and the weak evaporation assumption is valid, is too hard and we shall not try to solve it. There is, however, one obvious limitation: if the wormhole evaporates too violently it is conceivable that its horizons would approach each other and eventually intersect, thereby violating the assumption (10b) and rendering the model self-inconsistent. Let us check that such an intersection does not occur, at least when $\mathfrak{h}$ satisfies condition (9), because at such $\mathfrak{h}$

$$\hat{u}(\mu) < \hat{v}(\mu) \tag{50}$$

and, hence, the entire right/left horizon lies in the right/left half-plane. Indeed, this inequality is clearly true for all $\hat{u} \leqslant v_0$, i.e. for all $\mu \geqslant \mu_\star$. At the same time, $\mu \leqslant \mu_\star$ implies

$$\hat{u}/v \leqslant \hat{u}_\infty / v(\mu_\star) \leqslant \frac{e^{\frac{\mu_\star^3 - 1}{3\mathfrak{c}}}}{2 - \mathfrak{h}} \equiv \frac{e^{\frac{\mu_\star^2 - 1}{2\mathfrak{c}}}}{(2 - \mathfrak{h})\mu_\star} \mu_\star e^{\frac{1}{3\mathfrak{c}}(\mu_\star - 1)^2(\mu_\star + \frac{1}{2})} = \frac{1 - \mathfrak{h}^{-1}}{2 - \mathfrak{h}} \leqslant 1$$

[the second inequality follows from the combination of (49) and (27), and the last one—from (9). The last equality is obtained by applying (44) and (45)]. Thus, (50) is true and, correspondingly, the horizons do not meet.

The traversability time $\mathscr{T}_{\mathrm{L}}^{\mathrm{trav}}$ is macroscopic. It is small, however: even for a supermassive black hole like those inhabiting the centres of massive galaxies, it is of the order of minutes.

Remark 7 It does *not* follow from our analysis that the small traversability times are typical of empty (spherically symmetric) wormholes *in general*. The wormholes for which this may not be the case, merely cannot be considered within this simple model dealing exceptionally with 'almost Schwarzschild' spacetimes.

As a conveyance such quick-breaking wormholes do not look too promising.[8] Nonetheless, their existence—especially in the *intra*-universe version—may be of great importance. Let us dwell on this subject.

To adapt our model to description of intra-universe wormholes, first, enclose the throat in a surface

$$\mathcal{C} = \{q \in M_{\text{wh}}: \quad r(q) = \flat\}, \qquad \flat \gg 2m_0$$

(physically, the constant $\flat$ is interpreted as the radius of the mouth). This surface is a disjoint union of two cylinders $\mathbb{L}^1 \times \mathbb{S}^2$, one of which, $\mathcal{C}_L$, lies in the left asymptotically flat region and the other, $\mathcal{C}_R$,—in the right:

$$\mathcal{C} = \mathcal{C}_L \cup \mathcal{C}_R, \qquad \mathcal{C}_{L(R)} \subset \text{II(IV)}$$

(the Roman numerals refer to corresponding regions in Fig. 2). We shall consider the spacetime outside $\mathcal{C}$ (which is, correspondingly, a disjoint union of two asymptotically flat regions M_L and M_R) as flat. Hopefully, the resulting error is negligible—the space far enough from a gravitating body *is* more or less flat. Pick Cartesian coordinates in $M_{L(R)}$ so that the $t_{L(R)}$-axes are parallel to the generators of $\mathcal{C}_{L(R)}$, and $\mathcal{E} \cap M_{L(R)}$ are the surfaces $t_{L(R)} = 0$. Now an intra-universe wormhole is obtained by the standard surgery: one removes the half-spaces $x_L > d/2$ and $x_R < -d/2$ from, respectively, M_L and M_R [it is understood that $|x(\mathcal{C})| \leqslant \flat < d/2$] and identifies the points on their boundaries (i.e. on the three surfaces $x_L = d/2$ and $x_R = -d/2$) that have the same t-, y- and z- coordinates. The resulting spacetime, see Fig. 3, is a Minkowski space, in which the interiors of two cylinders (their boundaries are $\mathcal{C}_L$ and $\mathcal{C}_R$) are replaced by a *connected* region, so that, for example, a photon intersecting $\mathcal{C}_L$ at the moment $t_{\text{in}} \in (\mathcal{T}_{\text{L}}^{\text{op}}, \mathcal{T}_{\text{L}}^{\text{cl}})$, emerges from $\mathcal{C}_R$ at some t_{out}. The only difference between this spacetime and the wormhole considered in Sect. 2.1 in Chap. 3, is that in the former case $\Delta \leftrightharpoons t_{\text{out}} - t_{\text{in}}$ varies with time. Now note that it would take only time d for the photon to return to $\mathcal{C}_L$. So, the spacetime in question is causal only if

$$t_{\text{in}} < t_{\text{out}} + d \qquad \forall\, t_{\text{in}} \in (\mathcal{T}_{\text{L}}^{\text{op}}, \mathcal{T}_{\text{L}}^{\text{cl}}). \tag{$\star$}$$

By changing κ_L to κ'_L—all other parameters being fixed—we add $\approx 2m_0 \ln(\kappa'_L/\kappa_L)$ to each of the quantities t_{in}, $\mathcal{T}_{\text{L}}^{\text{op}}$ and $\mathcal{T}_{\text{L}}^{\text{cl}}$, see (8), without changing t_{out}. So, for

[8]Note, though, that owing to Lorentz contraction $\mathcal{T}_{\text{L}}^{\text{trav}}$ is *larger* for an observer moving towards the wormhole [100].

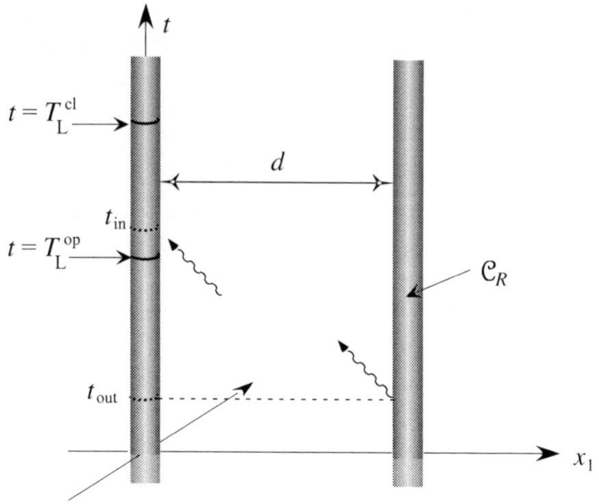

Fig. 3 The two wavy lines depict the world line of the same photon

sufficiently large κ'_L the inequality ($\star$) will fail. Thus, we conclude that irrespective of the values of m_0, d, $\mathfrak{h}$, and κ_R, traversable intra-universe evaporating wormholes with sufficiently large κ'_L are time machines with all the ensuing consequences.

Chapter 10
At and Beyond the Horizon

Но тормоза отказывают, – кода! –
Я горизонт промахиваю с хода!

В. С. Высоцкий[1]

As follows from the whole discussion above, there is nothing impossible in the existence of a traversable wormhole, when quantum corrections are taken into account. Likewise, classically there is nothing impossible in its transformation into a time machine. It remains to check that there are no known *quantum* effects that would protect causality after all. In fact, there *is* some reason for believing that such effects do exist. As we have seen, in the vicinity of any compactly determined Cauchy horizon, there are always null geodesics returning infinitely many times, each time more and more blue-shifted, to the same—arbitrarily small—region. This might make one suspect [74] that time machines with compactly generated Cauchy horizons are unstable: infinitely amplified quantum fluctuations will produce infinite energy density on the horizon. Partly, this idea is corroborated by the behaviour of the classical massless scalar field near the horizon in the Misner space, see Sect. 2.3 in Chap. 4. And, indeed, Yurtsever [183] argued that in the Misner space, the quantum version of this field has a state $|Q\rangle$ in which $\langle T_{ab}\rangle_Q$ diverges on the horizon. This, however, may simply mean that there is a flaw in $|Q\rangle$, while the spacetime *per se* is just as nice as, say, Minkowski space, see Example 7 in Chap. 7. It would be different if the divergence took place in *all* (or in 'all reasonable') states. Let us check, therefore, that for the Misner-type spaces with massless conformally coupled scalar field that is not the case [90].

We start with an auxiliary spacetime $(M, \mathring{g})$, which is a flat cylinder:

$$\mathring{g}: \qquad ds^2 = -d\tau^2 + d\psi^2 \qquad \tau \in \mathbb{R}, \quad \psi = \psi + \psi_0. \tag{1}$$

[1]"But the brakes refuse to work—coda! And I cross the horizon at full tilt." V. S. Vysotsky (H. William Tjalsma. Translation, 1982).

S. Krasnikov, *Back-in-Time and Faster-than-Light Travel in General Relativity*, Fundamental Theories of Physics 193, https://doi.org/10.1007/978-3-319-72754-7_10

The vacuum $|\mathring{\mathcal{U}}\rangle$ is defined by choosing the following set of modes $\mathring{\mathcal{U}} = \{u_k\}$ to be the basis of the one-particle Hilbert space:

$$u_0 \leftrightharpoons \frac{1}{\sqrt{2|\psi_0|}}(\varsigma\tau + i\varsigma^{-1}), \qquad u_k \leftrightharpoons \frac{1}{\sqrt{|4\pi k|}} e^{\frac{2\pi i}{\psi_0}(k\psi - |k|\tau)}, \quad k \in \mathbb{Z}, \quad k \neq 0,$$

where ς is a real constant. The Hadamard function in this state is of the form

$$G^{(1)}(\mathring{\mathcal{U}}; p, p') = |\psi_0|^{-1}(\varsigma^2\tau\tau' + \varsigma^{-2}) + D^{(1)}(p, p'),$$

see (Eq. 8 in Chap. 7), where

$$D^{(1)}(p, p') \leftrightharpoons \sum_{k\neq 0} u_k(p)u_k^*(p') + \text{complex conjugate}.$$

Now to find $\langle T_{ab}\rangle_{\mathring{\mathcal{U}}}$ one should apply (Eq. 9 in Chap. 7). Fortunately, we need not do so, because there is a well-known state $|0_{\psi_0}\rangle$ in which the Hadamard function is $D^{(1)}$ and the stress–energy tensor is already found [14, Sects. 4.1, 4.2]:

$$\langle T_{\tau\tau}\rangle_{0_{\psi_0}} = \langle T_{\psi\psi}\rangle_{0_{\psi_0}} = -\frac{\pi}{6\psi_0^2}, \qquad \langle T_{\tau\psi}\rangle_{0_{\psi_0}} = 0.$$

We shall not analyse the subtle problem of how $|0_{\psi_0}\rangle$ is defined (in fact, it is understood as the massless limit of some set of states of the *massive* fields), because irrespective of the answer (recall that T_{ab}^{div} does not depend on the state)

$$\langle T_{ab}\rangle_{\mathring{\mathcal{U}}} = \langle T_{ab}\rangle_{0_{\psi_0}} + \lim_{p\to p'} \frac{1}{2|\psi_0|}\mathfrak{D}_{ab}(\varsigma^2\tau\tau' - \varsigma^{-2}),$$

where $\mathfrak{D}_{ab}$ is defined by (Eq. 7 in Chap. 7) with $m = 0$. Thus, the expectation value of the stress–energy tensor in the state $|\mathring{\mathcal{U}}\rangle$ is

$$\langle T_{\tau\tau}\rangle_{\mathring{\mathcal{U}}} = \langle T_{\psi\psi}\rangle_{\mathring{\mathcal{U}}} = \frac{1}{4}\varsigma^2|\psi_0|^{-1} - \frac{1}{6}\pi\psi_0^{-2}, \quad \langle T_{\tau\psi}\rangle_{\mathring{\mathcal{U}}} = 0. \tag{2}$$

Remark 1 $\langle T_{ab}\rangle_{\mathring{\mathcal{U}}}$ violates the weak energy condition at $\varsigma^2 < \frac{2\pi}{3\psi_0}$.

The initial globally hyperbolic regions of the Misner-type time machines are spacetimes conformally related to the cylinders (1) with $|\psi_0| = 2|\ln\kappa|$ and the conformal factors Ω's listed in Remark 9 in Chap. 4. Accordingly, in the said regions the expectation of the stress–energy tensor in the state $|\mathcal{U}\rangle$, conformally related to $|\mathring{\mathcal{U}}\rangle$, can be found by substituting those Ω's and the expression (2) into (Eq. 14 in Chap. 7):

$$\langle T_{\tau\psi}\rangle_{\mathcal{U}} = 0, \quad \langle T_{\mu\mu}\rangle_{\mathcal{U}} = \frac{\varsigma^2}{8|\ln\kappa|} - \frac{\pi}{24\ln^2\kappa} - \frac{1}{96\pi} + \frac{R}{48\pi}g_{\mu\mu}, \tag{$\star$}$$

where $\mu \leftrightharpoons \tau, \psi$ and no summation over μ is implied. As a point p approaches the horizon, $\Omega(p)$ tends to 0, and we see that generally $\langle T_{ab}\rangle_{\mathcal{U}}\langle T^{ab}\rangle_{\mathcal{U}}$ diverges. But this quantity is a scalar, so we conclude that in any frame at least *some* components of $\boldsymbol{T}$ diverge too. To summarize, for a general ς in the case at hand, some components of the stress–energy tensor will diverge in the proper frame of an observer who crosses the horizon.

However, in the special case where $\varsigma = \sqrt{4\pi/|\ln\kappa| + |\ln\kappa|/\sqrt{12\pi}}$ the diverging terms in ($\star$) cancel each other:

$$\langle T_{ac}\rangle_{\mathcal{U}} = \frac{R}{48\pi} g_{ac}.$$

Thus, for each of the three time machines, we have found a state in which the stress–energy tensor remains bounded throughout the causal region.

Examples of equally nice *non-vacuum* states and states of the *automorphic* scalar field, see Remark 11 in Chap. 7, are adduced in [90]. Yet another 'regular' state (for the automorphic field in Misner space) was found in a more direct way [162], but this result must be taken with some caution, because the Fock space used there is non-conventional.[2]

So far, the behaviour of the stress–energy tensor has been studied only in the regions *preceding* the Cauchy horizon. One may suspect, however, that it behaves pathologically in the *causality violating* region and/or on the horizon [30, 31]. Were this the case, one might infer that the back reaction does protect causality. To verify this suspicion one would have to calculate $\langle T_{ab}\rangle_{Q}$ in the causality violating region. It is, however, absolutely unclear, how exactly this could be done:

1. The quantization of a field theory is a notoriously hard problem. As of today, it is solved satisfactorily only in the simplest toy cases like the free scalar field. However, even in these cases, the procedure leans heavily upon the global hyperbolicity of the background spacetime. Dropping this condition gives rise to severe problems, which differ depending on how we quantize the field,[3] but are of the same origin—there is no surface any more at which data could be fixed to determine uniquely the solution of the field equation.

A natural way out would be to develop a quantization scheme little sensitive to non-local properties of the background spacetime. The idea is that the laws in the small should coincide with the 'usual laws for quantum field theory on globally hyperbolic spacetimes' [83]. As a mathematical implementation of that principle, Kay formulated the 'F-locality condition' and proposed to dismiss as impossible the spacetimes which do not admit field algebras obeying that condition [83]. Among those ruled out spacetimes are, in particular, all time machines with compactly generated

---

[2]In particular, the vacuum modes, as it seems [163], do not satisfy the completeness condition, see Sect. 1 in Chap. 7.

[3]In the approach outlined in Sect. 1 in Chap. 7, such a problem is the lack of adequate substitutes for y and the canonical commutation relations.

Cauchy horizons [84]. This program,[4] however, has never been realized: it was shown that the F-locality condition includes as its part a strong extraneous (i.e. not implied by the mentioned principle) *non-local* requirement on the geometry of spacetime [94]. So, the prohibition of the time machines on the ground of the F-locality condition does not differ much from their prohibition by fiat.

2. The typical quantum mechanical problem is this: given a system is initially in a state $|\mathcal{A}\rangle$, what is the probability that its final state is $|\mathcal{B}\rangle$? Suppose now that the system in its final state is located in the causality violating part of a spacetime. Then some extraordinary effort is necessary even to *give meaning* to the above formulated problem (let alone to solve it):

a. If the system in its initial state is prepared in the causal region M^{r}, then in the course of its evolution it has to cross the Cauchy horizon. After which it is exposed to the *uncontrollable* environment (Cauchy demons, in terms of Sect. 3 in Chap. 2). In other words, the system ceases to be closed. Yet another (more quantum sounding) way to say the same is: the evolution of the system ceases to be unitary (cf. [34, 53]).
b. The situation is even worse, if the system finds itself on the same causal loop both in the initial and in the final state. It is hard (if possible at all) to assign in a meaningful way the probabilities to different results of a future measurement, if the measurement actually had place in the past and one can, in principle, just *recall* the outcome.

Taking into account the aforesaid, one would not expect quantum mechanics to be generalized to non-causal spaces in the near future.

[4]For another possible approach see [184].

Appendix A
Details

A.1 Externally Flat Wormhole as Charge

In this section, we use a simple model (classical electrostatics in an externally flat wormhole) to explore Wheeler's idea of 'charge without charge'. Our main task is to find the force acting on a (sole) pointlike charge in this space, see the expression (A.28) below, and to compare it with that which would act on the test charge be the wormhole replaced by another pointlike charge.

A.1.1 The Formulation of the Problem

The spacetime under consideration is a static inter-universe wormhole M_{X} with throat of length $2\mathfrak{z}$, see Example 7 in Chap. 3. The metric of this spacetime is

$$\mathrm{d}s^2 = -\mathrm{d}t^2 + \mathrm{d}x^2 + r^2(x)(\mathrm{d}\vartheta^2 + \sin^2\vartheta\,\mathrm{d}\varphi^2), \qquad x \in \mathbb{R}^1 \tag{A.1a}$$

where r is a smooth positive even function, satisfying the condition

$$r(x)\big|_{x>\mathfrak{z}} = x + 1. \tag{A.1b}$$

The consideration of just such spacetimes is justified by the fact that our subject is macroscopic—not cosmological—wormholes. This implies that moving away from the throat one gets eventually in an 'approximately Minkowski' region. We take it to be *strictly* flat, which considerably simplifies, as we shall see, some really hard problems encountered in the general case.

Remark 1 Striving for the maximal simplicity we, in particular, do not consider wormholes of a bit more general type, where $r = x + \mathfrak{p}$ at $x > \mathfrak{z}$, because their interaction with charge is qualitatively the same. Specific formulas for the case $\mathfrak{p} \neq 1$

S. Krasnikov, *Back-in-Time and Faster-than-Light Travel in General Relativity*, Fundamental Theories of Physics 193, https://doi.org/10.1007/978-3-319-72754-7

can be found in [102], or derived independently by noticing that the metric of such a wormhole is obtained from (A.1) by a conformal transformation with constant factor.

The spacetime M_{χ} is static and is represented as a product $M_{\chi} = \mathbb{L}^1 \times \mathcal{Z}_{\chi}$, where the second factor is a spacelike three-dimensional surface $\{p \in M_{\chi}\colon\ t(p) = const\}$. We split Z_{χ} into the throat and two flat regions—$\mathfrak{F}^-$ and $\mathfrak{F}^+$. The last two are defined as regions outside of the spheres $x = \mp\mathfrak{z}$:

$$\mathfrak{F}^- \leftrightharpoons \{p \in \mathcal{Z}_{\chi}\colon\ x(p) < -\mathfrak{z}\}, \qquad \mathfrak{F}^+ \leftrightharpoons \{p \in \mathcal{Z}_{\chi}\colon\ x(p) > \mathfrak{z}\}.$$

Now pick a point $p_* \in \mathfrak{F}^+$ and put in it a pointlike charge $\mathfrak{q}$ (note that p_* is a point of $\mathcal{Z}_{\chi}$, not of M_{χ}, because the charge is at rest). The Maxwell equations in the Lorentz gauge take the form

$$A^{a;b}{}_{;b} = -4\pi J^a,$$

see, for example, [119, Problem 14.16], and the fact that the charge under consideration is pointlike and stationary means that neither $\boldsymbol{A}$ nor $\boldsymbol{J}$ depends on t and

$$J^0 = \mathfrak{q}\delta(p - p_*), \qquad J^{\mu} = 0 \text{ at } \mu = 1, 2, 3, \quad p_* \in \mathfrak{F}^+. \tag{A.2}$$

Thus, the *scalar potential* $\Phi \leftrightharpoons A^0$ is subject to the (three-dimensional) equation

$$\Delta\Phi = -4\pi\mathfrak{q}\delta(p - p_*). \tag{A.3a}$$

We supplement this equation with the boundary condition [from now on the subscript $*$ means that the corresponding quantity is found in the point p_*, so, $\varphi_* \leftrightharpoons \varphi(p_*)$, $r_* \leftrightharpoons r(p_*)$, etc.]

$$\max_{\vartheta,\varphi} |\Phi(r, \vartheta, \varphi)| < const < \infty \qquad \forall r > 2r_*, \tag{A.3b}$$

implying that solutions diverging at infinity are rejected as unphysical.

Remark 2 The generalization of the Maxwell equations to curved spacetimes is not fixed uniquely by the rule 'comma-goes-to-semicolon' [119, Problem 14.16]. Fortunately, in the space under consideration, $R^0_a = 0$ and it turns out that for the current (A.2) the mentioned ambiguity does not affect Eq. (A.3).

By solving (A.3), one can find the potential and, hence, the electric field

$$\boldsymbol{E} = -\nabla\Phi$$

of a pointlike stationary charge. Then, it would remain to formulate accurately and to solve the inverse problem: how does a given electric field act on such a charge? Generally speaking, this problem is rather hard and the answer is partly *postulated* [153]. In our case, however, these difficulties can be avoided owing again to the exceptionally simple shape of the wormhole: it would suffice to find the action of the field on a charge placed in a point $\breve{p}$ of the *flat* region $\mathfrak{F}^+$. And this charge—let its value be $\breve{\mathfrak{q}}$—experiences a force $\boldsymbol{F}$ that can be found exactly as in $\mathbb{E}^3$. Namely, the potential in $\breve{p}$ is split into the potential 'generated by the test charge itself' (in each point s of the vicinity of $\breve{p}$ it is *defined* to be $\breve{\mathfrak{q}}/|\breve{p}, s|$, where $|\breve{p}, s|$ is the distance from $\breve{p}$ to s) and the potential of the 'external field'. The force $\boldsymbol{F}$ is associated only with the latter term:

$$\boldsymbol{F}(\breve{p}) = -\breve{\mathfrak{q}}\nabla_s \Phi^{ren}(s)\big|_{s=\breve{p}}, \qquad \Phi^{ren}(s) \leftrightharpoons \Phi(s) - \frac{\breve{\mathfrak{q}}}{|\breve{p}, s|}, \tag{A.4}$$

where $\Phi^{ren}(s)$ in $s = \breve{p}$ is defined by continuity, and the subscript s at ∇ shows that the differentiation is performed with respect to the coordinates of s.

To reproduce this approach in the case of the wormhole, we find Φ in what follows (this is the gist of the section) and define an analogue of 'distance' for pairs of points lying in the same flat region. Specifically, we extend $\mathfrak{F}^+$ considered as a space by itself, rather than a part of $\mathfrak{Z}_\chi$, to the three-dimensional Euclidean space (this is done by gluing a flat ball of radius $\mathfrak{z} + 1$ into $\mathfrak{F}^+$) and for all $a, b \in \mathfrak{F}^+$ define $|a, b|$ as the distance between a and b in this 'fictitious' Euclidean space. Obviously, when the whole segment from a to b lies in $\mathfrak{F}^+$, the value of $|a, b|$ is also the distance between a and b in $\mathfrak{Z}_\chi$.

A.1.2 Multipole Expansion

The solution of (A.3a) will be found by the standard method of separation of variables [2]. Since the problem is spherically symmetric, we begin with writing it down in the coordinates used in (A.1a). Then, the three-metric $\boldsymbol{\gamma}$ takes the following form in $\mathfrak{Z}_\chi$:

$$\gamma^{xx} = 1, \quad \gamma^{\vartheta\vartheta} = r^{-2}, \quad \gamma^{\varphi\varphi} = r^{-2}\sin^{-2}\vartheta, \qquad \sqrt{\gamma} = r^2 \sin\vartheta$$

(all other components are zero) and, correspondingly,

$$\Delta\Phi = \frac{1}{\sqrt{\gamma}}\,\partial_\alpha \sqrt{\gamma}\gamma^{\alpha\beta}\partial_\beta\Phi = \Big[\gamma^{\alpha\beta}\partial_\alpha\partial_\beta + \gamma^{\alpha\beta}{}_{,\alpha}\,\partial_\beta + \gamma^{\alpha\beta}(\ln\sqrt{\gamma})_{,\alpha}\,\partial_\beta\Big]\Phi = \\ = \Big[\partial_x^2 + \frac{2r'}{r}\partial_x + \frac{1}{r^2}\Delta_\Omega\Big]\Phi, \tag{A.5}$$

where the prime denotes the derivative with respect to x, and Δ_Ω is the angular part of the Laplacian:

$$\Delta_\Omega \leftrightharpoons \partial_\vartheta^2 + \mathrm{ctg}\vartheta\,\partial_\vartheta + \sin^{-2}\vartheta\,\partial_\varphi^2.$$

The sought solution is built (for the specific prescription see Proposition 5 below) from the spherical functions and solutions of the auxiliary equation

$$\left[\partial_x^2 - \left(\frac{r''}{r} + \frac{l(l+1)}{r^2}\right)\right]z = 0 \tag{A.6}$$

[from now on l is a natural number], which is related to the radial part of the previous one due to the identity

$$\left[\partial_x^2 - \left(\frac{r''}{r} + \frac{l(l+1)}{r^2}\right)\right]z \equiv r\left[\partial_x^2 + \frac{2r'}{r}\partial_x - \frac{l(l+1)}{r^2}\right]\frac{z}{r}. \tag{A.7}$$

In the remainder of this subsection, we explore Eq. (A.6). To begin with, let us rewrite it as

$$(z'r - r'z)' = l(l+1)z/r$$

and integrate over the interval (x_0, x), where x_0 is a constant. Dividing the result by rz, we get

$$\frac{z'(l,x)}{z(l,x)} - \frac{r'(x)}{r(x)} = \frac{(z'r - r'z)|_{x_0}}{rz} + \frac{l(l+1)}{rz}\int_{x_0}^{x} \frac{z(l,\breve{x})\,\mathrm{d}\breve{x}}{r(\breve{x})}. \tag{A.8}$$

Now note that in the flat region $|x| > \mathfrak{z}$ the term with r'' in Eq. (A.6) vanishes and the general solution becomes merely a linear combination of r^{-l} and r^{l+1}.

Notation By $z_-(l,x)$ and $z_+(l,x)$, we denote those solutions of equation (A.6) that are equal to r^{-l} at, respectively, $x < -\mathfrak{z}$ and $x > \mathfrak{z}$. Below, for the sake of definiteness, we mostly discuss $z_-(l,x)$, but all its properties hold also for $z_+(l,x)$, when $\mathfrak{F}^+$ is replaced with $\mathfrak{F}^-$.

At $x \leqslant -\mathfrak{z}$ both z_- and z'_- are non-negative (recall that l is natural, while $r' = -1$ there). It turns out that (with some refinements) the same is true for larger x too. Indeed, let us try to find the *smallest* value of x, denote it x_1, at which z_- ceases to be positive

$$z_-(l,x_1) = 0, \qquad z_-(l,x) > 0 \quad \forall x < x_1.$$

The smoothness of z_- implies that the ratio $z'_-(l,x)/z_-(l,x)$ would have to tend to $-\infty$ at $x \to x_1 - 0$, but this is impossible: substituting $z = z_-$ and $x_0 = -\mathfrak{z}$ in Eq. (A.8) gives

$$\frac{z'_-(l,x)}{z_-(l,x)} = \frac{r'(x)}{r(x)} + (l+1)\frac{r(-\mathfrak{z})^{-l}}{rz_-(l,x)} + \frac{l(l+1)}{r(x)z_-(l,x)}\int_{-\mathfrak{z}}^{x} \frac{z_-(l,\breve{x})\,\mathrm{d}\breve{x}}{r(\breve{x})} \tag{$*$}$$

and we see that the first term in the right-hand side is bounded, while the following two are positive—by hypothesis—up to x_1, which must be greater than $-\mathfrak{z}$. So, x_1 does not exist and $z_-(l,x)$ is positive on the whole x-axis. Consequently, at $x > \mathfrak{z}$, when *all* terms in the right-hand side of ($*$) are positive, $z'_-(l,x)$ is positive too. Thus, we have established that

$$\forall x \quad z_-(l,x) > 0, \qquad \forall x > \mathfrak{z} \quad z'_-(l,x) > 0. \tag{A.9}$$

Further, it follows immediately from (A.6) that z''_- is positive 1) at $x > \mathfrak{z}$ and 2) at *any* x, when $l > \hat{l}$, where the constant $\hat{l}$ is defined by the requirement that the term in the parentheses be positive everywhere. Hence, the first derivative z'_- monotonically increases and, therefore, is positive. Summarizing, we have proved the following.

Proposition 3 *The functions $z_-(l,x)$ are positive. Outside the interval $|x| \leqslant \mathfrak{z}$ and—when $l > \hat{l}$—on this interval also, they are increasing and concave.*

Corollary 4 *For $l \neq 0$ the ratio z/r, where z is a solution of Eq.* (A.6)*, is unbounded.*

At $x > \mathfrak{z}$, i.e. in the 'other universe', $z_-(l,x)$ is, generally speaking, a linear combination

$$z_-(l,x)\big|_{x>\mathfrak{z}} = C_l(r^{l+1} + \alpha_l r^{-l})$$

(this representation would be impossible for $z_-(l,x)$ which remains proportional to r^{-l} in $\mathfrak{F}^+$, but Proposition 3 says that there are no such solutions). The reverse formulas

$$C_l\alpha_l = -\frac{1}{2l+1}[z_-(l,x)r^{-(l+1)}]'r^{2l+2}, \qquad C_l = \frac{1}{2l+1}[z_-(l,x)r^l]'r^{-2l},$$

valid at $x \geqslant \mathfrak{z}$, are easily verified by direct calculation. The latter of them in combination with (A.9) proves that

$$C_l > 0. \tag{A.10}$$

The coefficient α_l also admits a simple (though rough, as we shall see) estimate. Indeed,

$$\begin{aligned}\alpha_l &= -\frac{[z_-(l,x)r^{-(l+1)}]'r^{2l+2}}{[z_-(l,x)r^l]'r^{-2l}} = -r^{4l+2}\frac{z'_-r^{-(l+1)} - (l+1)z_-r^{-(l+2)}}{z'_-r^l + lz_-r^{l-1}} = \\ &= r^{2l+1}\frac{(l+1)z_- - z'_-r}{z'_-r + lz_-} = r^{2l+1}(\mathfrak{z})\frac{(l+1) - rz'_-/z_-\big|_{x=\mathfrak{z}}}{l + rz'_-/z_-\big|_{x=\mathfrak{z}}} = \\ &= r^{2l+1}(\mathfrak{z})\left[-1 + \frac{(2l+1)}{l + rz'_-/z_-\big|_{x=\mathfrak{z}}}\right].\end{aligned} \tag{A.11}$$

By Proposition 3, $rz'_-/z_-\big|_{x=\mathfrak{z}}$ is positive beginning from some l and we conclude that

$$\exists \hat{l}:\ \forall l > \hat{l} \qquad -r^{2l+1}(\mathfrak{z}) < \alpha_l < 2r^{2l+1}(\mathfrak{z}). \tag{A.12}$$

A.1.3 The Solution of Equation (3) [102]

Proposition 5 *Specify the coordinate system by the requirement that $\varphi_* = \vartheta_* = 0$. Then the general solution of Eq. (A.3) is*

$$\Phi = \Phi_{inh} - Q\int_3^x \frac{\mathrm{d}\breve{x}}{r^2} + \Phi_0, \tag{A.13}$$

where

$$\Phi_{inh} \leftrightharpoons \frac{\mathfrak{q}}{r}\sum_{l=0}^{\infty} v_l(x)\mathscr{P}_l(\cos\vartheta), \tag{A.14}$$

Q and Φ_0 are arbitrary constants, and

$$\begin{aligned} &v_l(x) \leftrightharpoons (r_* C_l)^{-1}\big[H(x - x_*)z_+(l,x)z_-(l,x_*) + H(x_* - x)z_-(l,x)z_+(l,x_*)\big],\\ &H(x) \leftrightharpoons \tfrac{1}{2}(\operatorname{sign} x + 1),\\ &\quad \mathscr{P}_l \textit{ are Legendre polynomials.} \end{aligned} \tag{A.15}$$

Compendium The Legendre polynomials are defined by the equality

$$\mathscr{P}_l(\mu) \leftrightharpoons \frac{1}{2^l l!}\frac{\mathrm{d}^l}{\mathrm{d}\mu^l}(\mu^2 - 1)^l, \qquad \mu \in [-1, 1], \quad l = 0, 1 \ldots.$$

Below we shall use the following facts [190] (the prime temporarily denotes the differentiation with respect to μ):

$$(1 - \mu^2)\mathscr{P}_l'' - 2\mu\mathscr{P}_l' + l(l+1)\mathscr{P}_l = 0; \tag{A.16}$$

$$|\mathscr{P}_l| \leqslant 1, \quad \mathscr{P}_l(1) = 1; \tag{A.17}$$

$$\mathscr{P}_{l+1}' - \mu\mathscr{P}_l' = (l+1)\mathscr{P}_l. \tag{A.18}$$

By (A.16),

$$\Delta_\Omega\mathscr{P}_l(\cos\vartheta) = \Big[(1-\mu^2)\partial_\mu^2 - 2\mu\partial_\mu\Big]\mathscr{P}_l(\mu)\big|_{\mu=\cos\vartheta} = -l(l+1)\mathscr{P}_l(\cos\vartheta), \tag{A.19}$$

and, correspondingly, for z solving Eq. (A.6)

$$\begin{aligned} &\Big[\partial_x^2 + \frac{2r'}{r}\partial_x + \frac{\Delta_\Omega}{r^2}\Big]\mathscr{P}_l(\cos\vartheta)\frac{z(l,x)}{r} =\\ &= \mathscr{P}_l(\cos\vartheta)\Big[\partial_x^2 + \frac{2r'}{r}\partial_x - \frac{l(l+1)}{r^2}\Big]\frac{z(l,x)}{r} =\\ &\qquad = \mathscr{P}_l(\cos\vartheta)\frac{1}{r}\Big[\partial_x^2 - \frac{r''}{r} - \frac{l(l+1)}{r^2}\Big]z(l,x) = 0 \end{aligned} \tag{A.20}$$

[in transition to the last line we have used the identity (A.7)]. Hence, in particular,

$$\Delta[\mathscr{P}_l(\cos\vartheta)r^{-l-1}] = 0. \tag{A.21}$$

The Legendre polynomials are used [2] to define the *spherical functions*

$$Y_l^m(\varphi,\vartheta) \leftrightharpoons \sqrt{\frac{2l+1}{4\pi}\frac{(l-|m|)!}{(l+|m|)!}} e^{im\varphi}(1-\mu^2)^{\frac{|m|}{2}} \frac{\mathrm{d}^{|m|}}{\mathrm{d}\mu^{|m|}} \mathscr{P}_l(\mu)\big|_{\mu=\cos\vartheta},$$

which solve the equation

$$\Delta_\Omega Y_l^m = -l(l+1)Y_l^m \tag{A.22}$$

[for $m=0$ this follows from (A.19)]. These functions make up an orthonormal basis in the space $L_2(\mathbb{S}^2)$ and each function f of this space can be represented [175] as

$$f(\varphi,\vartheta) = \sum_{l=0}^{\infty}\sum_{m=-l}^{l} \tilde{f}_l^{(m)} Y_l^m(\varphi,\vartheta), \text{ where} \tag{A.23}$$

$$\tilde{f}_l^{(m)} \leftrightharpoons \int \mathrm{d}\varphi \mathrm{d}\vartheta \sin\vartheta\, \bar{Y}_l^m(\varphi,\vartheta) f(\varphi,\vartheta). \tag{A.24}$$

Finally, for any $p, p_* \in \mathbb{E}^3$ and, hence, for any $p, p_* \in \mathfrak{F}^+$ the following relation is true (see, for example, [2, (П 2.13)]):

$$\frac{1}{|p,p_*|} = \frac{1}{r}\sum_{l=0}^{\infty}[H(x-x_*)(r_*/r)^l + H(x_*-x)(r/r_*)^{l+1}]\mathscr{P}_l(\cos\theta), \tag{A.25}$$

where $|p, p_*|$ is defined in the end of A.1.1.

Proof We shall prove Proposition 5 in two steps. Namely, we shall show that, first, Φ (or, rather, Φ_{inh}, since the last two terms, as will be shown later, are harmonic functions) satisfies Eq. (A.3) and that, second, *any* solution of that equation has the same form, the only difference being the values of Q and Φ_0.

A Particular Solution to the Nonhomogeneous Equation

For a sufficiently large l, consider separately two cases:
(1) $x < x_{**} \leftrightharpoons \frac{1}{2}(x_* + \mathfrak{z})$. In this case taking into account the inequality $x < x_*$ one has by the definition of v, see (A.15),

$$\begin{aligned}|v_l(x)| =& |(C_l r_*)^{-1} z_-(l,x) z_+(l,x_*)| < |(C_l r_*)^{-1} z_-(l,x_{**}) z_+(l,x_*)| = \\ &= |r_{**}^{l+1} + \alpha_l r_{**}^{-l}| r_*^{-l-1} \leqslant (r_{**}/r_*)^{l+1} + 2(r(\mathfrak{z})/r_{**})^l (r(\mathfrak{z})/r_*)^{l+1}\end{aligned}$$

[the first inequality follows from the fact that z_- monotonically increases, see Proposition 3, and the second—from the inequality (A.12)]. Further, z'_- also monotonically increases. So, applying the same reasoning to the derivative one obtains the estimate

$$\begin{aligned}|v_l'(x)| < |(C_l r_*)^{-1} z'_-(l,x_{**}) z_+(l,x_*)| = |(l+1) r_{**}^l - l\alpha_l r_{**}^{-l-1}| r_*^{-l-1} \leqslant \\ \leqslant \frac{l}{r_*}\big[2(r_{**}/r_*)^l + 2(r(\mathfrak{z})/r_{**})^{l+1}(r(\mathfrak{z})/r_*)^l\big].\end{aligned}$$

Finally, using the relation (A.6) between z and z'' one finds that

$$|v_l''(x)| = |(C_l r_*)^{-1} z_-''(l,x) z_+(l,x_*)| < 2l^2 |(C_l r_*)^{-1} z_-(l,x) r^{-2} z_+(l,x_*)| <$$
$$< 2r_{\min}^{-2} l^2 |(C_l r_*)^{-1} z_-(l,x_{**}) z_+(l,x_*)| < 2r_{\min}^{-2} l^2 (C_l r_*)^{-1} r_*^{-l} C_l (r_{**}^{l+1} +$$
$$+ |\alpha_l| r_{**}^{-l}) < 6r_{\min}^{-2} l^2 (r_{**}/r_*)^{l+1}$$

[the last inequality is derived using the estimate (A.12) and the fact that r is monotone on the interval $(\mathfrak{z}, x_{**})$]. The three just derived bounds show that (by the Weierstrass M-test) the series (A.14) and the series obtained from it by one or two term by term differentiations converge uniformly. Hence, the last two converge to the, correspondingly, first and second derivatives of Φ_{inh}. Therefore,

$$\Delta\Phi_{inh} = q \sum_{l=0}^{\infty} \left[\partial_x^2 + \frac{2r'}{r} \partial_x + \frac{\Delta_\Omega}{r^2} \right] \frac{v_l(x)}{r} \mathscr{P}_l(\cos\vartheta).$$

But $v_l(x)$ in the domain under consideration is proportional to $z_-(l, x)$, which means, see (A.20), that

$$\Delta\Phi_{inh} = 0,$$

i.e. Eq. (A.3a) is fulfilled.

Further, at $\vartheta = 0$, differentiate the series (A.14) with respect to x. Since this can be done termwise, as we just have established, and since v_l at $x < -\mathfrak{z}$ is r^{-l} [up to a positive, see (A.10), factor] the result is positive. So, $\Phi_{inh}(\vartheta = 0)$ strictly decreases (remaining positive) at $x \to -\infty$. But each term of the series (A.14) for all x has the maximum at $\vartheta = 0$, see (A.17). Thus,

$$|\Phi_{inh}(\vartheta, x)| \leqslant \Phi_{inh}(0, x), \qquad \forall x, \vartheta,$$

and we conclude that the condition (A.3b) is fulfilled too.

(2) Now consider the case $x \geqslant x_{**}$. To the right-hand side of (A.14), add the Coulomb term and subtract the same term in the form of the series (A.25):

$$\Phi_{inh} = \frac{\mathfrak{q}}{|pp_*|} + \frac{\mathfrak{q}}{r} \sum_{l=0}^{\infty} \mathscr{P}_l(\cos\vartheta) \left\{ H(x - x_*) \left[\frac{C_l(r_*^{l+1} + \alpha_l r_*^{-l})}{r_* C_l r^l} - \left(\frac{r_*}{r}\right)^l \right] + \right.$$
$$\left. + H(x_* - x) \left[\frac{C_l(r^{l+1} + \alpha_l r^{-l}) r_*^{-l}}{r_* C_l} - \left(\frac{r}{r_*}\right)^{l+1} \right] \right\} =$$
$$= \frac{\mathfrak{q}}{|pp_*|} + \sum_{l=0}^{\infty} \frac{\mathfrak{q}}{r r_*} \mathscr{P}_l(\cos\vartheta) \frac{\alpha_l}{r^l r_*^l}. \quad \text{(A.26)}$$

The estimate (A.12) means that the series in the right-hand side of (A.26) is dominated by a convergent (at r's in the interval under consideration) geometric series and, therefore, uniformly converges. Moreover, for the same reason, the series remains uniformly convergent after being once or twice differentiated term by term. Hence, taking the Laplacian can also be done term by term. At the same time, each term is harmonic, as seen from (A.21). So,

$$\Delta\Phi_{inh} = \Delta\Big(\frac{\mathfrak{q}}{|pp_*|}\Big) = -4\pi\mathfrak{q}\delta(p - p_*).$$

Thus, Φ_{inh} solves Eq. (A.3a). And the validity of (A.3b) is proved in much the same way as in the previous case.

The General Solution to the Homogeneous Equation

We are left with the task of proving that all solutions of the equation

$$\Big[\partial_x^2 + \frac{2r'}{r}\partial_x + \frac{\Delta_\Omega}{r^2}\Big]\Phi_{hom} = 0, \tag{A.27}$$

see (A.3a) and (A.5), that obey the condition (A.3b) have the form

$$-Q\int_3^x r^{-2}\,\mathrm{d}\breve{x} + \Phi_0,$$

where Q and Φ_0 are arbitrary constants.

Let us first estimate the coefficients (A.24), we denote them Φ_l^m, of the expansion (A.23) for the potential Φ_{hom}. On the one hand, (A.3b) requires them to be bounded. But on the other hand, it follows from the chain of equalities

$$\begin{aligned}
0 &= \int \bar{Y}_l^m(\varphi,\vartheta)\Delta\Phi_{hom}(\varphi,\vartheta,x)\,\sin\vartheta\,\mathrm{d}\varphi\mathrm{d}\vartheta = \\
&= \Big[\partial_x^2 + \frac{2r'}{r}\partial_x\Big]\int \Phi_{hom}\bar{Y}_l^m\,\sin\vartheta\,\mathrm{d}\varphi\mathrm{d}\vartheta + \frac{1}{r^2}\int \Phi_{hom}\Delta_\Omega\bar{Y}_l^m\,\sin\vartheta\,\mathrm{d}\varphi\mathrm{d}\vartheta = \\
&= \Big[\partial_x^2 + \frac{2r'}{r}\partial_x - \frac{l(l+1)}{r^2}\Big]\int \Phi_{hom}\bar{Y}_l^m\,\sin\vartheta\,\mathrm{d}\varphi\mathrm{d}\vartheta = \\
&\qquad = \Big[\partial_x^2 + \frac{2r'}{r}\partial_x - \frac{l(l+1)}{r^2}\Big]\Phi_l^m(x)
\end{aligned}$$

[where we have used (A.22)] that the product $r\Phi_l^m(x)$ solves Eq. (A.6), see the identity (A.7). And this means, by Corollary 4, that Φ_l^m may be bounded only at $l = 0$. Thus, as easily verified by substitution,

$$\Phi_{hom} = \Phi_0^{(0)}(x)Y_0^0(\varphi,\vartheta) = z_0/r,$$

where z_0 is a solution of the equation $(\partial_x^2 - r''/r)z_0 = 0$, that is, an arbitrary linear combination of r and $r\int_3^x r^{-2}\mathrm{d}\breve{x}$.

□

A.1.4 The Action of a Wormhole on a Pointlike Charge

In the presence of a pointlike charge, the electric field outside the wormhole (i.e. in the region $\mathfrak{F}^+$) has the form

$$\boldsymbol{E} = -\nabla\Phi = -\nabla\left(\frac{\mathfrak{q}}{|pp_*|} + \frac{\tilde{Q}}{r} + \mathfrak{q}\sum_{l=1}^{\infty}\frac{\alpha_l \mathscr{P}_l(\cos\vartheta)}{(rr_*)^{l+1}}\right)$$

[this expression is obtained by substituting (A.26) into (A.13), taking explicitly the integral at Q, discarding the constant term, and, finally, replacing Q with $\tilde{Q}$ to absorb the zeroth term of the series]. Hence, the force exerted on the test charge $\mathfrak{q}$ with coordinate r is[1]

$$\begin{aligned}\boldsymbol{F} &= \frac{\tilde{Q}\mathfrak{q}}{r_*^3}\boldsymbol{r}_* - \mathfrak{q}^2\nabla\sum_{l=1}^{\infty}\mathscr{P}_l(\cos\vartheta)\frac{\alpha_l}{(rr_*)^{l+1}}\Big|_{\vartheta=0,r=r_*} = \\ &= \frac{\tilde{Q}\mathfrak{q}}{r_*^3}\boldsymbol{r}_* - \mathfrak{q}^2\frac{\boldsymbol{r}_*}{r_*}\frac{\mathrm{d}}{\mathrm{d}r}\left(\sum_{l=1}^{\infty}\frac{\alpha_l}{(rr_*)^{l+1}}\right)\Big|_{r=r_*} \\ &= \frac{\tilde{Q}\mathfrak{q}}{r_*^3}\boldsymbol{r}_* + \frac{\mathfrak{q}^2\boldsymbol{r}_*}{r_*^2}\sum_{l=1}^{\infty}\frac{(l+1)\alpha_l}{r_*^{2l+2}},\end{aligned} \tag{A.28}$$

as discussed in the beginning of the section, see (A.4). The first, Coulomb, term does coincide with the force that *would* act upon $\mathfrak{q}$, be $\mathfrak{F}^+$ a part of a Euclidean space—not of a wormhole—containing, in addition to $\mathfrak{q}$, a pointlike charge $\tilde{Q}$ in the origin of coordinates. In this sense, the wormhole, in agreement with the concept of 'charge without charge', resembles a charged body even though the matter filling it is electrically neutral, see (A.2). There are, however, significant differences too.

A.1.4.1 The wormhole in comparison with a charged body

So far $\tilde{Q}$ has been an absolutely arbitrary function of p_*. Correspondingly, at this stage expression (A.28) does not suffice to find the force $\boldsymbol{F}(p_*)$ experienced by a test charge located at p_*. Moreover, even if we somehow knew $\boldsymbol{F}(p_*)$ this would not enable us to find $\boldsymbol{F}(p'_*)$ for any $p'_* \neq p_*$. As a consequence, we cannot compare the law (A.28) with the Coulomb law.

The problem can be solved when the final state (in which the charge $\mathfrak{q}$ rests at p'_*) evolves from the initial one (the charge resting at p_*) in agreement with the Maxwell equations (this condition can be satisfied only *approximately*, the field in both states being static). In such a case the flux of $\boldsymbol{E}$ through a sufficiently large (not crossed

[1] Since the charge is in the flat region, the 'renormalization' by dropping the term $\nabla\frac{\mathfrak{q}^2}{|pp_*|}$ requires no special justification: by the electrodynamics in curved spacetime, we understand a local geometric generalization of the usual one, cf. Sect. 1 in Chap. 2.

by $\mathfrak{q}$ in the course of evolution) sphere remains constant [181]. Thus, $\tilde{Q}$ (not Q as argued in [15, 86][86][15]) deserves the name of the wormhole's charge, being a constant which generates the potential $\tilde{Q}/r$.

Now consider the second term of the expression (A.28). This term does not depend on $\tilde{Q}$ and is proportional to $\mathfrak{q}^2$. So, it is natural to think of it as a force exerted on the charge by itself ('self-force'). However, it also owes its existence to the presence of the wormhole, or, at least, of the curved regions—in a totally flat spacetime it is lacking. Note that the self-force does not have to be small: in fact, it may well constitute the greater part of the force acting on a test charge.

Finally, there is an important asymmetry: if the sphere $x = \mathfrak{z}$ bounded a charged body, instead of the wormhole's mouth, the body would move under the attraction of $\mathfrak{q}$. Whether the throat will move in this case is unclear.

A.1.5 The Short Throat Approximation

In this subsection, we consider the self-force in the limit $\mathfrak{z} \to 0$. The reason is that the result, as we shall see, does not depend on the shape of the wormhole. The only—quite innocuous—condition that we impose on it is

$$r \geqslant r^{\min}, \qquad |r'/r| \leqslant |r'/r|^{\max}, \qquad \forall x \in (-1, 1), \tag{A.29}$$

where the right-hand sides of the inequalities are some positive constants (i.e. quantities that are the same for all r *and* $\mathfrak{z}$).

Proposition 6 *Let $\mathfrak{z}$ tend to 0 and let l be fixed. Then the derivative of the function $\lambda \leftrightharpoons \ln' z_- - \ln' r$ is bounded on the interval* $(-1, 1)$ *uniformly over* $\mathfrak{z}$.

Proof By definition, λ solves the equations

$$\lambda' = -\lambda^2 - 2r'r^{-1}\lambda + l(l+1)r^{-2}, \tag{A.30a}$$

$$\lambda\big|_{x \leqslant -\mathfrak{z}} = (l+1)/r, \tag{A.30b}$$

see (A.6). Consequently, λ is positive. Indeed, to the left of $-\mathfrak{z}$ this follows from (A.30b), while to change its sign at larger x, λ would have first to become small, which [due to the last term in the right-hand side of (A.30a)] would make the derivative λ' positive and, correspondingly, would make λ increasing.

The right-hand side (RHS) of Eq. (A.30a) admits the estimate

$$\text{RHS} \leqslant -\lambda^2 + 2|r'/r|^{\max}\lambda + l(l+1)/(r^{\min})^2 \leqslant (|r'/r^{\max}|)^2 + l(l+1)/(r^{\min})^2$$

[the second inequality is obtained by simply finding the maximum of the square polynomial (in λ) in its left-hand side] and, hence, at any fixed l the derivative λ' on the interval $(-1, 1)$ is bounded *above*, uniformly over $\mathfrak{z}$. The same is also true for λ [since $\lambda(x) \leqslant \lambda(-1) + (x+1)(\lambda')^{\max}$]. On the other hand, λ, as we just have established, is positive. On the strength of Eq. (A.30a) combined with the hypothesis (A.29), the just proven boundedness of λ implies the boundedness—uniform over $\mathfrak{z}$, again —of λ', now from *below*. $\square$

This enables us to find the limit values of α_l. Indeed,

$$\ln' z_-(\mathfrak{z}) = \lambda(\mathfrak{z}) + 1 = \lambda(-\mathfrak{z}) + 1/(1+\mathfrak{z}) + 2\mathfrak{z}\lambda'(x_\mathfrak{z}) = l + 2 + O(\mathfrak{z}),$$

where $x_\mathfrak{z} \in (-\mathfrak{z}, \mathfrak{z})$ and the last equality follows from Proposition 6. Substituting this estimate into (A.11) yields

$$\alpha_l \to \frac{-(2l+2)+(2l+1)}{2l+2} = -\frac{1}{2(l+1)}.$$

Now that we know this limit we, at last, can find the force acting on the charge. As follows from (A.28), it is the sum of the 'Coulomb' field $\frac{\tilde{Q}\mathfrak{q}}{r_*^3}\boldsymbol{r}_*$, which may be interpreted as generated by the wormhole, and the self-force $\boldsymbol{F}^{s-i}$. The latter is directed radially and its value is

$$F_r^{s-i} \underset{\mathfrak{z}\to 0}{\longrightarrow} \frac{\mathfrak{q}^2}{r_*} \lim_{\mathfrak{z}\to 0} \sum_{l=1}^{\infty} \frac{\alpha_l(l+1)}{r_*^{2l+2}} =$$

$$= \frac{\mathfrak{q}^2}{r_*} \sum_{l=2}^{\infty} \frac{l \lim_{\mathfrak{z}\to 0} \alpha_{l-1}}{r_*^{2l}} = -\frac{\mathfrak{q}^2}{2r_*} \sum_{l=2}^{\infty} \frac{1}{r_*^{2l}} = \frac{-\mathfrak{q}^2}{2r_*^3(r_*^2-1)}$$

[lim and $\sum$ commute, because by (A.12) the series converges uniformly over $\mathfrak{z}$].

Conclusion A charge draws itself into a short wormhole with a finite force.

A.2 Shortcut in Curved Spacetime

In this section, we demonstrate by means of an explicit example that a spacetime *resembling* a shortcut may satisfy the weak energy condition. The difference between such a spacetime and a real shortcut is that outside U (see Definition 5 in Chap. 3) the former is the Schwarzschild rather than the Minkowski space. This makes one suspect that exotic matter is not at all necessary for faster-than-light travel, WEC violations resulting merely from the poor definition of shortcut.

The situation modelled in this section is similar to that considered in Sect. 2.5 in Chap. 2. in connection with gravitational signalling. At the moment $t = 0$, an earthling $\mathscr{E}$ decides to contact an alien $\mathscr{A}$. The space separating them is not any longer considered empty, in particular, it is not flat. Instead, it is filled with matter constituting a static globular cluster B with the centre located exactly between $\mathscr{E}$ and $\mathscr{A}$. The earthling considers two scenarios again:

1. They send a photon, which traverses the cluster and reaches $\mathscr{A}$, who—in the opposite direction—an answer signal, which arrives to the Earth at a moment[2] $t(w)$.
2. Instead of a test photon, the Earth sends a spaceship, which disturbs the metric—by exploding stars, emitting gravitational waves, etc.—in a way consistent with causality. This will not enable the spaceship to outrun a photon emitted at $t = 0$, but by the time the former arrives to $\mathscr{A}$ the metric in B will have *changed* (even though it will remain the same outside the cluster). It is obvious, therefore, that generally the spaceship will return to the Earth at $t(w')$ different from $t(w)$. And it may happen, in particular, that $t(w') < t(w)$. Then the return trip will be superluminal in the sense of Sect. 1 in Chap. 3.

In the second scenario, the spacetime is essentially a shortcut, but the role of the Minkowski space is played now by the Schwarzschild space. Our task is to verify that the WEC does not forbid such spacetimes. Specifically, we are going to build a spherically symmetric spacetime (M, g) with the following properties:

1. Outside of a cylinder $U \leftrightharpoons \{p\colon\ r(p) < r_0\}$, it is the Schwarzschild space, in which r is the standard radial coordinate;
2. Throughout M, the weak energy condition holds;
3. The minimal (Schwarzschild) time necessary for travelling through U from a point with $r > r_0$ to the diametrically opposite point *decreases with time.*

For this purpose, pick three positive constants: $\mathfrak{m}$, r_h and $r_0 > r_h$. Then, choose two smooth functions, ψ and Θ, obeying the following conditions:

$$\psi \geqslant 0, \qquad \Theta\big|_{r<r_h} = 10, \quad \Theta\big|_{r>r_0} = 1, \qquad \Theta' \leqslant 0.$$

Finally, use these quantities to define two more functions:

$$m(r) \leftrightharpoons \mathfrak{m} r^{-1/3} \exp \int_{r_h}^{r} \frac{\Theta(x)\,\mathrm{d}x}{3x}, \tag{A.31}$$

$$\epsilon(r) \leftrightharpoons \int_{r_0}^{r} (r - x)\frac{x^2[m'(x)x^{-2}]'}{x - 2m(x)}\psi(x)\,\mathrm{d}x. \tag{A.32}$$

For later use, let us examine these functions. First, note that

$$m(r)\big|_{r<r_h} = \mathfrak{m} r_h^{-10/3} r^3, \qquad m(r)\big|_{r>r_0} = const, \tag{A.33a}$$

i.e. m behaves as the mass of a non-relativistic ball with the density that is constant at $r < r_h$ and smoothly decays to zero as $r \to r_0$. Second, its derivative is non-negative

$$m'(r) = \frac{(\Theta - 1)m(r)}{3r} \geqslant 0, \tag{A.33b}$$

[2] Notation imitates that used in Chap. 3.

but

$$[m'(r)/r^2]' = \frac{m(r)}{9r^4}[3r\Theta' + (\Theta - 1)(\Theta - 10)] \leqslant 0, \tag{A.33c}$$

The equality here is attained at $r \notin (r_h, r_0)$. Substituting (A.33c) in the definition (A.32) gives

$$\epsilon\Big|_{r<r_h} = const, \qquad \epsilon\Big|_{r>r_0} = 0. \tag{A.33d}$$

Let the constant $\mathfrak{m}$ be so small that

$$r > 2m(r), \quad \text{and, hence, also } \epsilon' \geqslant 0 \tag{A.34}$$

[that such $\mathfrak{m}$ can be found, follows from (A.31) and (A.33a)]. The latter inequality is obtained from the former by differentiating (A.32) and using (A.33c) when the fact is taken into account that *inside* U the upper limit of the integral (A.32) is smaller than the lower, while *outside* it $\epsilon = 0$.

Now consider the (static and, hence, auxiliary) metric

$$\mathrm{d}s^2 = -e^{2\epsilon}(1 - 2m/r)\mathrm{d}t^2 + (1 - 2m/r)^{-1}\mathrm{d}r^2 + r^2(\mathrm{d}\vartheta^2 + \sin^2\vartheta\mathrm{d}\varphi^2). \tag{A.35}$$

on M. As already mentioned, outside of U, i.e. 'beyond the cluster', $\epsilon = 0$ and $m = const$. So, this region is simply a part of the Schwarzschild space lying outside the horizon.

Let us verify that by choosing a suitable function ψ (which has been almost arbitrary so far) one can force the metric (A.35) to satisfy the weak energy condition. A simple way to do this is to verify with the aid of equations [135, (14.43)] that in the orthonormal basis $\boldsymbol{e}_{(0)} \sim \partial_t$, $\boldsymbol{e}_{(1)} \sim \partial_r$, $\boldsymbol{e}_{(2,3)} \sim \partial_{\vartheta,\varphi}$

$$G^{\hat{0}\hat{0}} = 2r^{-2}m' \geqslant 0, \qquad G^{\hat{0}\hat{0}} + G^{\hat{1}\hat{1}} = 2\frac{r-2m}{r^2}\epsilon' \geqslant 0, \tag{A.36}$$

and for $i = 2, 3$

$$G^{\hat{0}\hat{0}} + G^{\hat{i}\hat{i}} = (1 - 2m/r)\Big[\epsilon'^2 + \frac{\epsilon'}{r}\Big(1 - 3\frac{m'r - m}{r - 2m}\Big) + \epsilon'' - \frac{r^2}{r - 2m}(m'/r^2)'\Big].$$

The first term in the square brackets is non-negative and—if $\mathfrak{m}$ is sufficiently small—the second term, too. So, by twice differentiating (A.32) we get

$$G^{\hat{0}\hat{0}} + G^{\hat{i}\hat{i}} \geqslant (\psi - 1)\frac{r^2}{r - 2m}(m'/r^2)'. \tag{A.37}$$

The right-hand side of the inequality is non-negative, see (A.33b), when $\psi < 1$.

Thus, when $\mathfrak{m}$ is sufficiently small, the metric (A.35) satisfies the WEC for *any* positive $\psi < 1$. Moreover, it is easy to see that in some interval $[r_1, r_2] \subset (r_h, r_i)$ the inequalities (A.36) and (A.37) are, in fact, *strict*. And hence they also hold for the metric (A.35), in which $\psi(r)$ is changed to $\psi(r) - \kappa(t)\psi_1(r)$, where κ and ψ_1 are non-negative, supp $\psi_1 \subset [r_1, r_2]$, and, finally, κ, $\dot{\kappa}$, and $\ddot{\kappa}$ are sufficiently small (the higher derivatives of κ do not enter the Einstein tensor). Consider such a 'deformed' metric in the case, where κ increases with t (and, correspondingly, ϵ—which, as follows from (A.33d) and (A.34), is non-negative and whose absolute value, by (A.32), decreases—grows too). In this metric, any causal curve $\gamma \subset U$ has the following property. Each $\tilde{\gamma}$, obtained from γ by translating every point the same $\Delta t > 0$ along the lines of t, is causal and, moreover, contains *timelike* segments (they lie in the regions where $|g_{00}|$ has increased). According to Proposition 1.23(d), $\tilde{\gamma}$ can be deformed so that its past end point remains fixed, the curve itself remains causal and the future end point is displaced in the past direction. Thus, even though the metric outside the cluster remains intact, it takes less and less time (by the Earth clock) to cross the cluster. Put differently, the return takes less time than the outbound trip.

A.3 Topology Evolution

Can a space change its topology in the course of evolution? The answer is simple in the two extreme cases:

(1) If we restrict our consideration to globally hyperbolic spacetimes, it becomes possible to formulate the question rigorously: the subject matter in this case is the topology of the Cauchy surfaces. The impossibility of the topology changes now follows from Geroch's splitting theorem, see Proposition 51(b) in Chap. 1.

(2) If, on the other hand, we consider *arbitrary* spacetimes, it is unclear even whose topology is discussed. It makes no sense, for example, to simply turn to slices[3] instead of Cauchy surfaces, because the former, in contrast to the latter, may have different topologies even if they intersect, as, for example, in the de Sitter space, cf. [64, Fig. 5.16].

The problem, it seems, can be solved by the following modification [89]: the topology of spacetime is regarded changing if there are causally related points p, q and an achronal surface $\mathcal{P} \ni p$ such that none of achronal surfaces through q is homeomorphic to $\mathcal{P}$. According to this definition, topology does not change either in the de Sitter, or in the Bardeen spaces (the metric of the latter can be found in [136] and the description of its geometry—in [76]). However, there are simple spacetimes where the topology does change, see Example 30 in Chap. 2 (for more refined transformations see [182]). The question now can be reformulated as follows: is a topology change, understood as above, possible? Or, put differently, what pathologies are *unavoidable* in spacetimes with changing topology? Important results in this direction were obtained in [60, 166], where it was proven that topology changes are impossible in spacetimes of a certain type. These are spacetimes in which some subsets (loosely speaking, the places where the changes occur) are compact and where (in the case

[3] A *slice* is a spacelike, three-dimensional closed submanifold of spacetime [64].

of [166]) some energy conditions hold, including the WEC. However, contrary to a widespread opinion, this fact does not forbid the topology changes outright. Perhaps, the compactness of the mentioned sets is a *too* strong requirement, because its lack does not necessarily imply that the spacetime is extendible or singular. At the same time a milder condition, which excludes WEC violations and singularities, but not *infinities*, does not rule out some topology changes. To prove this, we are going to present a procedure yielding spacetimes (M_w, g) with such changes.

The procedure consists in the following three steps: first, by cutting-and-gluing, see Sect. 6 in Chap. 1, we construct from a flat two-dimensional space (N, η) a (singular) spacetime (N_w, η) with the changing, in the above sense, topology. Then, we find a conformal factor w^{-2} which sends all the singularities to infinity. Now the desired spacetime (M_w, g) is the product $(N_w \times \mathbb{S}^2, g^{(w)} \times g^{(r_*)})$, where $g^{(w)} \leftrightharpoons w^{-2}\eta$ and $(\mathbb{S}^2, g^{(r_*)})$ is the standard sphere of radius r_*

$$g\colon \qquad ds^2 = w^{-2}(-dx_0^2 + dx_1^2) + r_*^2(dx_2^2 + \sin^2 x_2 dx_3^2) \qquad \text{(A.38)}$$

(x_2 and x_3 are just aliases for the usual spherical coordinates φ and ϑ). The topology of (M_w, g) changes with time, but this does not make it any 'worse' in other respects. In particular, M_w may contain a region isometric to the complement to a compact set in a maximal globally hyperbolic space. Such M_w describes a 'laboratory' (as opposed to 'cosmological') topology change. The same construction is used in building a non-singular time machine in Sect. 5 in Chap. 4.

Example 7 1. *Bridge.* Cut the Minkowski plane N along the segment $|x_0| \leqslant 1$, $x_1 = 1$, see Fig. A.1a, and along the same segment at $x_1 = -1$. Let N_w be the result of gluing the right bank of each cut to the left bank of the other (i.e. N_w is the spacetime M_2 from Example 30 in Chap. 2). Let, further,

$$\rho \leftrightharpoons (x_0^2 - 1)^2 + (x_1^2 - 1)^2$$

($x_{0,1}$ are the coordinates inherited from N, they cover the whole N_w except for the two segments) and w be a positive function equal to ρ at small ρ and constant at $\rho \geqslant 1$. The space M_w obtained from this N_w as described above is a universe two distant regions of which become connected for some time by two 'bridges'.
2. *Compactification.* Now let N be a flat cylinder

$$ds^2 = -dx_0^2 + dx_1^2, \qquad x_0 \in \mathbb{R}, \quad x_1 = x_1 + 3$$

and w be an arbitrary smooth function, constant at $x_0^2 + x_1^2 \geqslant 1$ and positive (only) in $N_w \leftrightharpoons \{p\colon\ p \in N,\ x_0^2 + x_1^2 > 1/2\}$. Then, the universe M_w is initially closed in the sense that the sections $x_0 = const$ (which are Cauchy surfaces for this initial region) are compact. With time, however, it becomes open and then compactifies again, see Fig. A.1b. All this occurs, as we shall see, without the appearance of any singularities.

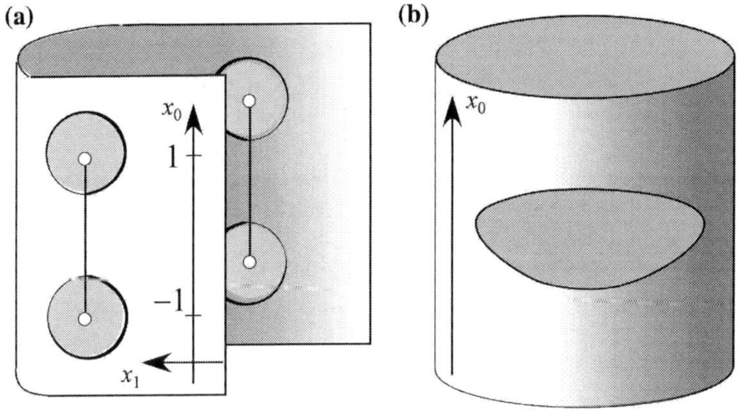

Fig. A.1 **a** To obtain N_w, identify the banks of the vertical slits crosswise. w is constant beyond the light grey regions (note that there are *two*—not four—such regions). **b** The space N_w is a cylinder from which a disc is removed. In the course of evolution the universe, starts from being closed, becomes open and then closes again

The spacetimes M_w built above are not globally hyperbolic, but are 'almost as nice'. In particular, they are strongly (and stably [76]) causal. Let us check that for a suitable r_* they also satisfy the weak energy condition and are free from singularities.

The Riemann tensor for the metrics under consideration can be easily found with the aid of the formula [135, (14.50)]. In the orthonormal basis

$$\omega^{(0)} \leftrightharpoons w^{-1}\mathrm{d}x_0, \quad \omega^{(1)} \leftrightharpoons w^{-1}\,\mathrm{d}x_1, \quad \omega^{(2)} \leftrightharpoons r_*\,\mathrm{d}x_2, \quad \omega^{(3)} \leftrightharpoons r_*\sin x_2\,\mathrm{d}x_3$$

its non-zero components (up to exchange of the indices) are

$$R_{\hat{2}\hat{3}\hat{2}\hat{3}} = r_*^{-2}, \quad R_{\hat{0}\hat{1}\hat{0}\hat{1}} = -E \leftrightharpoons w,_{x_0}^2 - w,_{x_1}^2 + w(w,_{x_1x_1} - w,_{x_0x_0})$$

(the hats over the indices stand to suit the notation of [135]). Thus, all curvature scalars are bounded and so are tidal forces acting on a free falling (or moving with a finite acceleration for a finite time) observer [89]. The Einstein tensor in the same basis is $G_{\hat{a}\hat{b}} = \mathrm{diag}(r_*^{-2}, -r_*^{-2}, -E, -E)$. So, the weak energy condition holds, if

$$r_*^{-2} > w(w,_{x_1x_1} - w,_{x_0x_0}) + w,_{x_0}^2 - w,_{x_1}^2\,. \tag{A.39}$$

It remains to prove that there are no singularities in M_w. This assertion needs some refinement, because there are different understandings of what is a singularity, see [62, 76]. Roughly speaking, they differ in how the length of a curve is defined (the problem is that the metric is *pseudo*-Riemannian) and which inextendible curves are required to be infinitely long [62]. To regard a spacetime singularity-free, one could, for example, require that all inextendible timelike geodesics be complete or that the proper time be unbounded on any timelike curve with bounded acceleration lasting for a finite time. We adopt the weakest of common definitions (i.e. such that a spacetime is singular in any other—usual—sense, if it is singular by our definition). Specifically, we identify singularity with *b*-incompleteness, see below. Since we are interested not in *all* singularities, but only in those associated with our procedure, we consider only curves along which $w \to 0$, i.e. we verify, loosely speaking, that the conformal transformation does rid the spacetime of the singularities $w = 0$ transforming them into infinities.

Thus, consider a curve $\gamma(\zeta)\colon [0, 1) \to M_w$. Denote by $\boldsymbol{l}$ its velocity

$$l^i = \mathrm{d}x^i(\zeta)/\mathrm{d}\zeta, \qquad \text{where} \quad x^i(\zeta) \leftrightharpoons x^i\big(\gamma(\zeta)\big),$$

and by γ_{N_w} and $\gamma_{\mathbb{S}}$—the projections of γ to, respectively, N_w and $\mathbb{S}^2$. Next, pick an orthonormal basis $\{\boldsymbol{e}_{(i)}(0)\}$ in $\gamma(0)$.[4] Then, in all points of γ the bases $\{\boldsymbol{e}_{(i)}(\zeta)\}$ are defined to be the parallel transport of $\{\boldsymbol{e}_{(i)}(0)\}$ along γ:

$$e^a_{(i);b} l^b = 0. \tag{A.40}$$

The *generalized affine length* of a segment $\gamma\Big([0, \zeta)\Big)$ is defined as

$$L_\gamma(\zeta) \leftrightharpoons \int_0^\zeta \Big(\sum_i g(\boldsymbol{l}, \boldsymbol{e}_{(i)})^2\Big)^{1/2} \mathrm{d}\zeta', \tag{A.41}$$

see [159]. Specific values of $L_\gamma(\zeta)$ depend on the choice of the initial basis, but its *boundedness* does not, which makes the following definition correct: γ is called *b-complete*, if $L_\gamma(\zeta)$ is unbounded.

To find out whether γ is *b*-complete, it is convenient to choose the vectors of the initial basis to be directed along the coordinate lines $\boldsymbol{e}_{(i)}(0) \sim \partial_{x^i}$. We also introduce the coordinates $\alpha \leftrightharpoons x_1 + x_0$, $\beta \leftrightharpoons x_1 - x_0$, in which the metric takes the form

$$g^{(w)}\colon \qquad \mathrm{d}s^2 = w^{-2}\mathrm{d}\alpha\mathrm{d}\beta,$$

and the corresponding vectors $\boldsymbol{e}_{(\alpha)} \leftrightharpoons \frac{1}{2}(\boldsymbol{e}_{(1)} + \boldsymbol{e}_{(0)})$, $\boldsymbol{e}_{(\beta)} \leftrightharpoons \frac{1}{2}(\boldsymbol{e}_{(1)} - \boldsymbol{e}_{(0)})$ (they are, of course, also parallel along γ). The only non-zero Christoffel symbols with the indices α or β are

$$\Gamma^\mu_{\mu\mu} = \tfrac{1}{2}g^{\mu\nu}2g_{\mu\nu},_\mu = (\ln g_{\mu\nu}),_\mu = -2\sigma,_\mu \qquad \sigma \leftrightharpoons \ln w, \quad \mu,\nu \leftrightharpoons \alpha,\beta, \quad \mu \neq \nu \tag{A.42}$$

(no summation over repeated indices is implied). It is seen that when the tetrad $\{\boldsymbol{e}_{(i)}(0)\}$ is parallel transported along γ, the dyads $\{\boldsymbol{e}_{(m)}\}$, $m = 0, 1$ and $\{\boldsymbol{e}_{(j)}\}$, $j = 2, 3$ are parallel transported along, respectively, γ_{N_w} and $\gamma_{\mathbb{S}}$. Define the lengths $L_{\gamma_{N_w}}$ and $L_{\gamma_{\mathbb{S}}}$ by changing in (A.41) i to, respectively, m and j. Clearly,

[4] To be precise, in $T_{\gamma(0)}$.

$$L_\gamma \geqslant L_{\gamma_{N_w}}, L_{\gamma_S}. \tag{A.43}$$

Proposition 8 *If $w[\gamma(\zeta)] \to 0$ as $\zeta \to 1$, then γ is b-complete.*

Proof In the coordinates α, β, the parallel transport Eq. (A.40) takes the form

$$e^{\mu}_{(i),\zeta} = -\Gamma^{\mu}_{\mu\mu}\,\dot{\mu} e^{\mu}_{(i)}$$

(still no summation over μ is implied, the dot denotes differentiation with respect to ζ) and is easily solved after (A.42) is substituted in it:

$$e^{\mu}_{(\mu)} = e^{\mu}_{(\mu)}(0)\exp\{2\int_0^\zeta \sigma,_\mu\,\dot{\mu}\,\mathrm{d}\zeta\}, \qquad e^{\mu}_{(\nu)} = 0, \qquad \mu \neq \nu; \tag{A.44}$$

$$e^{m}_{(j)} = e^{j}_{(m)} = 0 \qquad m = 0, 1, \quad j = 2, 3.$$

Pick a specific value of ζ and assume that

$$\int_0^\zeta \sigma,_\beta\,\dot{\beta}\,\mathrm{d}\zeta' \geqslant \int_0^\zeta \sigma,_\alpha\,\dot{\alpha}\,\mathrm{d}\zeta'. \tag{$*$}$$

(this does not lead to any loss of generality since the sense of the inequality changes with the coordinate transformation $\alpha \leftrightarrow \beta$, i.e. $x_0 \leftrightarrow -x_0$). Assume also that $\ln[w/w(0)] \leqslant 0$ (since w tends to zero along γ, this assumption involves no loss of generality either). Now it follows from the evident identity

$$\big[w(\zeta)\big]^{-2} = \big[w(0)\big]^{-2}\exp\{-2\int_0^\zeta (\sigma,_\alpha\,\dot{\alpha} + \sigma,_\beta\,\dot{\beta})\,\mathrm{d}\zeta'\} \tag{A.45}$$

that ($*$) implies the inequalities

$$\int_0^\zeta \sigma,_\alpha\,\dot{\alpha}\,\mathrm{d}\zeta' < 0 \quad \text{and} \quad \left|\int_0^\zeta \sigma,_\alpha\,\dot{\alpha}\,\mathrm{d}\zeta'\right| > \left|\int_0^\zeta \sigma,_\beta\,\dot{\beta}\,\mathrm{d}\zeta'\right|,$$

whence, in particular,

$$\int_0^\zeta |\sigma,_\alpha\,\dot{\alpha}|\,\mathrm{d}\zeta' > -\frac{1}{2}\int_0^\zeta (\sigma,_\alpha\,\dot{\alpha} + \sigma,_\beta\,\dot{\beta})\,\mathrm{d}\zeta' = \tfrac{1}{2}\ln[w(0)/w(\zeta)]. \tag{A.46}$$

In order to estimate $L_{\gamma_{N_w}}$, substitute in its definition the inequality

$$\big[g(\boldsymbol{l}, \boldsymbol{e}_{(0)})^2 + g(\boldsymbol{l}, \boldsymbol{e}_{(1)})^2\big]^{1/2} = \big[2g(\boldsymbol{l}, \boldsymbol{e}_{(\alpha)})^2 + 2g(\boldsymbol{l}, \boldsymbol{e}_{(\beta)})^2\big]^{1/2} \geqslant$$
$$\geqslant |g(\boldsymbol{l}, \boldsymbol{e}_{(\alpha)})| + |g(\boldsymbol{l}, \boldsymbol{e}_{(\beta)})| = w^{-2}\Big[|\dot{\alpha} e^{\beta}_{(\beta)}| + |\dot{\beta} e^{\alpha}_{(\alpha)}|\Big]$$

combined with the expression (A.44). Then setting $e^{\alpha}_{(\alpha)}(0) = e^{\beta}_{(\beta)}(0) = w(0)$ gives

$$L_{\gamma_{N_w}}(\zeta) \geqslant w(0) \int\limits_0^{\zeta} \Big(|\dot\alpha| \exp\{2 \int_0^{\zeta'} \sigma_{,\beta}\, \dot\beta \,\mathrm{d}\zeta''\} + |\dot\beta| \exp\{2 \int_0^{\zeta'} \sigma_{,\alpha}\, \dot\alpha \,\mathrm{d}\zeta''\}\Big) w^{-2} \,\mathrm{d}\zeta'.$$

Hence, dropping the second—obviously positive—term, substituting the identity (A.45), and, finally, using the inequality (A.46) we get

$$L_{\gamma_{N_w}}(\zeta) \geqslant w(0) \int\limits_0^{\zeta} \Big(|\dot\alpha| \exp\{2 \int_0^{\zeta'} \sigma_{,\beta}\, \dot\beta \,\mathrm{d}\zeta''\}\Big) w^{-2} \,\mathrm{d}\zeta' =$$

$$= w^{-1}(0) \int\limits_0^{\zeta} \Big(|\dot\alpha| \exp\{-2 \int_0^{\zeta'} \sigma_{,\alpha}\, \dot\alpha \,\mathrm{d}\zeta''\}\Big) \,\mathrm{d}\zeta' \geqslant \int\limits_0^{\zeta} |\dot\alpha / w| \,\mathrm{d}\zeta' \geqslant$$

$$\geqslant \frac{1}{\max |w_{,\alpha}|} \int\limits_0^{\zeta} |\sigma_{,\alpha}\, \dot\alpha| \,\mathrm{d}\zeta' \geqslant \frac{\ln[w(0)/w(\zeta)]}{2 \max |w_{,\alpha}|},$$

where the last inequality is again obtained by application of (A.46). □

A.4 The Metric of a 'Portal'

Our task in this section is to find an explicit analytic expression for the metric of the spacetime described in the beginning of Sect. 2.3 in Chap. 8. Specifically, we build a static 'portal' $M = \mathbb{L}^1 \times \mathcal{P}_{8}$, where the Riemannian space $\mathcal{P}_{8}$ has the structure shown in Fig. 2 in Chap. 8. To simplify the task, we write down the metric of *another* space, $\mathcal{P}_{X}$, from which $\mathcal{P}_{8}$ is obtained by a simple cut-and-paste surgery. By analogy with wormholes, see Sect. 2.1 in Chap. 3, $\mathcal{P}_{X}$ may be called 'inter-universe portal' and it is to $\mathcal{P}_{8}$ as W_{X} to W_{8}. The spacetime under discussion is axially symmetric, i.e. $\mathcal{P}_{8(X)}$ is obtained by revolution of some *two*-dimensional surface. It is this surface that is our main subject.

Denote by $\bar{\mathcal{W}}$ the space obtained by removing the solid torus $\bar{\mathcal{H}} \leftrightharpoons \{(\bar\rho - \rho_0)^2 + \bar z^2 \leqslant h^2\}$ from Euclidean space $\mathbb{E}^3$. Here, $(\bar z, \bar\rho, \bar\varphi)$ are the standard cylindrical coordinates, in which the metric of the Euclidean space takes the form

$$\mathrm{d}s^2 = \mathrm{d}\bar z^2 + \mathrm{d}\bar\rho^2 + \bar\rho^2 \mathrm{d}\bar\varphi^2, \qquad (\star)$$

and

$$h, \rho_0 \quad \text{are constants such that } 0 < h < \rho_0. \qquad (*)$$

Our plan is to build a double covering $\mathcal{W}$ of $\bar{\mathcal{W}}$ (we have excluded $h = 0$ lest $\mathcal{W}$ be a dihedral wormhole with their inherent singularities) and glue something—it will be denoted by $\mathcal{H}$—in place of $\bar{\mathcal{H}}$ so as to 'fill the hole' left by removing $\bar{\mathcal{H}}$, without forming an above-mentioned singularity.

Assertion The spacetime $\mathcal{P}_{\mathrm{X}}$, whose metric is[5]

$$\begin{aligned} \mathrm{d}s^2 = 4[\chi_\varepsilon^2(\eta) + \eta^2](\mathrm{d}\eta^2 + \eta^2\mathrm{d}\psi^2) + \rho^2(\eta, \psi)\mathrm{d}\varphi^2, \\ \rho(\eta, \psi) \leftrightharpoons \rho_0 - \eta^2 \cos 2\psi, \end{aligned} \qquad (\star\star)$$

has the just described geometrical structure. Here, the following *conventions and notations* are adopted: $\chi_\varepsilon(\eta)$ is a 'hat of radius $\sqrt{h}$', i.e. a smooth even function that is non-zero at and only at $|\eta| < \sqrt{h}$, while (η, ψ) and (ρ, φ) are two sets of 'polar coordinates' in the sense that in $\mathcal{P}_{\mathrm{X}}$

$$\psi = \psi + 2\pi, \qquad \varphi = \varphi + 2\pi, \qquad \eta, \rho \geqslant 0,$$

and all points with $\eta = 0$, differing only by ψ, are identified, as well as all points with $\rho = 0$, differing only by φ.

To clarify the geometry of $\mathcal{P}_{\mathrm{X}}$, split it into

$$\mathcal{H} \leftrightharpoons \{\eta \leqslant \sqrt{h}\} \quad \text{and} \quad \mathcal{W} \leftrightharpoons \{\eta > \sqrt{h}\}.$$

$\mathcal{H}$ is, again, a solid torus—the product of the circle $\eta \leqslant \sqrt{h}$ lying in the plane (η, ψ) and the circumference of constant η, ψ; this circumference does not degenerate into a point, because $\rho \neq 0$ in $\mathcal{H}$ owing to condition $(\ast)$. The distinctive characteristic of $\mathcal{H}$ is that its interior is the set of all points of $\mathcal{P}_{\mathrm{X}}$ in which—in contrast to the region Int $\bar{\mathcal{H}}$— the metric is *non-flat*. It is $\mathcal{H}$ that plays the role of a 'hat' smoothing out the singularity of the initial dihedral wormhole. The metric in it—and even in a neighbourhood of it —is regular [first, as we have already mentioned, $\rho > 0$ in $\mathcal{H}$ and, second, $(\chi_\varepsilon^2 + \eta^2) \neq 0$; it is to make this factor non-vanishing—and, correspondingly, the metric non-singular—that we consider a *non-flat* case $\chi_\varepsilon \neq 0$].

It remains to verify that $\mathcal{W}$ is, indeed, a double covering of $\bar{\mathcal{W}}$. To this end, notice that the function $\varpi\colon p \mapsto \bar{p}$ defined by the equalities

$$\bar{\varphi}(\bar{p}) = \varphi(p), \qquad \bar{\rho}(\bar{p}) = \rho(p), \qquad \bar{z}(\bar{p}) = \eta^2 \sin 2\psi(p), \qquad \text{(A.47)}$$

is a (*local*) isometry [cf. $(\star)$ and $(\star\star)$]. Further, for each $p \in \mathcal{W}$, by definition,

$$h^2 < \eta^4(p) = [\bar{\rho}(\bar{p}) - \rho_0]^2 + \bar{z}^2(\bar{p})$$

and, hence, $\varpi(\mathcal{W}) \subset \bar{\mathcal{W}}$. And conversely, complementing (A.47) with the definition of ρ, see $(\star\star)$, we convince ourselves that for all $\bar{p} \in \bar{\mathcal{W}}$ the equation $\varpi(p) = \bar{p}$ has *two* solutions. These are the points p with the coordinates

---

[5]Note that $\bar{\rho}, \bar{\varphi}$ and ρ, φ refer to *different* spaces.

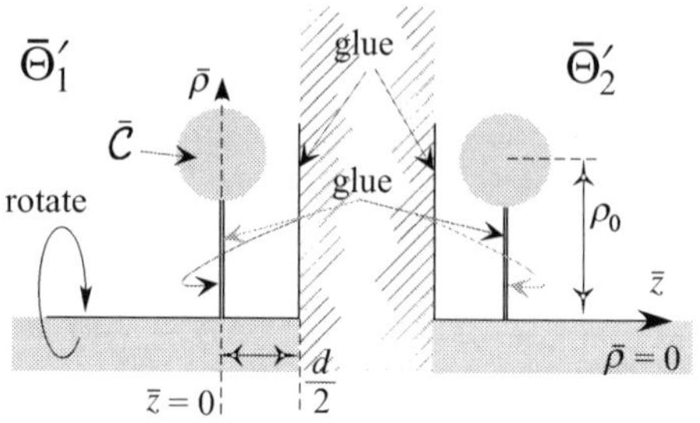

Fig. A.2 The space $\mathcal{P}_{\chi}$. It is obtained by (1) removing a circle from the upper half-plane (it is the left light grey circle in the picture), (2) cutting the remainder along a segment that connects the circle to the boundary of the half-plane, (3) gluing crosswise the edges of the resulting slit to the edges of the like slit in another (though isometric) space and (4) rotating the resulting space. To turn this wormhole into an *intra*-universe one, i.e. into $\mathcal{P}_{\gamma}$, the hatched regions must be removed and their boundaries identified

$$\eta(p) = \big([\bar\rho(\bar p) - \rho_0]^2 + \bar z^2(\bar p)\big)^{\frac14}, \quad \psi(p) = \sigma, \sigma + \pi, \quad \sigma \leftrightharpoons \frac12 \operatorname{arctg}\frac{\bar z(\bar p)}{\rho_0 - \bar\rho(\bar p)}.$$

They both satisfy the inequality $\eta(p) > \sqrt{h}$ and, hence, lie in $\mathcal{W}$.

To visualize the geometry of $\mathcal{P}_{\chi}$, it is instructive to construct it by the use of the 'cutting-and-pasting' method, see Sect. 6 in Chap. 1. To this end define $\bar\Theta$ to be the section $\bar\varphi = 0$ of $\bar{\mathcal{W}}$, which is the Euclidean half-space ($\bar z \in \mathbb{R}$, $\bar\rho \geqslant 0$) without the circle $\bar{\mathcal{C}} \leftrightharpoons \{(\bar\rho - \rho_0)^2 + \bar z^2 \leqslant h^2\}$. That circle is a section of $\bar{\mathcal{H}}$, see Fig. A.2. Further, take two copies—$\bar\Theta_1'$ and $\bar\Theta_2'$—of the space $\bar\Theta'$, which is obtained from $\bar\Theta$ by deleting the vertical segment from $\bar{\mathcal{C}}$ to the origin. Obviously, ϖ projects to $\bar\Theta_1'$ the 'right half' $-\pi/2 < \psi < \pi/2$ of $\Theta \leftrightharpoons \{x \in \mathcal{W}\colon \quad \varphi(x) = 0\}$ and to $\bar\Theta_2'$—the 'left half' $\pi/2 < \psi < 3\pi/2$ of Θ. Correspondingly, gluing crosswise the banks of the cuts we get the entire Θ. The space $\mathcal{P}_{\chi}$ is now obtained by rotating Θ and filling the appearing hole with the solid torus $\mathcal{H}$. $\mathcal{P}_{\chi}$ is not yet the shortcut we are after, because it still has *two* asymptotically flat regions, which correspond to $\bar\Theta_1'$ and $\bar\Theta_2'$, while our goal is an intra-universe wormhole $\mathcal{P}_{\gamma}$. To build it from $\mathcal{P}_{\chi}$ pick a constant $d > 2h$, remove the regions $\bar z > \frac12 d$ and $\bar z < -\frac12 d$ from $\bar\Theta_1'$ and $\bar\Theta_2'$, respectively, and glue together the boundaries of these removed regions (note that the surgery takes place in a part of $\mathcal{P}_{\chi}$ which is isometric to a region of Minkowski space, so it must not give birth to any new singularity). Rotating the thus obtained two-dimensional space around the axis $\rho = 0$ gives, finally, the sought $\mathcal{P}_{\gamma}$.

Outside some compact set, $\mathcal{P}_{\gamma}$ is merely a Euclidean space. Inside that set, it is also flat except in a region that *looks* like a pair of hoops, but is actually a *single* hoop $\mathcal{H}$. A traveller (the lion depicted in Fig. 2 in Chap. 8, say) jumping through one of these hoops instantaneously finds themself flying out of the other. Remarkably, throughout the whole journey the vicinity of the traveller remains flat and empty enabling them to avoid plunging into the Planck-density matter.

A.5 The Functions τ_i

In this section, we extract the estimates on τ_i needed in Chap. 9 from the results obtained in [24, 45]. In doing so, we neither reproduce the derivation of these results nor comment them.

Notation The quantities denoted in [24] by t, r, M, $T_{\alpha\beta}$ and T^{α}_{α} in this book are denoted by, respectively, t_S, $\mathring{r}$, m_0, $\mathring{T}_{\alpha\beta}$ and $\mathring{T}$. Without the risk of confusion, we use both notations as equivalent. The double brackets refer to the numbering in [24]. The index ξ used in [24] to label quantities associated with the Unruh vacuum will be dropped since we consider no other vacuum. It is also worth noting (even though we shall not use this right now) that our coordinates u and v *differ* from those denoted so in [24]. The latter—let us denote them $\bar{u}$ and $\bar{v}$, cf. Remark 6 in Chap. 9,—are related to the former as follows:

$$\bar{u} = -4m_0 \ln(-u), \qquad \bar{v} = 4m_0 \ln v.$$

1. The components $\mathring{T}^{\vartheta}_{\vartheta}$ and $\mathring{T}^{\varphi}_{\varphi}$ at large $\mathring{r}$ are given (implicitly) in Eq. ((5.5)), which says (when the Definition ((2.6)) and the expression

$$\mathring{T} = \frac{1}{60\pi^2}\frac{M^2}{r^6} \tag{(4.8)}$$

for the anomalous trace are taken into account)

$$\mathring{T}^{\vartheta}_{\vartheta} = \mathring{T}^{\varphi}_{\varphi} \approx \lambda K(2m_0)^{-4}x^{-4}, \tag{A.48}$$

where λ and K are constants

$$0 < \lambda \leqslant 27, \quad K \leftrightharpoons \frac{9}{40 \cdot 8^4\pi^2} \tag{A.49}$$

[the last equality is, in fact, ((6.21))]. Accordingly,

$$\tau_4 = 16\pi m_0^4 x^2 \mathring{T}^{\vartheta}_{\vartheta} \approx \pi\lambda K x^{-2} \qquad \text{at large } x. \tag{A.50}$$

2. Near the horizon the component $\mathring{T}^{\vartheta}_{\vartheta}$ was found, numerically, in [45] (it is denoted by p_t there):

$$0 < \mathring{T}^{\vartheta}_{\vartheta} \lesssim 2 \cdot 10^{-6} m_0^{-4}, \qquad \tfrac{\mathrm{d}}{\mathrm{d}x}\mathring{T}^{\vartheta}_{\vartheta} \approx -1,3 \cdot 10^{-6} m_0^{-4} \tag{A.51}$$

[the value of the derivative will be needed in (A.53)], whence

$$\tau_4(1) = \frac{m_0^4}{\pi}\mathring{T}^{\vartheta}_{\vartheta} \lesssim 10^{-5}. \tag{A.52}$$

3. It follows from ((4.8)) that

$$\mathring{T} = \frac{m_0^{-4}}{3840\pi^2}\, x^{-6} \approx 2,6 \times 10^{-5} m_0^{-4} x^{-6}, \qquad \mathring{T}'\big|_{x=1} \approx -1,6 \times 10^{-4} m_0^{-4}$$

(the prime, as usual, denotes differentiation with respect to x). Hence, for the quantity

$$Y \leftrightharpoons \mathring{T} - \mathring{T}^{\vartheta}_{\vartheta} - \mathring{T}^{\varphi}_{\varphi} = \mathring{T} - 2\mathring{T}^{\vartheta}_{\vartheta},$$

we have

$$Y\big|_{x=1} \approx 3{\cdot}10^{-5} m_0^{-4}, \qquad Y'\big|_{x=1} = \big(\mathring{T}' - 2\tfrac{\mathrm{d}}{\mathrm{d}x}\mathring{T}^{\vartheta}_{\vartheta}\big)\big|_{x=1} \approx -1,5{\times}10^{-4} m_0^{-4}. \quad \text{(A.53)}$$

Now it is convenient to introduce yet another pair of coordinates

$$r^* \leftrightharpoons 2m_0 \ln(-vu), \qquad t_S \leftrightharpoons 2m_0 \ln(-v/u),$$

which diagonalizes the Schwarzschild metric (9.1):

$$\mathrm{d}s^2 = \tfrac{x-1}{x}\big(-\,\mathrm{d}t_S^2 + \mathrm{d}r^{*2}\big) + \mathring{r}^2(\mathrm{d}\vartheta^2 + \cos^2\vartheta\,\mathrm{d}\varphi)$$

and yields

$$\mathring{T}_{uv} = \tfrac{4m_0^2}{vu}(\mathring{T}_{r^*r^*} - \mathring{T}_{t_St_S}) = -\tfrac{4m_0^2 e^{-x}}{x}(\mathring{T}^{r^*}_{r^*} + \mathring{T}^{t_S}_{t_S}) = -\tfrac{4m_0^2 e^{-x}}{x} Y\big|_{x=1} \approx -4\cdot 10^{-5} m_0^{-2},$$

whence

$$|\tau_3(1)| = |4\pi m_0^2 \mathring{T}_{uv}| \approx 5 \cdot 10^{-4}. \quad \text{(A.54)}$$

4. In the case of the Unruh vacuum, the quantity denoted in [24] by Q vanishes, see the passage between ((5.2)) and ((5.3)). Substituting this into ((2.5)) and using the fact that the component $T^{\vartheta}_{\vartheta}(x)$ and the trace $\mathring{T}(x)$ are bounded at $x \to 1$, see, e. g., (A.51) and ((4.8)), one gets

$$T^r_r = \frac{1}{r^2}\left(1 - \frac{2M}{r}\right)^{-1}\left(-\frac{K}{M^2} + \int_{2M}^{r}[M\mathring{T}(\acute{r}) + 2(\acute{r} - 3M)T^{\vartheta}_{\vartheta}(\acute{r})]\mathrm{d}\acute{r}\right) \quad \text{(A.55)}$$

and, thus,

$$T^{r^*}_{r^*} = T^{\mathring{r}}_{\mathring{r}} \to -\frac{K}{M^2 r^2}\left(1 - \frac{2M}{\mathring{r}}\right)^{-1} = -\frac{K}{4m_0^4 x(x-1)} \qquad \text{at } x \to 1.$$

This means, in particular, that one can neglect two last terms in ((2.4))

$$T^{t_S}_{t_S} = -T^{\mathring{r}}_{\mathring{r}} - 2T^{\vartheta}_{\vartheta} + \mathring{T} \to -T^{\mathring{r}}_{\mathring{r}} \qquad \text{at } x \to 1.$$

Further, substituting ((2.3)) into the formula above ((2.6)) gives

$$T_{ts}^{r^*} = \frac{x}{x-1}\left[-\frac{K}{m_0^2\mathring{r}^2}\right] = -\frac{K}{4m_0^4 x(x-1)}. \tag{A.56}$$

Gathering the expressions above, we have

$$\mathring{T}_{vv} = \tfrac{4m_0^2}{v^2}(\mathring{T}_{tsts} + \mathring{T}_{r^*r^*} + 2\mathring{T}_{tsr^*}) = \tfrac{(x-1)}{x}\left(\tfrac{4m_0^2}{xe^x}\right)^2 \mathring{r},_u^{-2}(-\mathring{T}_{ts}^{ts} + \mathring{T}_{r^*}^{r^*} + 2\mathring{T}_{ts}^{r^*}) \to$$
$$\to -\frac{(x-1)}{x}\left(\frac{4m_0^2}{xe^x}\right)^2 \mathring{r},_u^{-2}\frac{4K}{4m_0^4 x(x-1)}\Big|_{x=1} = -16e^{-2}K\mathring{r},_u^{-2} =$$
$$= -\frac{\mathring{F}^4(1)K}{16m_0^4}\mathring{r},_u^{-2},$$

which implies, among another things, the estimate

$$\tau_1(1) = 4\pi\mathring{r},_u^2 T_{vv} = -4\pi\mathring{r},_u^2 \times 16e^{-2}K\mathring{r},_u^{-2} = -\frac{64\pi e^{-2}\cdot 9}{40\cdot 8^4\pi^2} \approx -1,5\cdot 10^{-3}. \tag{A.57}$$

5. Similarly, by the use of (9.5b), we find

$$\mathring{T}_{uu} = \tfrac{4m_0^2}{u^2}(\mathring{T}_{tsts} - 2\mathring{T}_{r^*ts} + \mathring{T}_{r^*r^*}) = \tfrac{x}{x-1}\mathring{r},_u^2(2T_\vartheta^\vartheta - \mathring{T} + 2\mathring{T}_{r^*}^{r^*} - 2\mathring{T}_{ts}^{r^*}).$$

Substituting in this equality the definition of Y and expressions (A.55) and (A.56), one obtains

$$\mathring{T}_{uu} = \frac{x\mathring{r},_u^2}{x-1}\left[-Y + 2\frac{K}{4m_0^4 x(x-1)} + 2\mathring{T}_{r^*}^{r^*}\right] =$$
$$= \frac{x\mathring{r},_u^2}{x-1}\left[-Y + \frac{2}{x(x-1)}\int_1^x [\tfrac{1}{2}\mathring{T}(\acute{x}) + 2(\acute{x} - \tfrac{3}{2})T_\vartheta^\vartheta(\acute{x})]d\acute{x}\right] =$$
$$= \frac{x\mathring{r},_u^2}{x-1}\left[-Y + \frac{2}{x(x-1)}\int_1^x \left(\tfrac{1}{2}Y + 2(\acute{x}-1)T_\vartheta^\vartheta\right)d\acute{x}\right].$$

Finally, we introduce the quantity $\varepsilon \leftrightharpoons x - 1$ and check that

$$\mathring{T}_{uu}(x=1) = \lim_{\varepsilon\to 0}\frac{x\mathring{r},_u^2}{\varepsilon}\left[-Y + \frac{2}{\varepsilon(\varepsilon+1)}\int_0^\varepsilon \left(\tfrac{1}{2}Y + 2\acute{\varepsilon}T_\vartheta^\vartheta\right)d\acute{\varepsilon}\right] =$$
$$= \lim_{\varepsilon\to 0}\frac{x\mathring{r},_u^2}{\varepsilon}\left[-Y(0) - \varepsilon Y'(0) + \frac{2}{(\varepsilon+1)}\left(\tfrac{1}{2}Y(0) + \tfrac{1}{4}\varepsilon Y'(0) + \varepsilon T_\vartheta^\vartheta(0)\right)\right] =$$
$$= \lim_{\varepsilon\to 0}\mathring{r},_u^2\left(-Y(0) - \tfrac{1}{2}Y'(0) + 2T_\vartheta^\vartheta(0)\right) \approx -2\cdot 10^{-5}m_0^{-4}\mathring{r},_u^2,$$

whence

$$\tau_2(1) \approx -2,5\cdot 10^{-4}. \tag{A.58}$$

References

1. I. Asimov, *Escape!, I, Robot* (Bantam Dell, NY, 2004)
2. V.V. Batygin, I.N. Toptygin, *Problems in Electrodynamics* (Academic Press Inc., New York, 1978)
3. M. Alcubierre, The warp drive: hyperfast travel within general relativity. Class. Quantum Gravity **11**, L73 (1994)
4. L. Andersson, in *50 Years of the Cauchy Problem in General Relativity*, ed. by P. Chrusciel, H. Friedrich (Birkhauser, 2004)
5. K.V. Anisovich, *Problems of High Energy Physics and Field Theory (Proceedings of the XIV Workshop)* (Moscow, Nauka, 1992)
6. I.Ya. Aref'eva, I.V. Volovich, T. Ishiwatari, Cauchy problem on non-globally hyperbolic space-times. Theor. Math. Phys. **157**, 1646 (2008)
7. R. Banach, J.S. Dowker, Automorphic field theory-some mathematical issues. J. Phys. A **12**, 2527 (1979)
8. C. Barceló, M. Visser, Traversable wormholes from massless conformally coupled scalar fields. Phys. Lett. B **466**, 127 (1999)
9. J.M. Bardeen, Black holes do evaporate thermally. Phys. Rev. Lett. **46**, 382 (1981)
10. S. Beckett, *Act Without Words II in Collected Shorter Plays of Samuel Beckett* (Faber and Faber, London, 1984)
11. J. Beem, P. Ehrlich, *Global Lorentzian Geometry* (Marcel Dekker. Ink, New York, 1981)
12. A.N. Bernal, M. Sánchez, On smooth Cauchy hypersurfaces and Geroch's splitting theorem. Commun. Math. Phys. **243**, 461 (2003)
13. O.M. Bilaniuk, E.C.G. Sudarshan, Particles beyond the light barrier. Phys. Today **22**, 43 (1969)
14. N.D. Birrel, P.C.V. Davies, *Quantum Fields in Curved Space* (Cambridge University Press, Cambridge, 1982)
15. B. Boisseau, B. Linet, Electrostatics in a simple wormhole revisited. Gen. Relativ. Gravit. **45**, 845 (2013)
16. R. Bradbury, *A Sound of Thunder in The Stories of Ray Bradbury* (Alfred A. Knopf, New York, 1980)
17. F. Brickell, R.S. Clark, *Differentiable Manifolds: An Introduction* (Van Nostrand Reinhold company, London, 1970)
18. K.A. Bronnikov, Scalar-tensor theory and scalar charge. Acta Phys. Pol. B **4**, 251 (1973)
19. K.A. Bronnikov, S.-W. Kim, Possible wormholes in a brane world. Phys. Rev. D **67**, 064027 (2003)

S. Krasnikov, *Back-in-Time and Faster-than-Light Travel in General Relativity*, Fundamental Theories of Physics 193, https://doi.org/10.1007/978-3-319-72754-7

20. R. Brout, S. Massar, R. Parentani, Ph Spindel, Primer for black hole quantum physics. Phys. Rep. **260**, 329 (1995)
21. Y. Bruhat, in *Gravitation: An Introduction to Current Research*, ed. by L. Witten (Wiley, New York, 1962)
22. E. Capelas de Oliveira, W.A. Rodrigues, Finite energy superluminal solutions of Maxwell equations. Phys. Lett. A **291**, 367 (2001)
23. Y. Choquet-Bruhat, R. Geroch, Global aspects of the Cauchy problem in general relativity. Commun. Math. Phys. **14**, 329 (1969)
24. S.M. Christensen, S.A. Fulling, Trace anomalies and the Hawking effect. Phys. Rev. D **15**, 2088 (1977)
25. P.T. Chruściel, A remark on differentiability of Cauchy horizons. Class. Quantum Gravity **15**, 3845 (1998)
26. P.T. Chruściel, J. Isenberg, *On the Dynamics of Generators of Cauchy Horizons* (1993). arXiv:gr-qc/9401015
27. C.J.S. Clarke, Singularities in globally hyperbolic space-time. Commun. Math. Phys. **41**, 65 (1975)
28. C.J.S. Clarke, *The Analysis of Space-Time Singularities* (Cambridge University Press, Cambridge, 1993)
29. D.H. Coule, No warp drive. Class. Quantum Gravity **15**, 2523 (1998)
30. C.R. Cramer, B.S. Kay, Stress-energy must be singular on the misner space horizon even for automorphic fields. Class. Quantum Gravity **13**, L143 (1996)
31. C.R. Cramer, B.S. Kay, The thermal and two-particle stress-energy must be ill-defined on the 2-d Misner space chronology horizon. Phys. Rev. D **57**, 1052 (1998)
32. J.G. Cramer et al., Natural wormholes as gravitational lenses. Phys. Rev. D **51**, 3117 (1995)
33. P.L. Csonka, Causality and faster than light particles. Nucl. Phys. B **21**, 436 (1970)
34. D. Deutsch, Quantum mechanics near closed timelike lines. Phys. Rev. D **44**, 3197 (1991)
35. B.S. DeWitt, C.F. Hart, C.J. Isham, Topology and quantum field-theory. Physica A **96**, 197 (1979)
36. B.S. DeWitt, *The Global Approach to Quantum Field Theory* (Clarendon Press, Oxford, 2003)
37. D. Eardley, V. Moncrief, The global existence problem and cosmic censorship in general relativity. Gen. Relativ. Gravit. **13**, 887 (1981)
38. J. Earman, C. Wüthrich, J. Manchak, in *Time Machines*, The stanford encyclopedia of philosophy (Winter 2016 Edition), ed. by E.N. Zalta. https://plato.stanford.edu/archives/win2016/entries/time-machine/
39. F. Echeverria, G. Klinkhammer, K.S. Thorne, Billiard balls in wormhole spacetimes with closed timelike curves: Classical theory. Phys. Rev. D **44**, 1077 (1991)
40. A. Einstein in *Albert Einstein: Philosopher-Scientist*, Ed. P. A. Schilpp (Cambridge, Cambridge University Press, 1949)
41. A. Einstein, Üiber die vom Relativitätsprinzip geforderte Trägheit der Energie. Ann. Phys. **23**, 371 (1907)
42. A. Einstein, N. Rosen, The particle problem in the general theory of relativity. Phys. Rev. D **48**, 73 (1935)
43. G.F.R. Ellis, B.G. Schmidt, Singular space-times. Gen. Relativ. Gravit. **8**, 915 (1977)
44. H.G. Ellis, Ether flow through a drainhole: a particle model in general relativity. J. Math. Phys **14**, 104 (1971)
45. T. Elster, Vacuum polarization near a black hole creating particles. Phys. Lett. A **94**, 205 (1983)
46. A. Everett, Warp drive and causality. Phys. Rev. D **53**, 7365 (1996)
47. A.E. Everett, T.A. Roman, Superluminal subway: the Krasnikov tube. Phys. Rev. D **56**, 2100 (1997)
48. É.É. Flanagan, Quantum inequalities in two dimensional curved spacetimes. Phys. Rev. D **66**, 104007 (2002)
49. V. Fock, *The Theory of Space Time and Gravitation* (Pergamon Press, New York, 1959)

50. L.H. Ford, M.J. Pfenning, T.A. Roman, Quantum inequalities and singular negative energy densities. Phys. Rev. D **57**, 4839 (1998)
51. L.H. Ford, T.A. Roman, Averaged energy conditions and quantum inequalities. Phys. Rev. D **51**, 4277 (1995)
52. J.L. Friedman, M.S. Morris, Existence and uniqueness theorems for massless fields on a class of spacetimes with closed timelike curves. Commun. Math. Phys. **186**, 495 (1997)
53. J.L. Friedman, N.J. Papastamatiou, J.Z. Simon, Failure of unitarity for interacting fields on spacetimes with closed timelike curves. Phys. Rev. D **46**, 4456 (1992)
54. J.L. Friedman, K. Schleich, D.M. Witt, Topological censorship. Phys. Rev. Lett. **71**, 1486 (1993)
55. V.P. Frolov, Vacuum polarization in a locally static multiply connected spacetime and time machine problem. Phys. Rev. D **43**, 3878 (1991)
56. V. Frolov, I. Novikov, *Physics of Black Holes (Fundamental Theories of Physics)* (Springer, Berlin, 1989)
57. V.P. Frolov, I.D. Novikov, Physical effects in wormholes and time machines. Phys. Rev. D **42**, 1057 (1990)
58. R.W. Fuller, J.A. Wheeler, Causality and multiply connected space-time. Phys. Rev. **128**, 919 (1962)
59. S. Gao, R.M. Wald, Theorems on gravitational time delay and related issues. Class. Quantum Gravity **17**, 4999 (2000)
60. R. Geroch, Topology in general relativity. J. Math. Phys. **8**, 782 (1967)
61. R. Geroch, Spinor structure of space-times in general relativity I. J. Math. Phys. **9**, 1739 (1968)
62. R. Geroch, in *Relativity, Proceedings of Relativity Conference, Cincinnati*, June 2–6, 1969, New York–London, 1970
63. R. Geroch, *Foundations of Space-Time Theories, Minnesota studies in the philosophy of science*, vol. VIII (University of Minnesota Press, Minneapolis, 1977)
64. R. Geroch, G.T. Horowitz, in *General Relativity: An Einstein Centenary Survey*, ed. by S.W. Hawking, W. Israel (Cambridge University Press, Cambridge, 1979)
65. G.W. Gibbons, S.W. Hawking, Selection rules for topology change. Commun. Math. Phys. **148**, 345 (1992)
66. K. Gödel, An example of a new type of Cosmological solutions of Einstein's field equations of gravitation. Rev. Mod. Phys. **21**, 447 (1949)
67. J.R. Gott, Closed timelike curves produced by pairs of moving cosmic strings: exact solutions. Phys. Rev. Lett. **66**, 1126 (1991)
68. J.D. Grant, Cosmic strings and chronology protection. Phys. Rev. D **47**, 2388 (1993)
69. P. Gravel, J.-L. Plante, Simple and double walled Krasnikov tubes: I. Tubes with low masses. Class. Quantum Gravity **21**, L7 (2004)
70. A.A. Grib, S.G. Mamayev, V.M. Mostepanenko, *Vacuum Quantum Effects in Strong Fields* (Friedmann Laboratory Publishing, St.Petersburg, 1994)
71. J.B. Griffiths, J. Podolsky, *Exact Space-Times in Einstein's General Relativity* (Cambridge University Press, Cambridge, 2009)
72. H. Harrison, *The Technicolor Time Machine* (Doubleday, New York, 1967)
73. P. Hajicek, Origin of Hawking radiation. Phys. Rev. D **36**, 1065 (1987)
74. S.W. Hawking, Chronology protection conjecture. Phys. Rev. D **46**, 603 (1992)
75. S.W. Hawking, *A Brief History of Time* (Bantam, New York, 1988)
76. S.W. Hawking, G.F.R. Ellis, *The Large Scale Structure of Spacetime* (Cambridge University Press, Cambridge, 1973)
77. S. Hayward, H. Koyama, How to make a traversable wormhole from a Schwarzschild black hole. Phys. Rev. D **70**, 101502 (2004)
78. W.A. Hiscock, D.A. Konkowski, Quantum vacuum energy in Taub-NUT-type cosmologies. Phys. Rev. D **26**, 1225 (1982)
79. D. Hochberg, A. Popov, S.V. Sushkov, Self- consistent wormhole solutions of semiclassical gravity. Phys. Rev. Lett. **78**, 2050 (1997)

80. D. Hochberg, M. Visser, Geometric structure of the generic static traversable wormhole throat. Phys. Rev. D **56**, 4745 (1997)
81. A. Ishibashi, R.M. Wald, Dynamics in non-globally-hyperbolic static spacetimes II: general analysis of prescriptions for dynamics. Class. Quantum Gravity **20**, 3815 (2003)
82. A.J. Janca, *So You Want to Stop Time* (2007). arXiv:gr-qc/0701084
83. B.S. Kay, The principle of locality and quantum field theory on (non globally hyperbolic) curved spacetimes. Rev. Math. Phys. **4**, 167 (1992). Special Issue
84. B.S. Kay, M. Radzikowski, R.M. Wald, Quantum field theory on spacetims with a compactly generated Cauchy horizon. Commun. Math. Phys. **183**, 533 (1997)
85. V. Khatsymovsky, Rotating vacuum wormhole. Phys. Lett. B **429**, 254 (1998)
86. N.R. Khusnutdinov, I.V. Bakhmatov, Self-force of a point charge in the space-time of a symmetric wormhole. Phys. Rev. D **76**, 124015 (2007)
87. A.N. Kolmogorov, S.V. Fomin, *Elements of the Theory of Functions and Functional Analysis* (Martino Fine Books, 2012)
88. S.V. Krasnikov, On the classical stability of a time machine. Class. Quantum Gravity **11**, 2755 (1994)
89. S.V. Krasnikov, Topology changes without any pathology. Gen. Relativ. Gravit. **27**, 529 (1995)
90. S.V. Krasnikov, Quantum stability of the time machine. Phys. Rev. D **54**, 7322 (1996)
91. S. Krasnikov, Causality violations and paradoxes. Phys. Rev. D **55**, 3427 (1997)
92. S. Krasnikov, A singularity free WEC respecting time machine. Class. Quantum Gravity **15**, 997 (1998)
93. S. Krasnikov, Hyper-fast travel in general relativity. Phys. Rev. D **57**, 4760 (1998)
94. S. Krasnikov, Quantum field theory and time machines. Phys. Rev. D **59**, 024010 (1999)
95. S. Krasnikov, Traversable wormhole. Phys. Rev. D **62**, 084028 (2000), **76**, 109902 (2007). Erratum ibid
96. S. Krasnikov, Time travel paradox. Phys. Rev. D **65**, 064013 (2002)
97. S. Krasnikov, No time machines in classical general relativity. Class. Quantum Gravity **19**, 4109 (2002). *Corrigendum: No time machines in classical general relativity* ibid. **31** (2014) 079503
98. S. Krasnikov, The quantum inequalities do not forbid spacetime shortcuts. Phys. Rev. D **67**, 104013 (2003)
99. S. Krasnikov, Quantum inequalities and their applications. Gravit. Cosmol. **46**, 195 (2006)
100. S. Krasnikov, Evaporation induced traversability of the Einstein-Rosen wormhole. Phys. Rev. D **73**, 084006 (2006)
101. S. Krasnikov, Unconventional stringlike singularities in flat spacetime. Phys. Rev. D **76**, 024010 (2007)
102. S. Krasnikov, Electrostatic interaction of a pointlike charge with a wormhole. Class. Quantum Gravity **25**, 245018 (2008)
103. S. Krasnikov, Falling into the Schwarzschild black hole. Important details. Gravit. Cosmol. **14**, 362 (2008)
104. S. Krasnikov, Even the Minkowski space is holed. Phys. Rev. D **79**, 124041 (2009)
105. S. Krasnikov, Topological censorship is not proven. Gravit. Cosmol. **19**, 195 (2013)
106. S. Krasnikov, Time machines with the compactly determined Cauchy horizon. Phys. Rev. D **90**, 024067 (2014)
107. S. Krasnikov, Yet another proof of Hawking and Ellis's Lemma 8.5.5. Class. Quantum Gravity **31**, 227001 (2014)
108. S. Krasnikov, What is faster - light or gravity? Class. Quantum Gravity **32**, 075002 (2015)
109. M. Kriele, The structure of chronology violating sets with compact closure. Class. Quantum Gravity **6**, 1607 (1989)
110. S.K. Lamoreaux, Demonstration of the Casimir force in the 0.6 to 6 μ m range. Phys. Rev. Lett. **78**, 5 (1997)
111. L.D. Landau, E.M. Lifshitz, *The Classical Theory of Fields*, vol. 2 (Pergamon Press, USA, 1971)
112. S. Lem, *The Cyberiad* (Houghton Mifflin Harcourt, USA, 2002)

113. J. Leray, *Hyperbolic differential equations* (Princeton Institute for Advanced Studies, 1952). duplicated notes
114. D. Levanony, A. Ori, Extended time-travelling objects in Misner space. Phys. Rev. D **83**, 044043 (2011)
115. L.-X. Li, Must time machine be unstable against vacuum fluctuations? Class. Quant. Gravity **13**, 2563 (1996)
116. L.-X. Li, Time machines constructed from Anti-de sitter space. Phys. Rev. D **59**, 084016 (1999)
117. L.-X. Li, Two open universes connected by a wormhole: exact solutions. J. Geom. Phys. **49**, 254 (2001)
118. M. Li, X. Li, S. Wang, Y. Wang, Dark Energy: A Brief Rev. Front. Phys. **8**, 828 (2013)
119. A.P. Lightman, W.H. Press, R.H. Price, S.A. Teukolsky, *Problem Book in Relativity and Gravitation* (Princeton University Press, Princeton, 1975)
120. A. Lossev, I.D. Novikov, The Jinn of the time machine: non-trivial selfconsistent solutions. Class. Quantum Gravity **9**, 2309 (1992)
121. R.J. Low, Speed limits in general relativity. Class. Quantum Gravity **16**, 543 (1999)
122. R.J. Low, Time machines, maximal extensions and Zorn's lemma. Class. Quantum Gravity **29**, 097001 (2012)
123. H. Maeda, T. Harada, B.J. Carr, Cosmological wormholes. Phys. Rev. D **79**, 044034 (2009)
124. K. Maeda, A. Ishibashi, M. Narita, Chronology protection and non-naked singularity. Class. Quantum Gravity **15**, 1637 (1998)
125. J.B. Manchak, Is spacetime hole-free? Gen. Rel. Gravity **41**, 1639 (2009)
126. J.B. Manchak, No no-go: a remark on time machines. Stud. Hist. Philos. Mod. Phys. **42**, 74 (2011)
127. M.B. Mensky, I.D. Novikov, Three-dimensional billiards with time machine. Int. J. Mod. Phys. D **5**, 179 (1996)
128. A.A. Milne, *Winnie-The-Pooh* (Methuen Children's Books Ltd, 1998)
129. E. Minguzzi, Causally simple inextendible spacetimes are hole-free. J. Math. Phys. **53**, 062501 (2012)
130. E. Minguzzi, Completeness of Cauchy horizon generators. J. Math. Phys. **55**, 082503 (2014)
131. E. Minguzzi, Area theorem and smoothness of compact Cauchy horizons. Commun. Math. Phys. **339**, 57 (2015)
132. E. Minguzzi, The boundary of the chronology violating set. Class. Quantum Gravity **33**, 225004 (2016)
133. C.W. Misner, in *Lectures in applied mathematics, vol. 8*, ed. by J. Ehlers (American Mathematical Society, 1967)
134. C.W. Misner, J.A. Wheeler, Classical physics as geometry. Ann. Phys. **2**, 525 (1957)
135. C.W. Misner, K.S. Thorne, J.A. Wheeler, *Gravitation* (Freeman, San Francisco, 1973)
136. M.S. Morris, K.S. Thorne, Wormholes in spacetime and their use for interstellar travel: a tool for teaching general relativity. Am. J. Phys. **56**, 395 (1988)
137. M.S. Morris, K.S. Thorne, U. Yurtsever, Wormholes, time machines, and the weak energy condition. Phys. Rev. Lett. **61**, 1446 (1988)
138. P.J. Nahin, *Time Machines* (Springer, New York, 1999)
139. I.D. Novikov, An analysis of the operation of a time machine. Zh. Eksp. Theor. Fiz. **95** (1989). 769 = (1989 Sov. Phys. - JETP 68 439)
140. K. Olum, Superluminal travel requires negative energies. Phys. Rev. Lett. **81**, 3567 (1998)
141. B. O'Neill, *Semi-Riemannian Geometry* (Academic Press, New York, 1983)
142. A. Ori, Formation of closed timelike curves in a composite vacuum/dust asymptotically-flat spacetime. Phys. Rev. D **76**, 044002 (2007)
143. D. Page, Particle emission rates from a black hole: massless particles from an uncharged, nonrotating hole. Phys. Rev. D **13**, 198 (1976)
144. D. Page, Thermal stress tensor in Einstein spaces. Phys. Rev. D **25**, 1499 (1982)
145. R. Penrose, *Techniques of Differential Topology in Relativity* (SIAM, 1983)

146. R. Penrose, in *General Relativity: An Einstein Centenary Survey*, ed. by S.W. Hawking, W. Isreal (Cambridge University Press, Cambridge, 1992)
147. R. Penrose, The question of cosmic censorship. J. Astrophys. Astr. **20**, 233 (1999)
148. M.J. Pfenning, L.H. Ford, The unphysical nature of Warp Drive. Class. Quantum Gravity **14**, 1743 (1997)
149. M.J. Pfenning, L.H. Ford, Quantum inequality restrictions on negative energy densities in curved spacetimes. arXiv:gr-qc/9805037
150. H.D. Politzer, Simple quantum systems in spacetimes with closed timelike curves. Phys. Rev. D **46**, 4470 (1992)
151. M.M. Postnikov, *Geometry VI: Riemannian Geometry* (Springer Science & Business Media, New York, 2013)
152. K. Prutkov, Q. Blake, *Thoughts and Aphorisms from the Fruits of meditation of Kozma Prutkov* (Royal College of Art, London, 1975)
153. T.C. Quinn, R.M. Wald, Axiomatic approach to electromagnetic and gravitational radiation reaction of particles in curved spacetime. Phys. Rev. D **56**, 3381 (1997)
154. S.K. Rama, S. Sen, *Inconsistent Physics in the Presence of Time Machines*. arXiv:gr-qc/9410031
155. E. Recami, On localized X-shaped superluminal solutions to Maxwell equations. Physica A **252**, 586 (1998)
156. A.D. Rendall, Theorems on existence and global dynamics for the Einstein equations. Living Rev. Relativ. **8**, 6 (2005)
157. T.A. Roman, in *Proceedings of the Tenth Marcel Grossmann Meeting on General Relativity*, ed. by M. Novello, S. Perez-Bergliaffa, R. Ruffini (Rio de Janeiro, Brazil, 2003)
158. J.K. Rowling, *Harry Potter and the Prisoner of Azkaban* (Bloomsbury, UK, 1999)
159. B.G. Schmidt, A new definition of singular points in general relativity. Gen. Relativ. Gravit. **1**, 269 (1971)
160. M. Sparnaay, Measurements of attractive forces between flat plates. Physica **24**, 751 (1958)
161. S.V. Sushkov, A selfconsistent semiclassical solution with a throat in the theory of gravity. Phys. Lett. A **164**, 33 (1992)
162. S.V. Sushkov, Quantum complex scalar field in two-dimensional spacetime with closed time-like curves and a time-machine problem. Class. Quantum Gravity **12**, 1685 (1995)
163. S.V. Sushkov, Privat Communication
164. S.V. Sushkov, Chronology protection and quantized fields: complex automorphic scalar field in Misner space. Class. Quantum Gravity **14**, 523 (1997)
165. R. Takahashi, H. Asada, Observational upper bound on the cosmic abundances of negative-mass compact objects and Ellis wormholes from the Sloan digital sky survey quasar lens search. Ap. J. **768L**, 16 (2013)
166. F. Tipler, Singularities and causality violation. Ann. Phys. **108**, 1 (1977)
167. R.C. Tolman, *The Theory of Relativity of Motion* (California Press, Berkeley Univ, 1917)
168. D.F. Torres, G.E. Romero, L.A. Anchordoqui, Might some gamma ray bursts be the observable signature of natural wormholes? Phys. Rev. D **58**, 123001 (1998)
169. V.A. Ugarov, *Special Theory of Relativity* (Mir Publishers, Moscow, 1979)
170. C. Van Den Broeck, A warp drive with more reasonable total energy requirements. Class. Quantum Gravity **16**, 3973 (1999)
171. W.J. van Stockum, The gravitational field of a distribution of particles rotating around an axis of symmetry. Proc. Roy. Soc. Edin. **57**, 135 (1937)
172. M. Visser, *Lorentzian Wormholes-from Einstein to Hawking* (AIP Press, New York, 1995)
173. M. Visser, The reliability horizon for semi-classical quantum gravity: Metric fluctuations are often more important than back-reaction. Phys. Lett. B. **415**, 8 (1997)
174. D.N. Vollick, How to produce exotic matter using classical fields. Phys. Rev. D **56**, 4720 (1997)
175. V.S. Vladimirov, *Equations of Mathematical Physics* (Mir Publishers, Moscow, 1983)
176. R.M. Wald, *Gravitational Collapse and Cosmic Censorship*. arXiv:gr-qc/9710068
177. R.M. Wald, *General Relativity* (University of Chicago Press, Chicago, 1984)

178. R.M. Wald, *Quantum Field Theory in Curved Spacetime and Black Hole Thermodynamics* (University of Chicago Press, Chicago, 1994)
179. Y. Wang, e. a. Current Observational Constraints on Cosmic Doomsday. JCAP **12**, 006 (2004)
180. J.A. Wheeler, *Neutrinos, Gravitation and Geometry* (Bologna, 1960)
181. J.A. Wheeler, *Geometrodynamics* (Academic Press, New York, 1992)
182. P. Yodzis, Lorentz cobordism. Commun. Math. Phys. **26**, 39 (1972)
183. U. Yurtsever, Classical and quantum instability of compact Cauchy horizons in two dimensions. Class. Quantum Gravity **8**, 1127 (1991)
184. U. Yurtsever, Algebraic approach to quantum field theory on non-globally hyperbolic spacetimes. Class. Quantum Grav. **11**, 999 (1994)
185. Zeno's paradoxes in *Wikipedia*. https://en.wikipedia.org
186. Р. А. Александрян и Э. А. Мирзаханян *Общая топология* (Москва, Высшая школа, 1979)
187. С. В. Красников *Некоторые вопросы причинности в ОТО: <<машины времени>> и <<сверхсветовые перемещениая>>* (Москва, Ленанд, 2015)
188. Козьма Прутков *Сочинения Козьмы Пруткова* (Москво, Художествен-ная Литература, 1976)
189. А. Н. Стругацкий, Б. Н. Стругацкий *Понедельник начинается в субботу* (Москва, Моладая Гвадия, 1966)
190. П. К. Суетин *Классические ортогональные многочлены* Изд. 2-е (Москва, Наука, 1979)
191. М. В. Федорюк *Асимптотика: интегралы и ряды* (Москва, Наука, 1987)
192. Г. М. Фихтенгольц *Курс дифференциального и интегрального исчисления. Том I* (Москва, Госудапственное издательстьо техникотеоретической литературы, 1947)
193. Г. М. Фихтенгольц *Курс дифференциального и интегрального исчисления. Том I* (Москва, Госудапственное издательстьо техникотеоретической литературы, 1948)

Index

S. Krasnikov, *Back-in-Time and Faster-than-Light Travel in General Relativity*, Fundamental Theories of Physics 193, https://doi.org/10.1007/978-3-319-72754-7

Manufactured by Amazon.ca
Bolton, ON

39405721R00149